W9-CHW-230

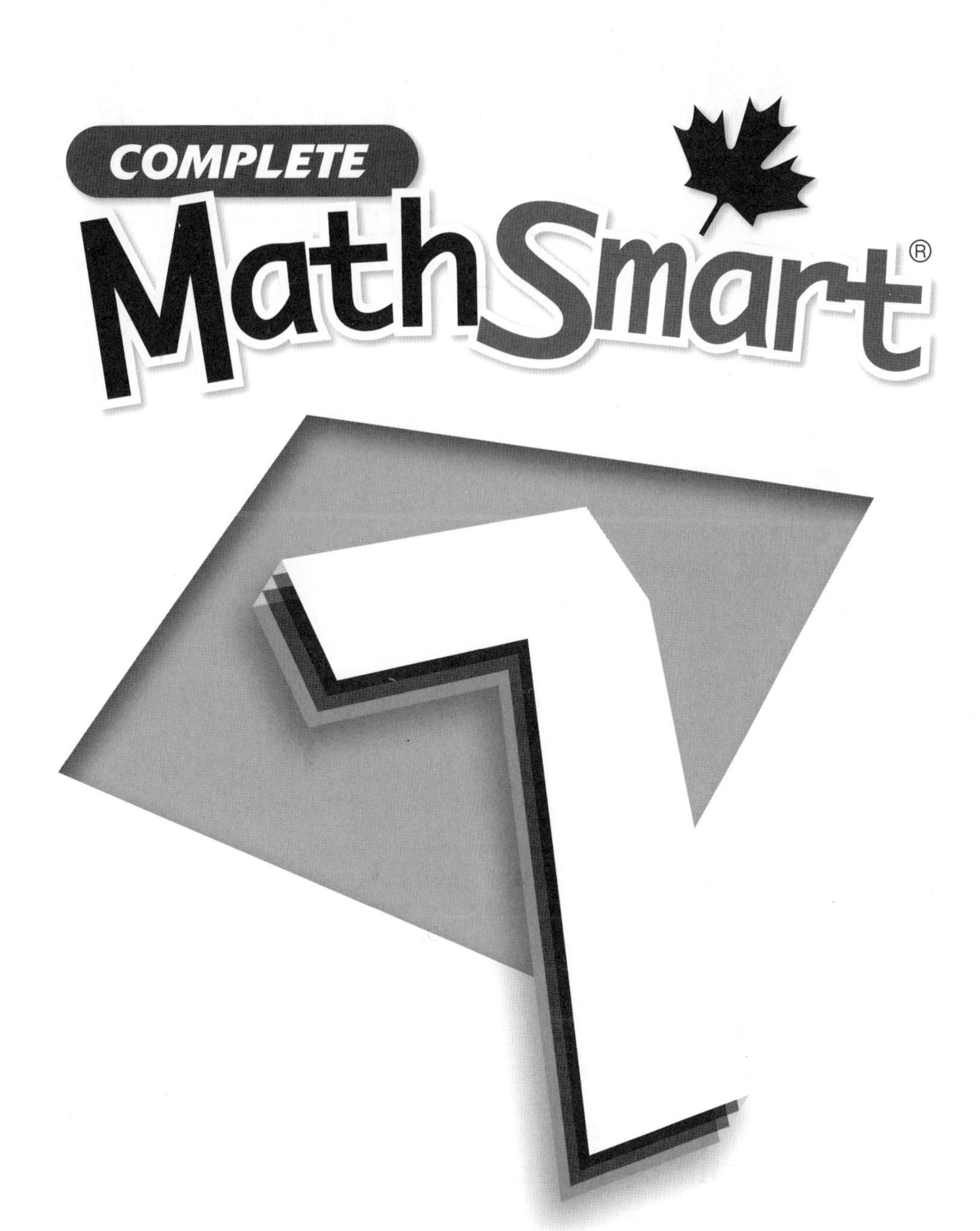
COMPLETE
MathSmart®
7

Contents

Level 1 – Basic Skills

Level 2 – Further Your Understanding

Dear Parent or Guardian,

Thank you for choosing our book to help sharpen your child's math skills. Our primary goal is to provide a learning experience that is both fun and rewarding. This aim has guided the development of the series in a few key ways.

Our ***Complete MathSmart*** series has been designed to help children achieve mathematical excellence. Each grade has 3 levels. In level 1, your child learns all the basic math concepts necessary for success in his or her grade. Key concepts are accompanied by helpful three-part introductions: "Read This" explains the concept, "Example" demonstrates the concept, and "Try It" lets your child put the concept to use. In level 2, and to a greater extent in level 3, these concepts are worked into relatable problem-solving questions. These offer a greater challenge and point children to the every-day usefulness of math skills.

Fun activities, lively illustrations, and real-world scenarios throughout the book help bring the concepts to life and engage your child. Additionally, the QR codes in the book link to motion graphics that explain key ideas in a fun and active way. After your child has completed the core content, they will find two assessment tests. These will test your child's general ability to apply the concepts learned, and prepare them for standardized testing. Finally, your child can use the answer key in the back of the book to improve by comparing his or her results and methods.

With the help of these features, we hope to provide an enriching learning experience for your child. We would love to hear your feedback, and encourage you to share any stories of how ***Complete MathSmart*** has helped your child improve his or her math skills and gain confidence in the classroom.

Your Partner in Education,
Popular Book Company (Canada) Limited

LEVEL 1

BASIC SKILLS

1 Multiples

- understanding multiples

A multiple of a number is the product of that number and another whole number.

Example Write the first 5 multiples of 2.

multiplied by the first 5 whole numbers to get the first 5 multiples of 2

2 x 1 = 2

2 x 2 = 4

2 x 3 = 6

2 x 4 = 8

2 x 5 = 10

The first 5 multiples of 2:

2, 4, 6, 8, 10

3 x 1 = ______

3 x 2 = ______

3 x 3 = ______

3 x 4 = ______

3 x 5 = ______

The first 5 multiples of 3:

____

Write the first 5 multiples of each number.

① 4 x 1 = ______
4 x 2 = ______
4 x 3 = ______
4 x 4 = ______
4 x 5 = ______

Multiples of 4

② 5 x 1 = ______
5 x 2 = ______
5 x 3 = ______
5 x 4 = ______
5 x 5 = ______

Multiples of 5

③ 6 x 1 = ______
6 x 2 = ______
6 x 3 = ______
6 x 4 = ______
6 x 5 = ______

Multiples of 6

④ 7 x 1 = ______
7 x 2 = ______
7 x 3 = ______
7 x 4 = ______
7 x 5 = ______

Multiples of 7

⑤ 8 x 1 = ______
8 x 2 = ______
8 x 3 = ______
8 x 4 = ______
8 x 5 = ______

Multiples of 8

⑥ 9 x 1 = ______
9 x 2 = ______
9 x 3 = ______
9 x 4 = ______
9 x 5 = ______

Multiples of 9

Mark the numbers on the hundreds chart as specified. Then list the multiples and find the common multiples.

⑦

1	2	3	4	5	6	7	8	9	10
11	12	13	14	15	16	17	18	19	20
21	22	23	24	25	26	27	28	29	30
31	32	33	34	35	36	37	38	39	40
41	42	43	44	45	46	47	48	49	50
51	52	53	54	55	56	57	58	59	60
61	62	63	64	65	66	67	68	69	70
71	72	73	74	75	76	77	78	79	80
81	82	83	84	85	86	87	88	89	90
91	92	93	94	95	96	97	98	99	100

○ : multiples of 2

△ : multiples of 3

□ : multiples of 5

✕ : multiples of 7

a. List the first 10 multiples of each number.

- 2: ______
- 3: ______
- 5: ______
- 7: ______

b. List the common multiples of each pair of numbers.

- 2 and 3: ______
- 2 and 5: ______
- 3 and 5: ______
- 3 and 7: ______
- 5 and 7: ______

Hints

A common multiple is a multiple that two or more numbers share.

Cross out the numbers that are not multiples of each given number.

⑧

2

4	8	13
16	25	30
33	38	44
48	51	60

3

6	9	15
19	29	36
45	48	72
81	90	95

4

4	16	20
26	38	44
56	60	74
84	94	100

5

15	20	38
40	45	55
66	70	85
90	93	95

6

14	18	28
30	36	45
48	56	64
66	78	84

7

14	21	35
45	55	63
77	80	84
91	98	99

Look at each set of multiples. Circle and write the correct numbers.

⑨

a.	**15, 30, 45, 60**	Multiples of **2 , 3 , 5 , 6 , 10**
		Common multiples of ______ and ______
b.	**12, 24, 36, 48**	Multiples of **3 , 4 , 5 , 7 , 10**
		Common multiples of ______ and ______
c.	**20, 40, 60, 80**	Multiples of **3 , 4 , 5 , 6 , 9**
		Common multiples of ______ and ______
d.	**16, 24, 32, 40**	Multiples of **3 , 4 , 5 , 8 , 10**
		Common multiples of ______ and ______

Complete the Venn diagrams with the given numbers.

⑩

12 10 6 40 30

35 45 20 15 50

Multiples of 2

Multiples of 5

⑪

18 27 72 36 63

66 54 81 90 45

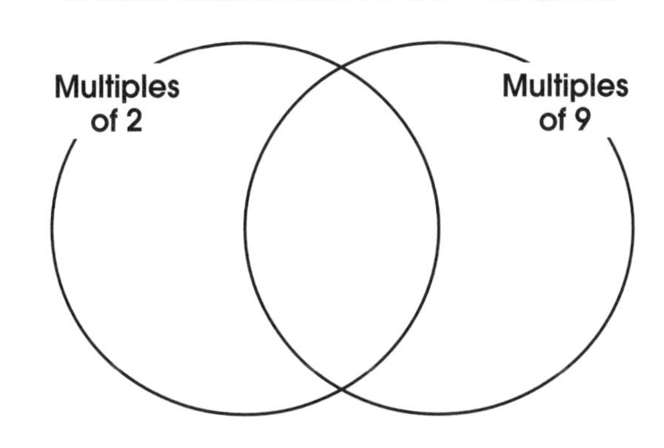

⑫

28 35 56 72 48

14 42 84 63 32

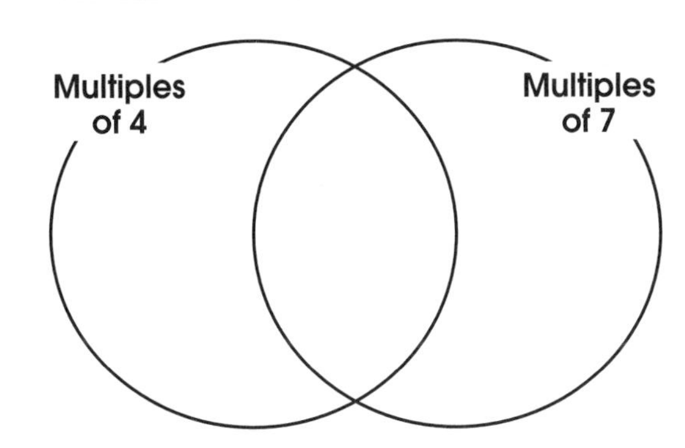

⑬

20 32 80 15 55

16 40 64 72 35

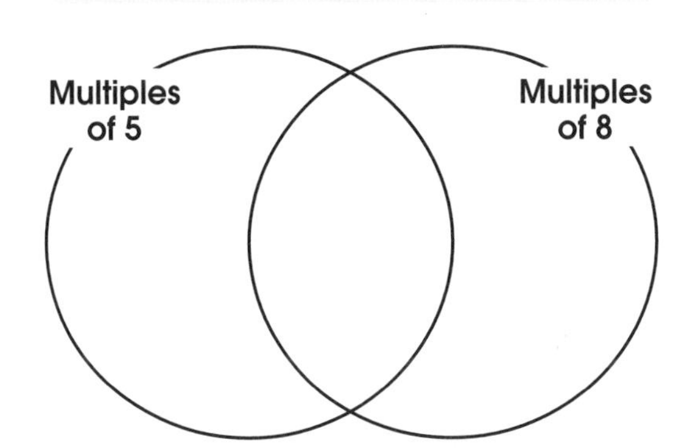

Circle "T" for the true statements and "F" for the false ones.

⑭ All multiples of even numbers are even. T / F

⑮ All multiples of odd numbers are odd. T / F

⑯ There is a limit to how many multiples a number has. T / F

⑰ If Number A is a multiple of Number B, then all the multiples of A are also multiples of B. T / F

⑱ The smallest multiple of a number is the number itself. T / F

⑲ The product of two numbers is always a common multiple of those two numbers. T / F

2 Factors

- understanding factors

Whole numbers that multiply to give a product are called the factors of the product.

Example Find the factors of 6.

List all the multiplication facts with a product of 6.

Factors		Product
1 x 6	=	**6**
2 x 3	=	**6**

← 1 and the number itself are always factors of the number.

Factors of 6: 1, 2, 3, 6

Find the factors of 4.

1 x ☐ = 4

☐ x ☐ = 4

Factors of 4:

____, ____, ____

Complete and list the multiplication sentences to find the factors.

① 1 x ____ = 3

Factors of 3:

1, ____

② ____ x 5 = 5

Factors of 5:

____, ____

③ ____ x ____ = 7

Factors of 7:

____, ____

④ ____ = 8

____ = 8

Factors of 8:

____

⑤ ____ = 10

____ = 10

Factors of 10:

____

⑥ ____ = 9

____ = 9

Factors of 9:

____

⑦ ____ = 12

____ = 12

____ = 12

Factors of 12:

____

⑧ ____ = 16

____ = 16

____ = 16

Factors of 16:

____

⑨ ____ = 20

____ = 20

____ = 20

Factors of 20:

____

List the multiplication facts and find the factors for each number.

⑩ 

Factors of 24: ____________

⑪ 

Factors of 27: ____________

When finding the factors of a number, remember to list the multiplication facts in order. Start with 1 and go up, testing each number to see if it is a factor. Then stop when a factor pair repeats.

e.g. $1 \times 10 = 10$

$2 \times 5 = 10$ ← no factors between 2 and 5

Factors of 10: 1, 2, 5, 10

⑫ 

Factors of 30: ____________

⑬ 32

Factors of 32: ____________

Cross out the numbers that are not factors of each given number.

⑭ Factors of 36

2	9	12	1
8	7	4	3
11	36	6	18

⑮ Factors of 18

3	1	5	18
12	9	2	8
6	7	16	4

⑯ Factors of 42

6	2	42	4
10	1	5	7
21	8	3	14

Draw all the possible rectangles with each given area. Then list the factors.

⑰ **14 square units**

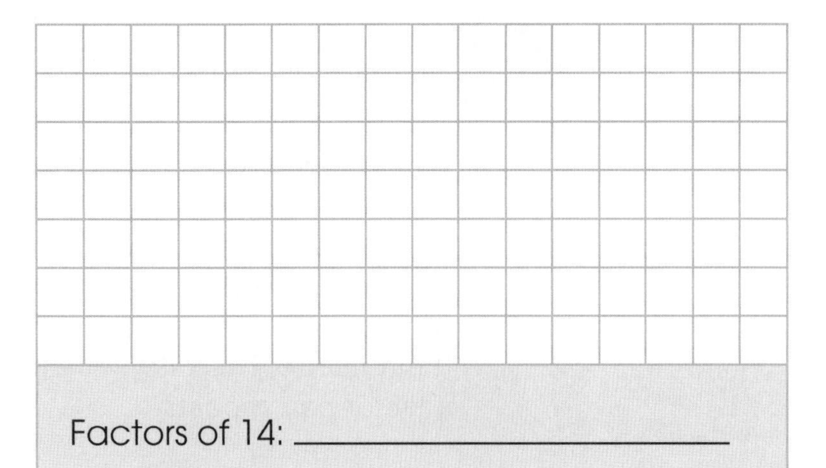

Tips

Factors of a number can be found by drawing rectangles. The number is the area of the rectangle and the factors are the dimensions.

⑱ **15 square units**

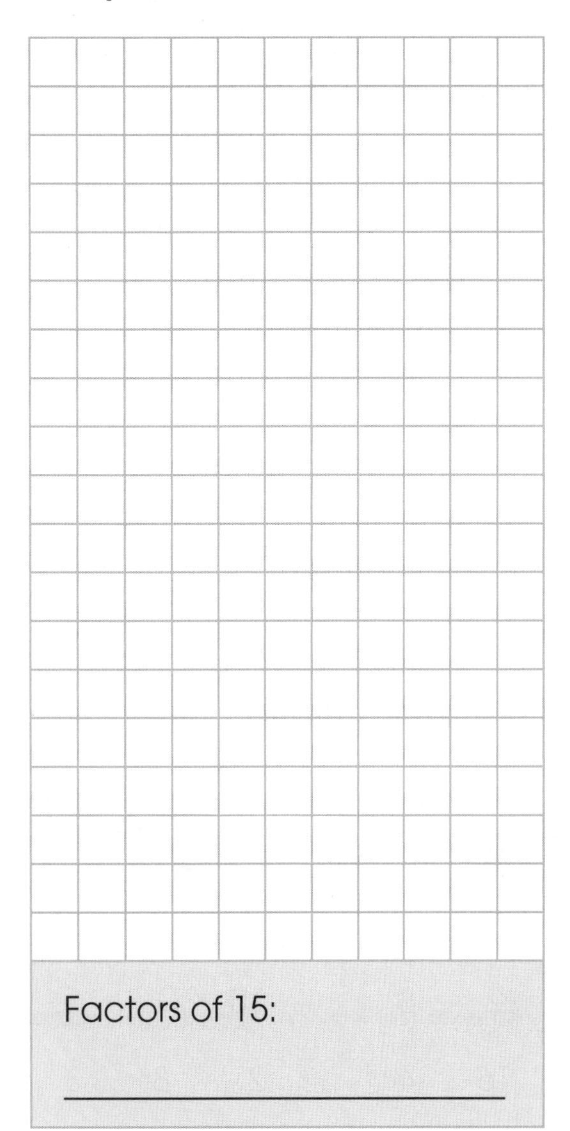

⑲ **16 square units**

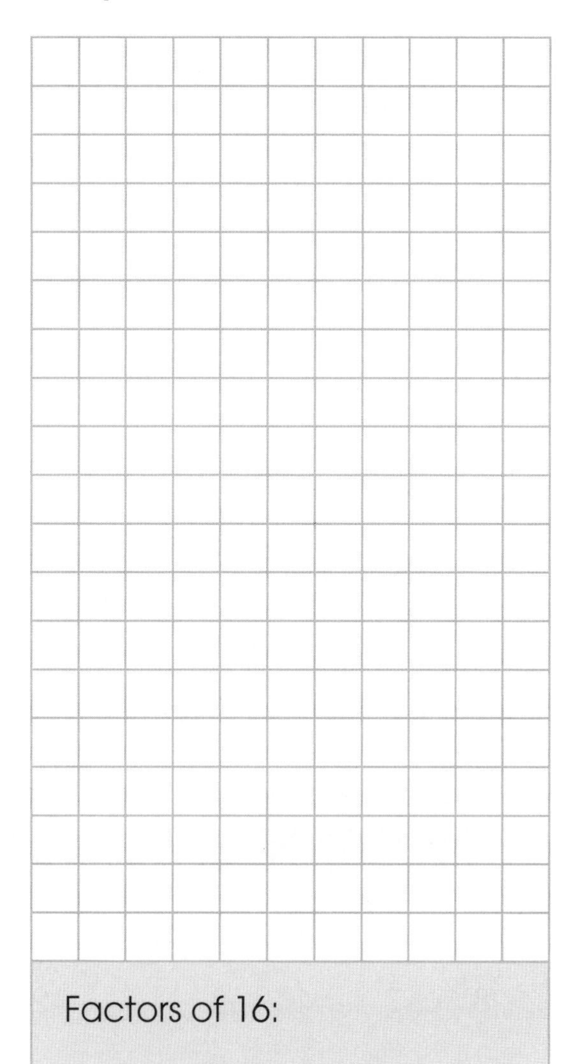

Write the factors for each number. Circle the common factors. Then list them.

⑳ **21**: ____________________

35: ____________________

Common factors: __________

㉑ **42**: ____________________

60: ____________________

Common factors: __________

A common factor is a factor that two or more numbers share.

e.g. Factors of 6: ①, 2, ③, 6

Factors of 9: ①, ③, 9

1 and 3 are the common factors of 6 and 9.

㉒ **32**: ____________________

40: ____________________

Common factors: __________

㉓ **27**: ____________________

63: ____________________

Common factors: __________

Complete the Venn diagrams. List the common factors.

㉔

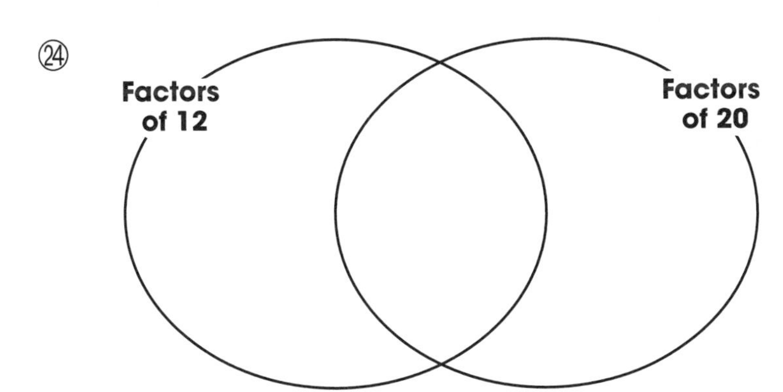

Common factors of 12 and 20:

____

㉕

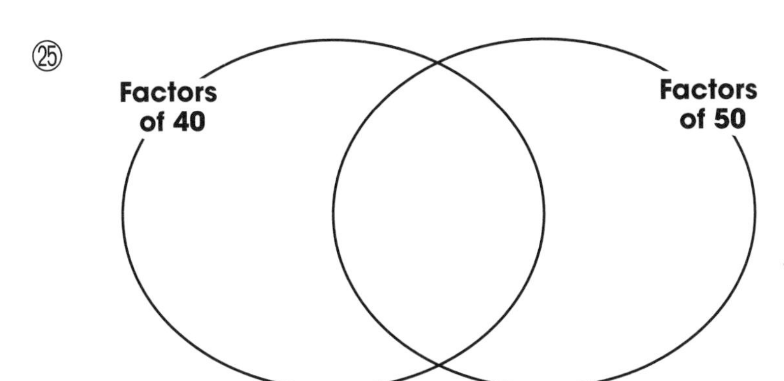

Common factors of 40 and 50:

____

3 Exponents

- understanding exponents

An exponent tells how many times a number is multiplied by itself. A number that is in exponential notation is also called a power.

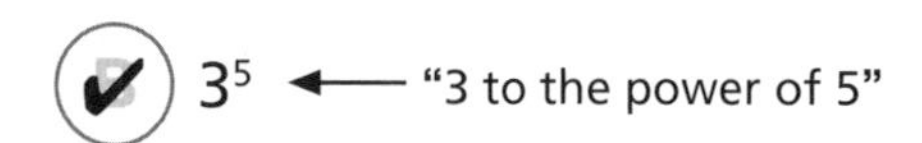

e.g. $2 \times 2 \times 2 = 2^3$ ← exponent (2 multiplied by itself 3 times)
3 twos — base

Example Check the correct exponential form.

$3 \times 3 \times 3 \times 3 \times 3$ ← 3 multiplied 5 times

(A) 5^3 (✔) 3^5 ← "3 to the power of 5"

$4 \times 4 \times 4$

(A) 4^3

(B) 3^4

Check the correct exponential form.

① $2 \times 2 \times 2 \times 2$
(A) 4^2
(B) 2^4

② 3×3
(A) 2^3
(B) 3^2

③ $4 \times 4 \times 4 \times 4$
(A) 4^4
(B) 4^2

④ $7 \times 7 \times 7$
(A) 7^3
(B) 3^7

⑤ $9 \times 9 \times 9 \times 9$
(A) 4^9
(B) 9^4

⑥ $5 \times 5 \times 5 \times 5 \times 5$
(A) 5^6
(B) 5^5

⑦ 3 to the power of 6
(A) 3^6
(B) 6^3

⑧ 8 to the power of 4
(A) 8^4
(B) 4^8

⑨ 6 to the power of 7
(A) 6^7
(B) 7^6

Match.

⑩

$3 \times 3 \times 3 \times 3$ •	• 3×4
$4 \times 4 \times 4$ •	• 5^3
$3 + 3 + 3 + 3$ •	• 3^4
$5 \times 5 \times 5$ •	• 5×3
$5 + 5 + 5$ •	• 4^3
$3 \times 3 \times 3 \times 3 \times 3$ •	• 3^5
$4 + 4 + 4 + 4 + 4$ •	• 4×5

Write each multiplication sentence in exponential form. Identify the base and the exponent.

⑪ 8 x 8 x 8

Base: ______

Exponent: ______

⑫ 9 x 9 x 9 x 9

Base: ______

Exponent: ______

⑬ 2 x 2 x 2 x 2 x 2

Base: ______

Exponent: ______

⑭ 4 x 4 x 4 x 4 x 4

Base: ______

Exponent: ______

⑮ 3 x 3 x 3 x 3

Base: ______

Exponent: ______

⑯ 6 x 6 x 6

Base: ______

Exponent: ______

The base is the number that is repeatedly multiplied in a power; the exponent is the number of times it is multiplied.

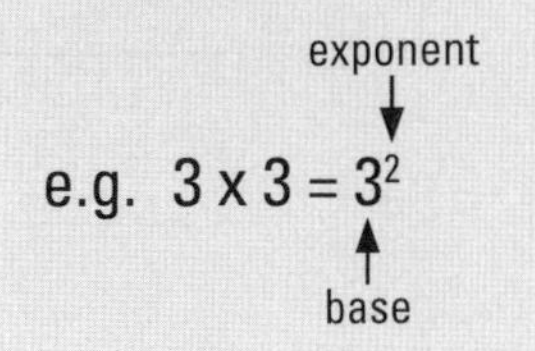

Evaluate.

⑰ 2^3

= 2 x ______ x ______

= ______

⑱ 3^3

= ______

= ______

⑲ 5^3

= ______

= ______

⑳ 6^3

= ______

= ______

㉑ 4^4

= ______

= ______

㉒ 2^5

= ______

= ______

㉓ 4^3

= ______

= ______

㉔ 5^4

= ______

= ______

㉕ 8^2

= ______

= ______

㉖ 1^6

= ______

= ______

㉗ 7^3

= ______

= ______

㉘ 12^2

= ______

= ______

Find the values.

㉙ 4^2 = ______

㉚ 2^4 = ______

㉛ 5^0 = ______

㉜ 6^1 = ______

㉝ 7^2 = ______

㉞ 3^4 = ______

㉟ 10^4 = ______

㊱ 16^1 = ______

㊲ 20^0 = ______

㊳ 8^3 = ______

㊴ 45^1 = ______

㊵ 2^6 = ______

㊶ 119^1 = ______

㊷ 9^2 = ______

㊸ 37^1 = ______

㊹ 89^0 = ______

㊺ 533^1 = ______

㊻ 741^0 = ______

Hints

A number raised to the power of 0 is always 1.

e.g. $7^0 = 1$

A number raised to the power of 1 is always the number itself.

e.g. $8^1 = 8$

Match.

㊼

6^3 •	• 256
4^2 •	• 1
2^8 •	• 216
6^2 •	• 16
6^0 •	• 36
26^1 •	• 26

㊽

11^2 •	• 81
3^4 •	• 361
11^1 •	• 121
121^0 •	• 21
19^2 •	• 11
21^1 •	• 1

㊾

2^6 •	
12^2 •	• 1
8^2 •	• 64
144^1 •	• 144
28^0 •	
1^8 •	

㊿

2^3 •	
8^1 •	• 8
2^7 •	• 128
8^3 •	• 512
2^9 •	
128^1 •	

Compare each pair and write ">", "<", or "=".

51 **Same Base, Different Exponents**

a. 4^3 ◯ 4^5

b. 2^4 ◯ 2^6

c. 3^1 ◯ 3^3

d. 6^5 ◯ 6^3

52 **Different Bases, Same Exponent**

a. 3^2 ◯ 4^2

b. 5^3 ◯ 3^3

c. 2^5 ◯ 3^5

d. 3^4 ◯ 1^4

53 **Different Bases and Different Exponents**

a. 2^2 ◯ 4^1

b. 2^3 ◯ 3^2

c. 6^2 ◯ 2^6

d. 3^3 ◯ 1^9

e. 3^2 ◯ 9^1

f. 7^3 ◯ 3^7

Tips

For powers that have the same base and different exponents, the one with the greater exponent is greater.

e.g. $4^3 > 4^2$

For powers that have different bases and the same exponent, the one with the greater base is greater.

e.g. $5^3 < 7^3$

Circle the correct answer in each group.

54 the greatest power

a. 4^3 6^3 2^3

b. 9^2 9^1 9^5

c. 5^3 5^5 4^5

55 the smallest power

a. 7^3 7^0 7^2

b. 8^2 2^2 5^2

c. 3^4 7^5 3^5

Put the powers in order from smallest to greatest.

56 2^3 2^2 2^1 2^4

____

57 3^3 3^1 3^9 3^0

____

58 4^3 7^3 1^3 8^3

____

59 5^2 8^2 0^2 1^2

____

60 5^2 4^3 5^3 4^2

____

61 9^8 8^9 8^8 9^9

____

4 Squares and Square Roots

- understanding squares and square roots

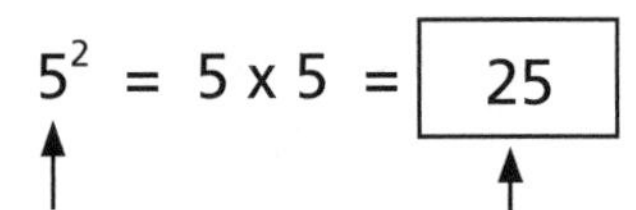

A square number is the product of a number multiplied by itself. The square root of a number is a value that, when multiplied by itself, gives the number.

Example Find the square of 5.

$5^2 = 5 \times 5 = \boxed{25}$

the square of 5 — a square number

Try It

Find the square root of 25.

$\sqrt{25} = \sqrt{5 \times 5}$

$= \boxed{}$

Find the first 20 square numbers. Then circle the correct answers.

Tips

The digit in the ones place of a number and the digit in the ones place of its square are related.

e.g. "9" in the ones place → "1" in the ones place

9 squared 81
19 → 361
29 → 841

① The squares of the first 20 whole numbers:

1^2 = ______ 11^2 = ______

2^2 = ______ 12^2 = ______

3^2 = ______ 13^2 = ______

4^2 = ______ 14^2 = ______

5^2 = ______ 15^2 = ______

6^2 = ______ 16^2 = ______

7^2 = ______ 17^2 = ______

8^2 = ______ 18^2 = ______

9^2 = ______ 19^2 = ______

10^2 = ______ 20^2 = ______

② A number ending in

a. 0 has a square ending in **0 / 5** .

b. 1 or 9 has a square ending in **1 / 6** .

c. 2 or 8 has a square ending in **4 / 8** .

d. 5 has a square ending in **0 / 5** .

e. 4 or 6 has a square ending in **4 / 6** .

f. 3 or 7 has square ending in **6 / 9** .

③ The ones digit of a number is 4, so the ones digit of its square **can / cannot** be 8.

④ The ones digit of a number is 9, so the ones digit of its square root **can / cannot** be 5.

SCAN IT!

Fill in the missing digits.

⑤ $15^2 = 22\_\_$ ⑥ $26^2 = 67\_\_$ ⑦ $43^2 = 184\_\_$

⑧ $30^2 = 90\_\_$ ⑨ $21^2 = 44\_\_$ ⑩ $28^2 = 78\_\_$

⑪ $12^2 = 14\_\_$ ⑫ $35^2 = 122\_\_$ ⑬ $49^2 = 240\_\_$

⑭ $10^2 = 10\_\_$ ⑮ $16^2 = 25\_\_$ ⑯ $14^2 = 19\_\_$

Match without doing any calculations.

⑰
- 24^2 • • 256
- 27^2 • • 576
- 16^2 • • 1089
- 33^2 • • 729

⑱
- 18^2 • • 841
- 21^2 • • 324
- 29^2 • • 1024
- 32^2 • • 441

⑲
- 35^2 • • 1296
- 36^2 • • 1225
- 45^2 • • 2116
- 46^2 • • 2025

⑳
- 23^2 • • 1936
- 37^2 • • 529
- 44^2 • • 1369
- 56^2 • • 3136

Tips

Look at the ones digits first. Then compare to match.

Find the areas of the squares using the given side lengths.

㉑

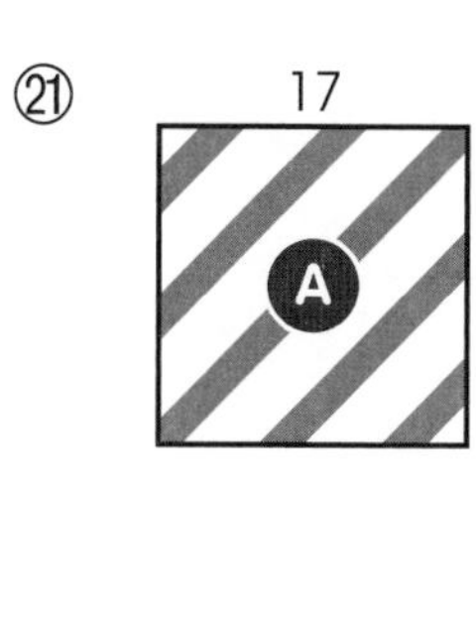

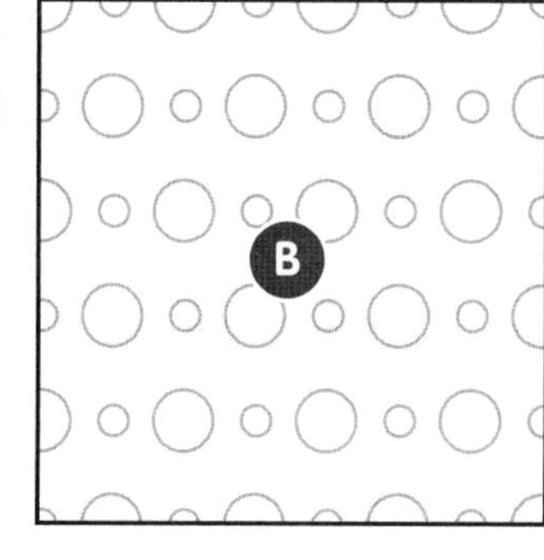

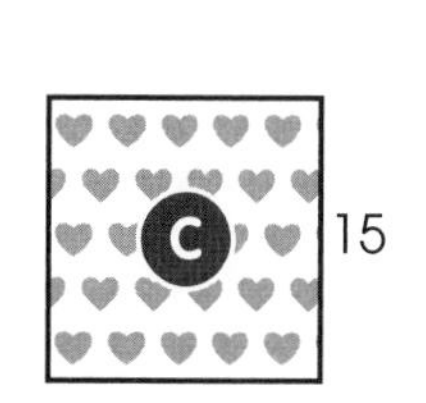

13
D

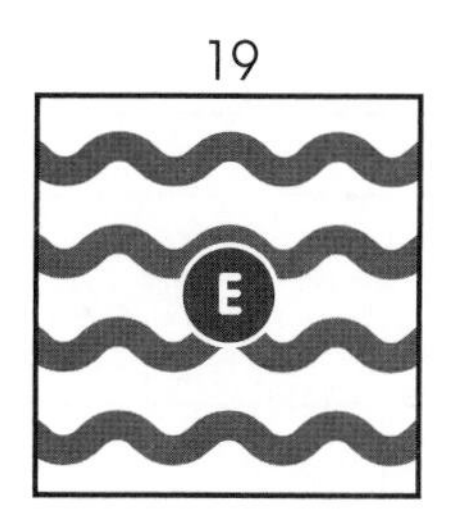

Area
(square units)

A $17^2 = \_\_\_\_\_\_$

B ______

C ______

D ______

E ______

Find the first 20 square roots.

㉒ **First 20 Square Roots**

$\sqrt{1}$ = ______	$\sqrt{36}$ = ______	$\sqrt{121}$ = ______	$\sqrt{256}$ = ______
$\sqrt{4}$ = ______	$\sqrt{49}$ = ______	$\sqrt{144}$ = ______	$\sqrt{289}$ = ______
$\sqrt{9}$ = ______	$\sqrt{64}$ = ______	$\sqrt{169}$ = ______	$\sqrt{324}$ = ______
$\sqrt{16}$ = ______	$\sqrt{81}$ = ______	$\sqrt{196}$ = ______	$\sqrt{361}$ = ______
$\sqrt{25}$ = ______	$\sqrt{100}$ = ______	$\sqrt{225}$ = ______	$\sqrt{400}$ = ______

Find the side lengths of the squares using the given areas.

㉓

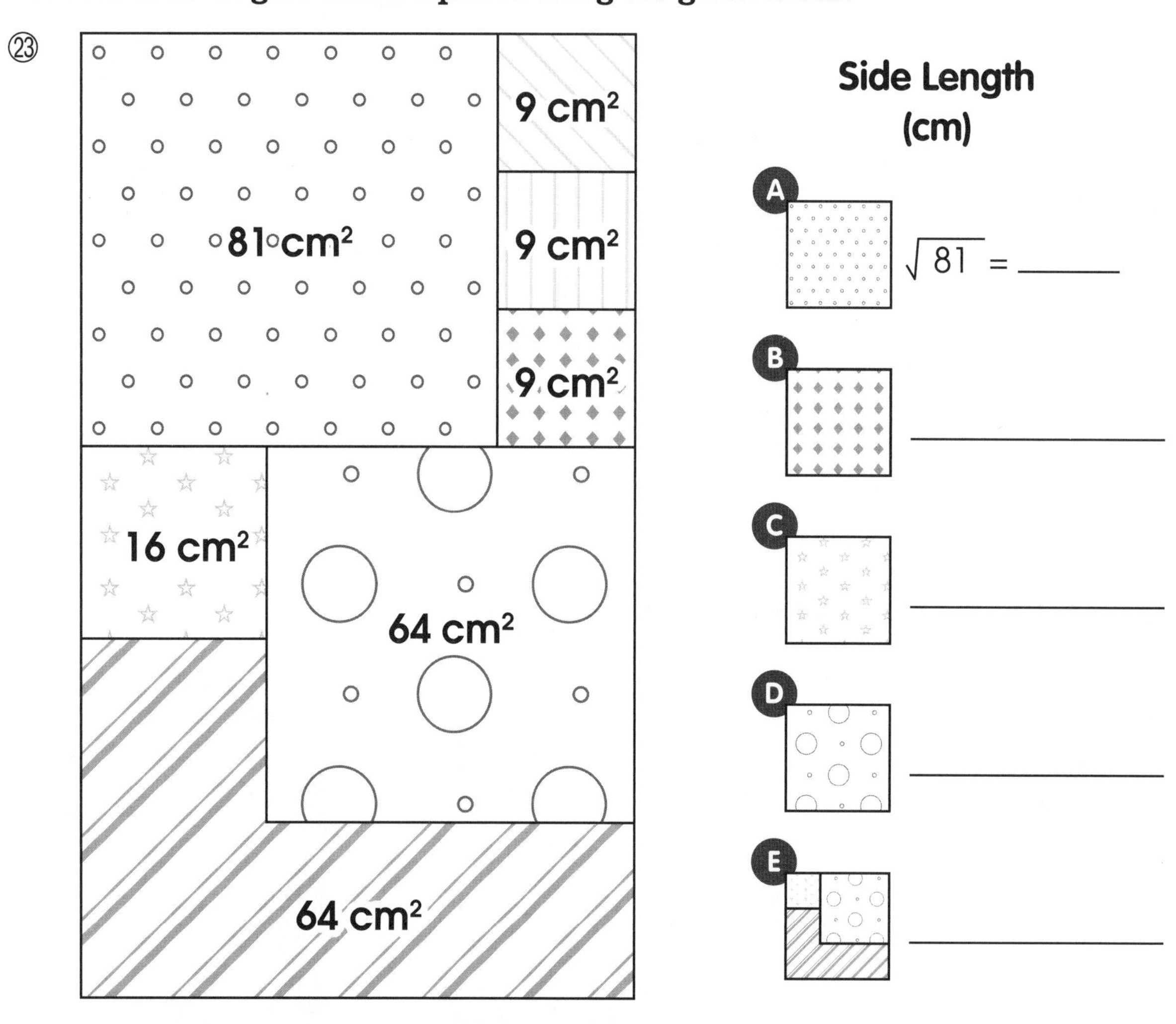

Side Length (cm)

A $\sqrt{81}$ = ______

B ______

C ______

D ______

E ______

㉔

Area of a Square (cm²)	25	49	121	144	225
Side Length (cm)					

Circle the perfect squares.

㉕ **Perfect Squares**

9	18	30	36	45	54
66	75	81	100	120	130
144	169	175	196	200	225
250	275	289	303	324	396
400	420	425	450	529	625

A perfect square has a whole number as its square root.

e.g. $\sqrt{144} = 12$ ← a whole number

$\sqrt{10} = 3.162...$

144 is a perfect square while 10 is not.

Locate each set of square roots on the number line. Then fill in the blanks with whole numbers.

㉖

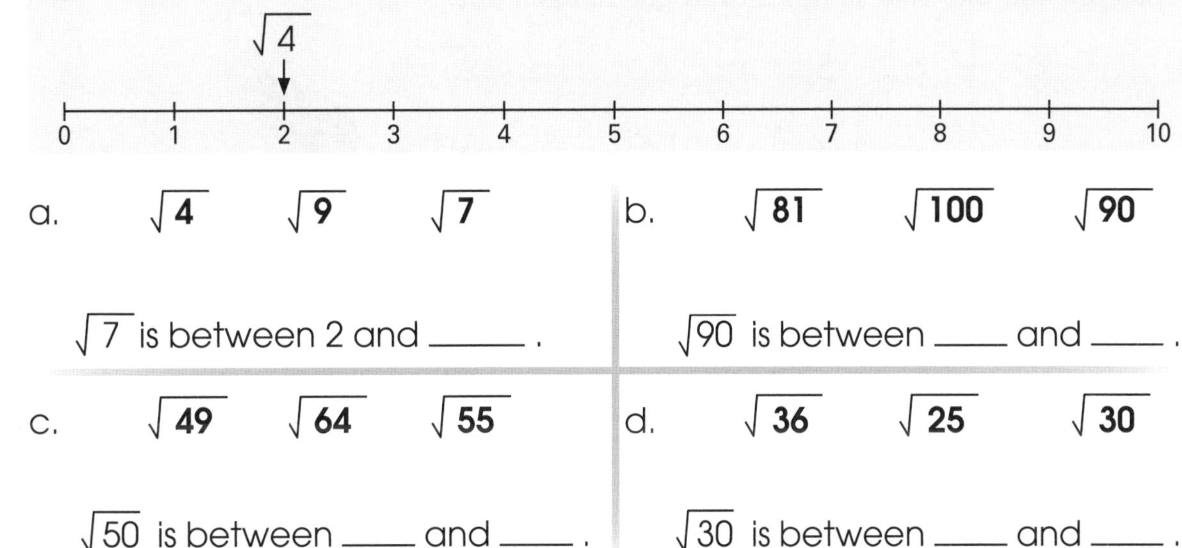

a. $\sqrt{4}$ $\sqrt{9}$ $\sqrt{7}$

$\sqrt{7}$ is between 2 and ____ .

b. $\sqrt{81}$ $\sqrt{100}$ $\sqrt{90}$

$\sqrt{90}$ is between ____ and ____ .

c. $\sqrt{49}$ $\sqrt{64}$ $\sqrt{55}$

$\sqrt{50}$ is between ____ and ____ .

d. $\sqrt{36}$ $\sqrt{25}$ $\sqrt{30}$

$\sqrt{30}$ is between ____ and ____ .

㉗

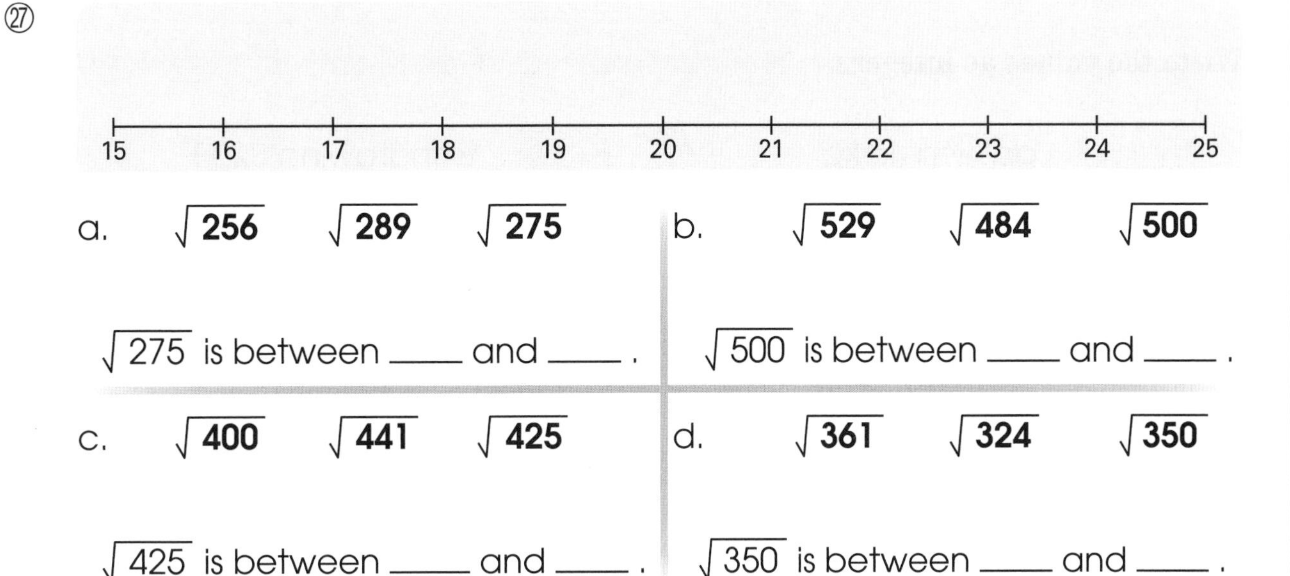

a. $\sqrt{256}$ $\sqrt{289}$ $\sqrt{275}$

$\sqrt{275}$ is between ____ and ____ .

b. $\sqrt{529}$ $\sqrt{484}$ $\sqrt{500}$

$\sqrt{500}$ is between ____ and ____ .

c. $\sqrt{400}$ $\sqrt{441}$ $\sqrt{425}$

$\sqrt{425}$ is between ____ and ____ .

d. $\sqrt{361}$ $\sqrt{324}$ $\sqrt{350}$

$\sqrt{350}$ is between ____ and ____ .

5 Integers

- understanding and comparing integers

Integers are all the whole numbers, their opposites, and zero. For example, 5 and -5 are opposite numbers. 5, -5, and 0 are integers.

Example Cross out the ones that are not integers.

+8 ~~1.5~~ -2 ~~$3\frac{4}{5}$~~ 0 12

- 1.5 ↑ a decimal, not a whole number
- $3\frac{4}{5}$ ↑ a fraction, not a whole number
- 12 ↑ The "+" sign can be omitted for positive numbers.

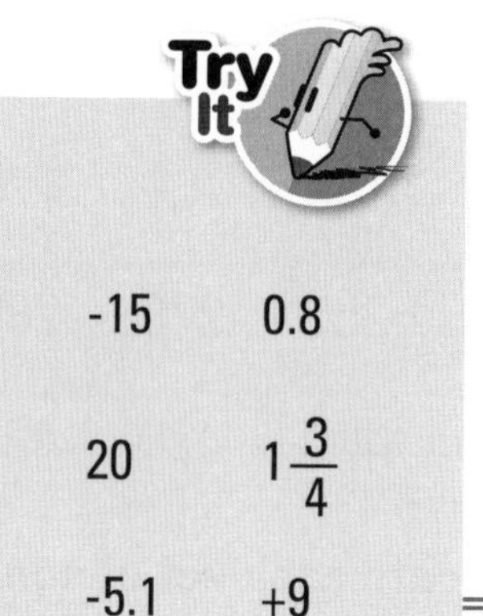

-15	0.8
20	$1\frac{3}{4}$
-5.1	+9

Cross out the one that is not an integer in each set.

①

a.	b.	c.	d.
0.9	-15	0	$\sqrt{25}$
+7	0.04	-120	-3.3
-12	1	30%	4

e.	f.	g.	h.
10^2	15%	$\sqrt{100}$	$-\frac{16}{8}$
$-3\frac{3}{4}$	-5	-2.5	500%
5	-11	-9	$\frac{22}{5}$

Evaluate to identify whether a number is an integer

e.g. $\sqrt{4}$ ← equals 2, an integer

Write the values as integers.

② ______ gaining 45 kg

③ ______ withdrawing $200

④ ______ 12° below 0°C

⑤ ______ 140 m above sea level

⑥ ______ depositing $35

⑦ ______ losing 2 kg

⑧ ______ cooling by 8°C

⑨ ______ 8 seconds longer than before

⑩ ______ climbing up 25 m

⑪ ______ 5 minutes before launch

Locate each integer on the number line. Then locate and write its opposite.

Opposite

⑫ 

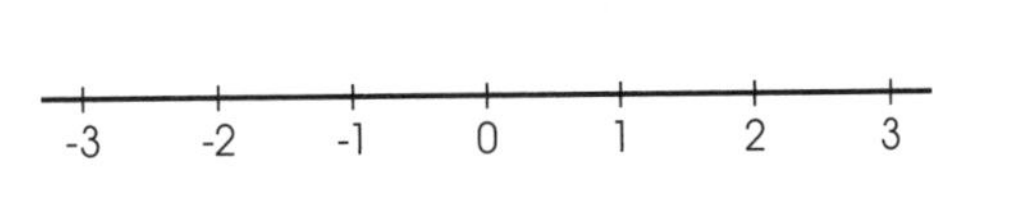

______

⑬ 

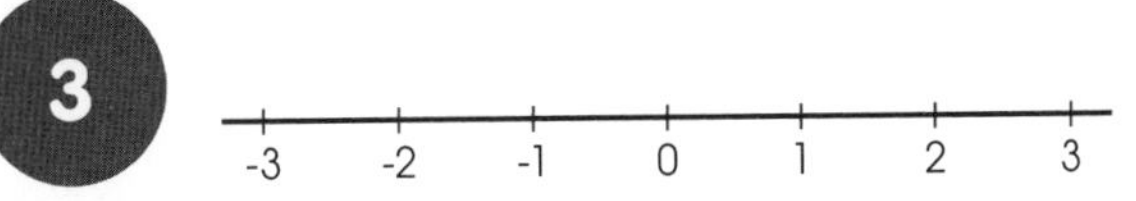

______

⑭ -5

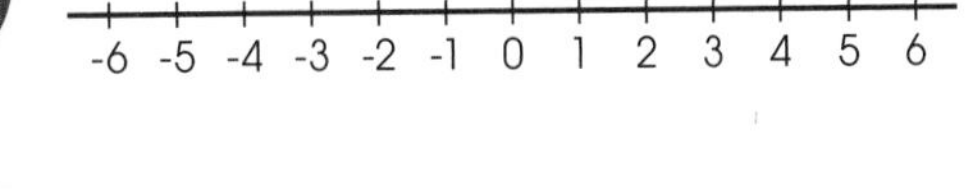

______

⑮ 

-6 -5 -4 -3 -2 -1 0 1 2 3 4 5 6

______

Hints

An integer has the same distance from 0 as its opposite.

e.g.

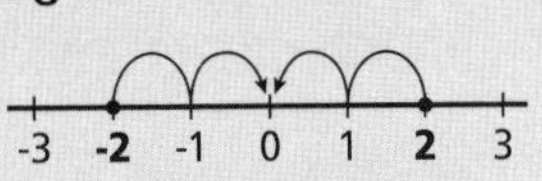

-2 and 2 are both 2 units from 0.

Locate the integers in each pair on the number line. Compare them and write "<" or ">" in the circle.

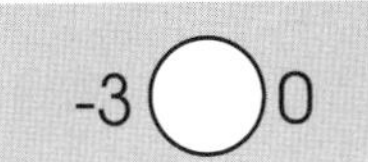

-4 -3 -2 -1 0 1 2

⑰ 

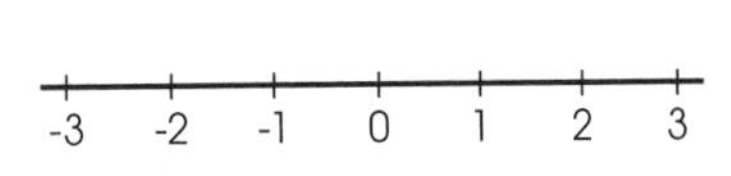

⑱ 

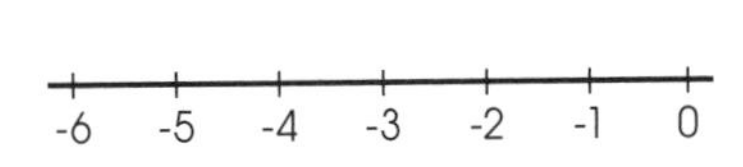

⑲ 

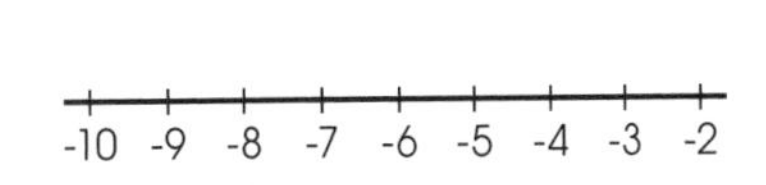

Tips

A number to the right on a number line is greater than one to the left.

e.g.

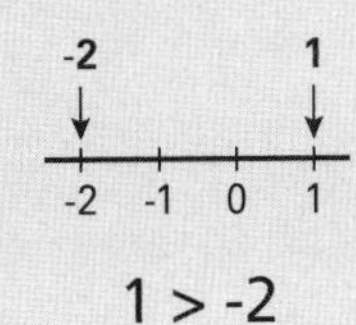

1 > -2

⑳

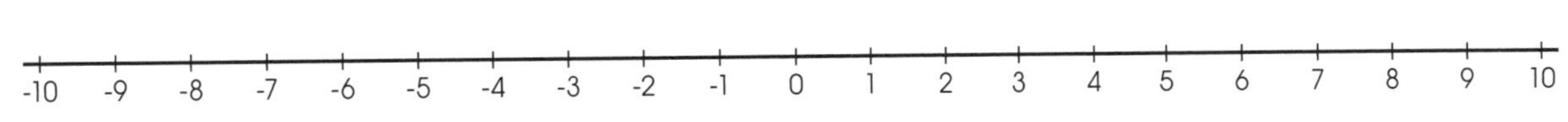

a. -6 ◯ 3　　b. 2 ◯ -4　　c. 3 ◯ 8

d. -10 ◯ -6　　e. 4 ◯ -3　　f. -9 ◯ 10

g. -6 ◯ -7　　h. -8 ◯ 5　　i. -1 ◯ 0

Locate the integers on the number lines. Then list the integers in the specified order.

㉑
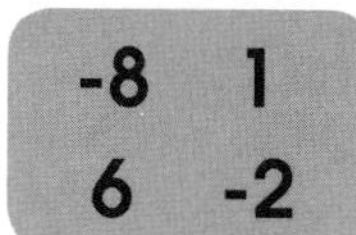

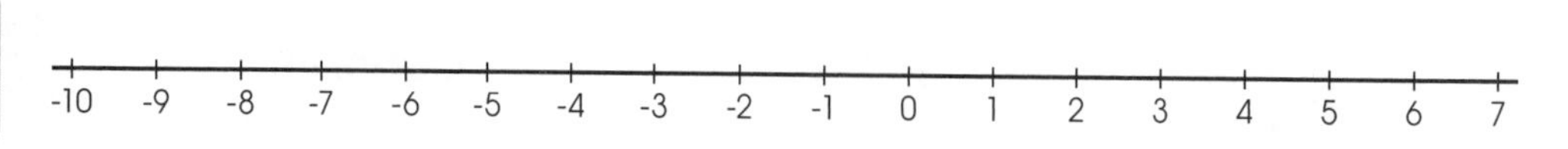

From smallest to greatest: ______________________

㉒
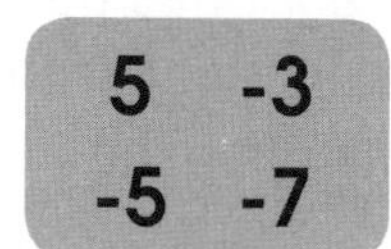

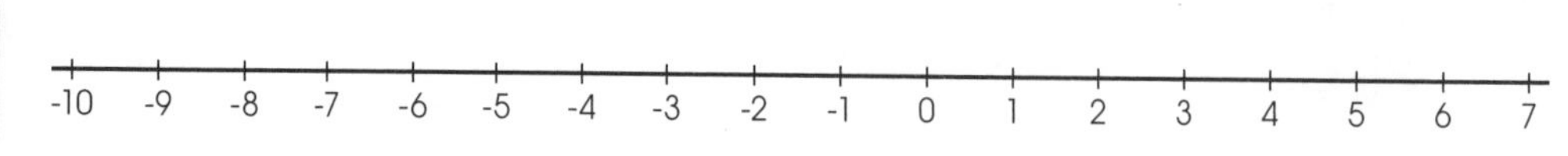

From greatest to smallest: ______________________

㉓
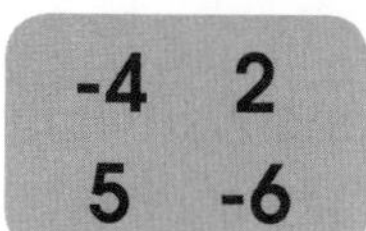

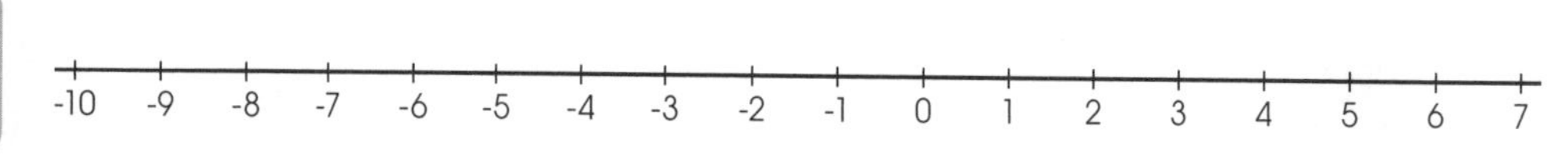

From smallest to greatest: ______________________

㉔
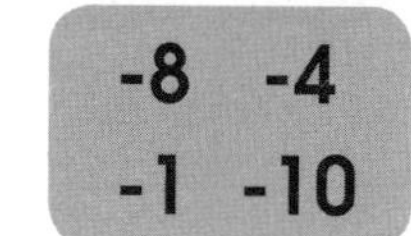

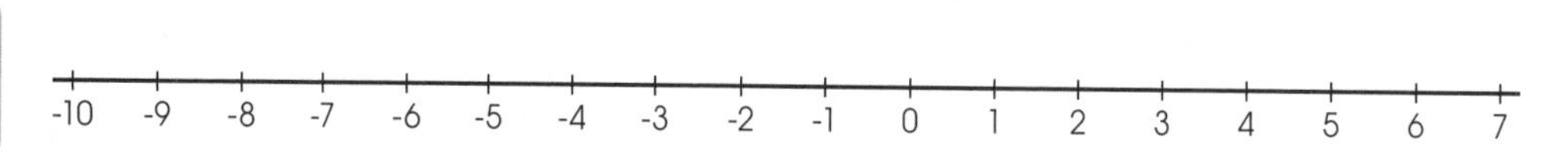

From greatest to smallest: ______________________

㉕ -3 -6 0 4

_____ < _____ < _____ < _____

㉖ 5 -4 -8 -7

_____ > _____ > _____ > _____

㉗ 9 -7 -2 3

_____ > _____ > _____ > _____

㉘ -6 -1 -5 2

_____ < _____ < _____ < _____

Circle "T" for the true statements and "F" for the false ones.

㉙ A positive integer is always greater than a negative integer. **T / F**

㉚ The difference between an integer and its opposite can be an odd or even number. **T / F**

㉛ 0 is the only integer that is neither positive nor negative. **T / F**

㉜ -1 is the smallest integer. **T / F**

㉝ -1 is the greatest negative integer. **T / F**

Look at each description. Answer the questions with the given numbers.

㉞

-10.2	-1
-6	-11
-0.5	-8
-2	5
0	15
-9	2
$3\frac{1}{2}$	9.2

a. the opposite of 6 ______

b. the opposite of -5 ______

c. integers greater than -2 ______________

d. numbers greater than -1 ______________

e. integers less than -5 ______________

f. numbers less than -9 ______________

g. integers between -3 and 2 ______________

Read and record the temperatures. Then fill in the blanks.

㉟

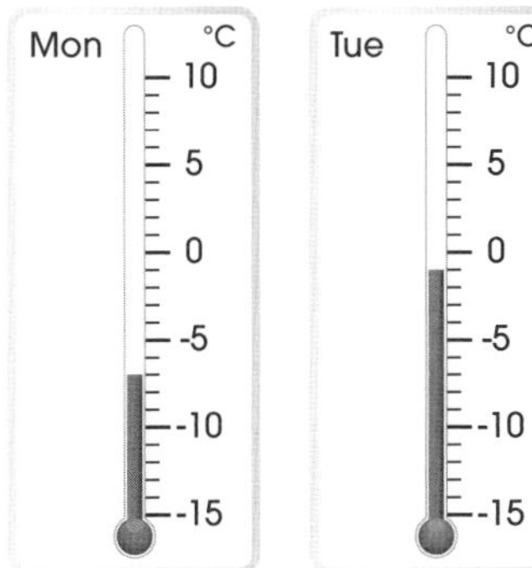

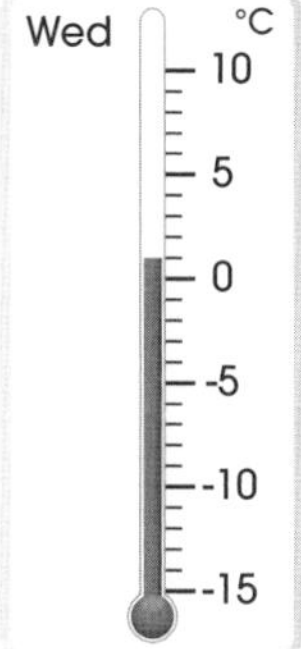

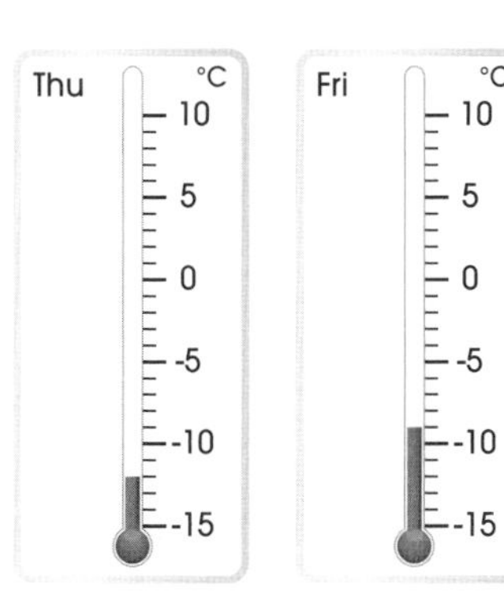

Mon: ________ °C

Tue: ________

Wed: ________

Thu: ________

Fri: ________

a. It was the hottest on ____________ .

b. It snowed when it was below -10°C on ____________ .

㊱

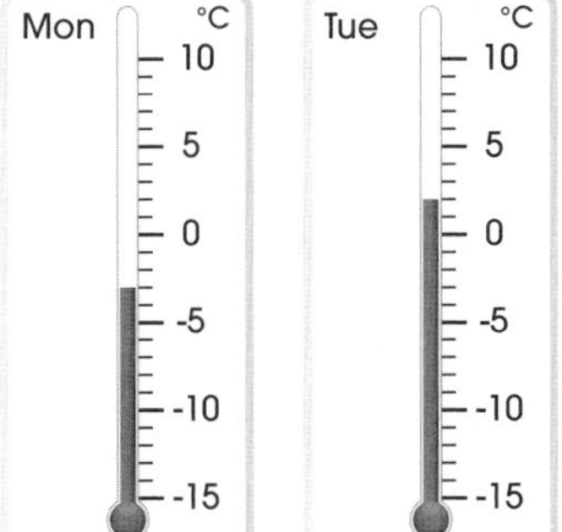

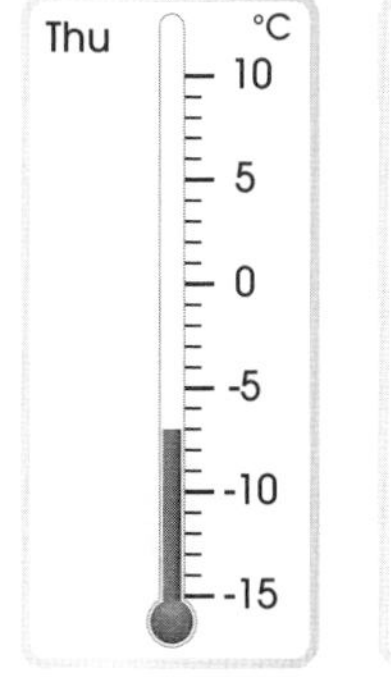

Mon: ________ °C

Tue: ________

Wed: ________

Thu: ________

Fri: ________

a. It was the coldest on ____________ .

b. Karen went skiing on the day that had a temperature between -3°C and -7°C. She went skiing on ____________ .

6 Fractions

- adding, subtracting, multiplying, and dividing fractions

To add or subtract fractions with different denominators, first rewrite the fractions as equivalent fractions with the same denominator.

Example Find the answer.

$\frac{1}{2} + \frac{3}{4}$ ← 2 and 4 have a common multiple of 4.

$= \frac{2}{4} + \frac{3}{4}$ ← Rewrite to make both fractions have a common denominator of 4.

$= \frac{5}{4}$ ← Add the numerators and keep the denominator the same.

$= 1\frac{1}{4}$ ← Write in simplest form.

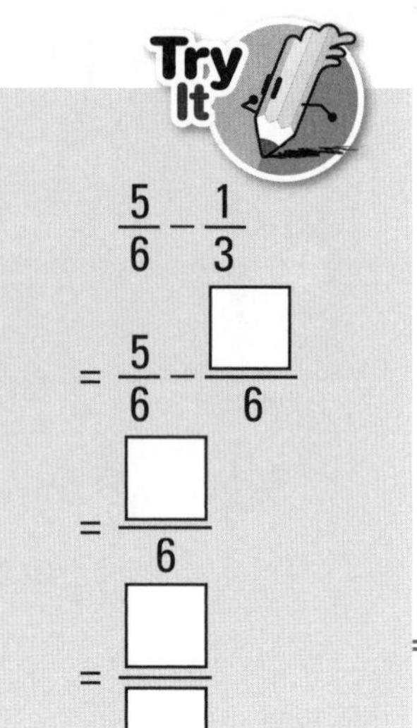

$\frac{5}{6} - \frac{1}{3}$

$= \frac{5}{6} - \frac{\square}{6}$

$= \frac{\square}{6}$

$= \frac{\square}{\square}$

Find the answers and write them in simplest form.

① $\frac{5}{8} + \frac{1}{16}$

② $\frac{7}{12} + \frac{7}{24}$

③ $\frac{3}{10} + \frac{4}{5}$

④ $\frac{3}{5} + \frac{6}{7}$

⑤ $1\frac{4}{9} + 8\frac{1}{18}$

⑥ $7\frac{4}{5} + 2\frac{2}{3}$

⑦ $\frac{1}{2} - \frac{1}{10}$

⑧ $\frac{14}{15} - \frac{1}{3}$

⑨ $3 - 1\frac{1}{3}$

⑩ $\frac{5}{6} - \frac{5}{9}$

⑪ $5\frac{1}{6} - 3\frac{2}{3}$

⑫ $6\frac{1}{2} - 5\frac{1}{4}$

Multiply. Write the answers in simplest form.

⑬ $\frac{2}{5} \times \frac{3}{4}$

⑭ $\frac{1}{3} \times \frac{5}{6}$

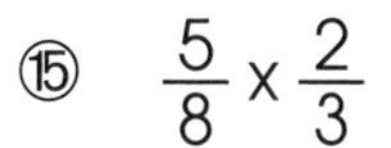

⑮ $\frac{5}{8} \times \frac{2}{3}$

⑯ $2\frac{1}{3} \times \frac{2}{14}$

⑰ $5\frac{1}{2} \times \frac{4}{11}$

⑱ $2\frac{5}{6} \times 7\frac{1}{2}$

⑲ $4 \times 3\frac{5}{8}$

⑳ $3\frac{3}{7} \times 14$

Hints

Follow these steps to multiply fractions.

1. Change any mixed numbers to improper fractions.
2. Reduce to make the calculation easier.
3. Multiply the numerators and denominators separately.
4. Reduce to simplest form.

e.g. $1\frac{1}{3} \times 2\frac{1}{2}$

$= \frac{^{2}\not{4}}{3} \times \frac{5}{\not{2}_{1}}$

$= \frac{10}{3}$ ← $2 \times 5 = 10$; ← $3 \times 1 = 3$

$= 3\frac{1}{3}$

Look at the recipe and solve the problems.

㉑

Egg Salad Sandwich

Time needed: $\frac{7}{12}$ hour

$\frac{2}{3}$ cartons of eggs

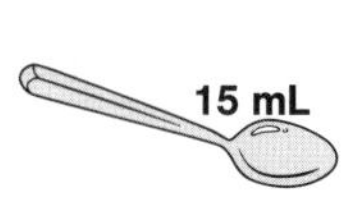

$1\frac{1}{2}$ tablespoons of mayonnaise

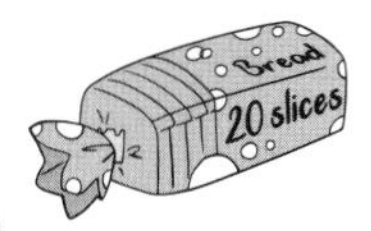

$\frac{2}{5}$ bag of bread

a. Time (min):

______ x ______ = ______

b. Number of eggs:

____

c. Amount of mayonnaise (mL):

____

d. Slices of bread:

____

Find the reciprocal of each fraction.

㉒ $\frac{3}{4}$ ______

㉓ $\frac{5}{2}$ ______

㉔ $\frac{6}{7}$ ______

㉕ $1\frac{2}{3}$ ______

㉖ $2\frac{4}{5}$ ______

㉗ $1\frac{7}{10}$ ______

㉘ $1\frac{5}{7}$ ______

㉙ $3\frac{1}{2}$ ______

To find the reciprocal of a fraction, swap its numerator and denominator. If it is a mixed number, rewrite it as an improper fraction before swapping to find the reciprocal.

e.g. $1\frac{2}{3} = \frac{5}{3} \xrightarrow{\text{swap}} \frac{3}{5}$

improper fraction

$\frac{3}{5}$ is the reciprocal of $1\frac{2}{3}$

Divide. Write the answers in simplest form.

㉚ $\frac{2}{3} \div \frac{1}{15}$

㉛ $\frac{5}{6} \div \frac{3}{12}$

㉜ $20 \div \frac{10}{19}$

㉝ $\frac{5}{14} \div \frac{5}{6}$

㉞ $\frac{3}{4} \div 4\frac{1}{2}$

㉟ $8 \div 3\frac{2}{3}$

㊱ $6\frac{1}{4} \div 1\frac{1}{2}$

㊲ $2\frac{1}{5} \div \frac{1}{10}$

To divide any number by a fraction, multiply the number by the reciprocal of the fraction.

e.g. $\frac{2}{3} \div \frac{4}{5}$

$= \frac{2}{3} \times \frac{5}{4}$ ← reciprocal of $\frac{4}{5}$

$= \frac{5}{6}$

㊳ $\frac{3}{5} \div \frac{3}{4}$ = ______

㊴ $\frac{1}{2} \div \frac{7}{8}$ = ______

㊵ $7\frac{1}{2} \div \frac{5}{6}$ = ______

㊶ $5\frac{5}{6} \div 15$ = ______

㊷ $5 \div 2\frac{1}{7}$ = ______

㊸ $2\frac{1}{2} \div 1\frac{1}{2}$ = ______

Find the answers in simplest form.

44. $\frac{2}{9}+\frac{5}{12}$ = ______
45. $\frac{5}{8}-\frac{7}{16}$ = ______
46. $\frac{2}{7}\times\frac{1}{4}$ = ______
47. $\frac{8}{9}\div\frac{4}{3}$ = ______
48. $2\frac{2}{5}+1\frac{1}{2}$ = ______
49. $\frac{5}{8}\times 12$ = ______
50. $\frac{3}{4}\times 6$ = ______
51. $1\frac{1}{5}+3\frac{1}{3}$ = ______
52. $3\frac{1}{8}-2\frac{1}{5}$ = ______
53. $9\frac{1}{3}\div 3\frac{1}{9}$ = ______
54. $3\frac{3}{4}\div 5\frac{5}{8}$ = ______
55. $\frac{5}{7}\times\frac{28}{45}$ = ______

Find the answers in simplest form following the order of operations.

Order of Operations:
Brackets
Exponents
Division
Multiplication
Addition
Subtraction

56. $12\frac{2}{5}-10\times\frac{1}{5}$
57. $3\frac{1}{2}\times\frac{6}{7}-3$
58. $4\div 1\frac{1}{7}+6\frac{7}{8}$
59. $1\frac{5}{12}+2\frac{4}{9}\div 1\frac{1}{3}$
60. $10-\frac{6}{13}\div\frac{3}{26}$
61. $5\frac{2}{5}-2\frac{1}{2}\div 5$ = ______
62. $1\frac{1}{10}+4\frac{1}{2}\times 3$ = ______
63. $\frac{3}{4}\times 6-2\frac{2}{3}$ = ______
64. $\frac{1}{10}+2\frac{7}{8}\times 2\frac{2}{5}$ = ______
65. $1\frac{7}{18}\div(3\frac{1}{3}-1\frac{1}{9})$ = ______
66. $(1\frac{1}{2}+5\frac{7}{8})\times 2\frac{2}{3}$ = ______
67. $1\frac{4}{5}\times(2-\frac{3}{5})\div 1\frac{1}{5}$ = ______
68. $\frac{1}{2}+(1\frac{2}{3}+2\frac{1}{6})\div 3$ = ______

7 Decimals

- adding, subtracting, multiplying, and dividing decimals

When adding or subtracting decimals, align the decimal points. Add or subtract as whole numbers. Then add the decimal point to the answer in the same place.

Example 6.89 + 2.2 = ?

Align.

$$\begin{array}{r} 6.89 \\ +\ 2.20 \\ \hline \boxed{9.09} \end{array}$$

Add "0" as a placeholder.

6.89 + 2.2 = 9.09

$$\begin{array}{r} 0.74 \\ -\ 0.13 \\ \hline \boxed{} \end{array}$$

Add or subtract.

① $\begin{array}{r} 70.86 \\ +\ 7.07 \\ \hline \end{array}$

② $\begin{array}{r} 98.04 \\ -\ 6.13 \\ \hline \end{array}$

③ $\begin{array}{r} 24.09 \\ +\ 15.98 \\ \hline \end{array}$

④ $\begin{array}{r} 45.73 \\ -\ 6.84 \\ \hline \end{array}$

⑤ 36.19 + 4.2 = ________

⑥ 100 – 6.23 = ________

⑦ 9.02 + 92.1 = ________

⑧ 6.71 + 68.9 = ________

⑨ 40.19 – 5.87 = ________

⑩ 8.96 – 4.16 = ________

⑪ 63.05 + 7.28 = ________

⑫ 71.04 – 2.13 = ________

Find the answers. Show your work.

⑬ 67.08 + 9.2 – 6.72
= ________ – 6.72
= ________

⑭ 21.42 – 16.9 + 7.65
= ________
= ________

⑮ 72.29 – 68.11 + 49.12
= ________
= ________

⑯ 26.99 – 18.12 – 8.66
= ________
= ________

⑰ 56.45 + 47.58 – 20.98
= ________
= ________

⑱ 60 – 24.8 – 19.91
= ________
= ________

Put the decimal point in the correct place for each product.

⑲ 6.9 x 3.1 = 2 1 3 9

⑳ 5.23 x 6 = 3 1 3 8

㉑ 1.07 x 2.3 = 2 4 6 1

㉒ 0.7 x 5.904 = 4 1 3 2 8

㉓ 2.25 x 1.35 = 3 0 3 7 5

㉔ 0.82 x 4.15 = 3 4 0 3 0

The number of decimal places in a product is the sum of the number of decimal places in the factors.

e.g.

```
     7.0 2  ← 2 decimal places
x      4.1  ← 1 decimal place
----------
       7 0 2
   2 8 0 8 0
----------
   2 8.7 8 2  ← 3 decimal places
```

Multiply.

㉕ 18.2 x 0.3

㉖ 0.096 x 9

㉗ 11.93 x 0.04

㉘ 8.29 x 0.6

㉙ 7.11 x 1.5

㉚ 0.05 x 84

㉛ 3.08 x 0.26

㉜ 2.53 x 7.1

㉝ 24.8 x 0.1 = ____________

㉞ 68.02 x 0.4 = ____________

㉟ 2.43 x 1.6 = ____________

㊱ 104.1 x 3.2 = ____________

㊲ 1.92 x 0.35 = ____________

㊳ 5.76 x 0.19 = ____________

㊴ 3.43 x 26 = ____________

㊵ 16.04 x 2.9 = ____________

㊶ 8.02 x 6.5 = ____________

㊷ 31.05 x 4.3 = ____________

Divide.

㊸

A 1.08 ÷ 0.6 = ____

10.8 ÷ ____

B 4.16 ÷ 0.8 = ____

____ ÷ ____

C 0.768 ÷ 0.08 = ____

____ ÷ ____

D 625 ÷ 2.5 = ____

____ ÷ ____

E 8.04 ÷ 0.04 = ____

____ ÷ ____

F 2.24 ÷ 0.07 = ____

____ ÷ ____

G 40.5 ÷ 0.9 = ____

____ ÷ ____

H 4.25 ÷ 1.7 = ____

____ ÷ ____

Hints

To divide a number by a decimal, change the decimal to a whole number and apply the same change to the dividend. Then divide.

e.g. 0.235 ÷ 0.5 = ___
(x10) (x10)

2.35 ÷ 5

$$\begin{array}{r} 0.47 \\ 5\overline{)2.35} \\ \underline{2\,0} \\ 35 \\ \underline{35} \end{array}$$

0.235 ÷ 0.5 = 0.47

A ☐)10.8	B	C	D
E	F	G	H

㊹ 0.84 ÷ 0.03 = ________

㊺ 32.2 ÷ 1.4 = ________

㊻ 19.53 ÷ 0.31 = ________

㊼ 64.48 ÷ 0.52 = ________

㊽ 2.684 ÷ 8.8 = ________

㊾ 164.08 ÷ 5.6 = ________

Find the answers. Show your work.

㊿ 16.39 – 6.2 x 0.7

(51) 4.2 + 1.6 x 5

(52) 67.5 ÷ 0.05 ÷ 0.2

(53) 53.9 – 12.8 ÷ 3.2

(54) 20 – 4.4 x 2.05

(55) 36.45 ÷ 2.5 x 1.25

Find the mean of each set of decimals.

(56) 8.4 3.7 6.38

Mean:

(______ + ______ + ______) ÷ 3 = ______

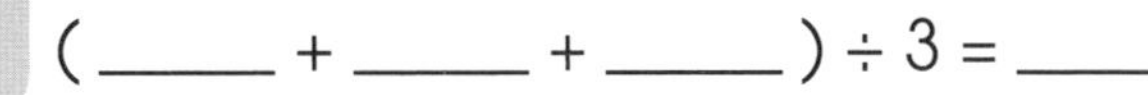

(57) 4.27 5.62 9.15 13.28

Mean:

______________________________ = ______

(58) 5.49 4.91 6.5 3.7 1.35

Mean:

______________________________ = ______

To find the mean, add all the decimals and divide the sum by the number of decimals added.

Use the distributive property to find the answers.

(59) 15.8 x 0.5

= (15 + 0.8) x 0.5 ← 15.8 = 15 + 0.8

= 15 x ______ + 0.8 x ______ ← Apply the distributive property.

= ______ + ______

= ______

10.1 x 0.5
= (10 + 0.1) x 0.5
= 10 x 0.5 + 0.1 x 0.5
= 5 + 0.05
= 5.05

(60) 20.4 ÷ 0.2

(61) 4.25 x 0.4

(62) 2.98 ÷ 0.2

8 Rates

- understanding rates

A rate is a ratio that compares two quantities with different units. A unit rate is a rate in which the second quantity is 1.

Example Find the unit rates.

150 km in 2 h

$= \dfrac{150 \text{ km}}{2 \text{ h}}$

= 75 km/h

↑

75 km in 1 h

$200 for 4 boxes

$= \dfrac{\$200}{4 \text{ boxes}}$

= $50/box

↑

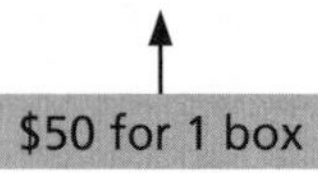

$10 for 5 kg

$ ____ /kg

320 g for 4 bags

____ g/bag

Find the unit rates.

①

840 ÷ 6 = ________

________ g/orange

②

2 ÷ 8 = ________

$________ /crayon

③

980 ÷ 5 = ________

________ mL/serving

④

15 ÷ 30 = ________

$________ /m

⑤

4.2 ÷ 12 = ________

$________ /egg

⑥ $5 for 4 kg

$________ /kg

⑦ 525 words in 15 min

____

⑧ 372 pages in 4 days

____

⑨ $272 for 400 mL

____

⑩ 58 claps in 29 s

____

⑪ 15 push-ups in 50 s

____

⑫ 92 rotations in 4 s

____

⑬ 30 bracelets in 4 h

____

Find the unit prices.

⑭

Ⓐ 6 apples

$3.60

Ⓑ $11.97

3 toothbrushes

Ⓒ 8 cups

$3.36

Ⓓ $12

400 mL

Ⓔ $8.95

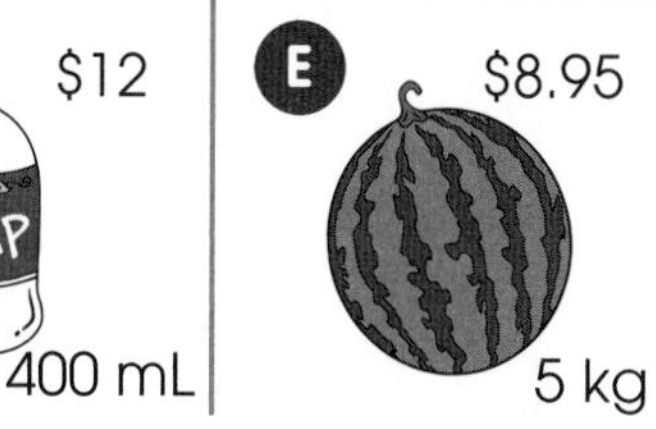

5 kg

Ⓕ 50 m

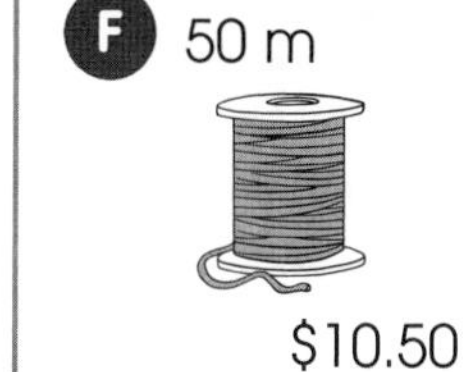

$10.50

Hints

The unit price is the price of one item.

e.g. 10 candies for $3.50.

Unit price:
$3.5 \div 10 = 0.35$
$0.35/candy

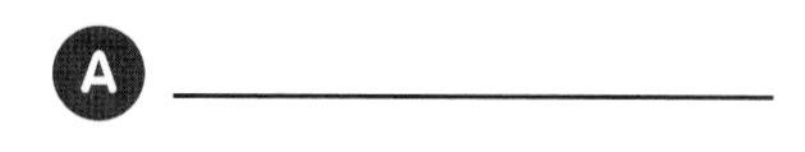

Ⓓ ____________

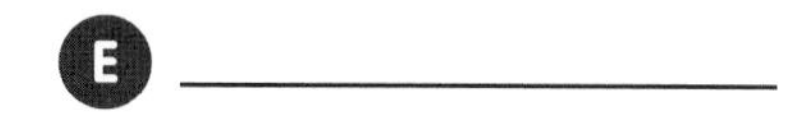

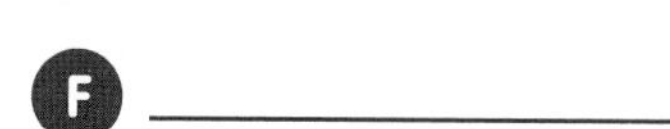

Find and compare the unit prices. Then check the better buy.

⑮

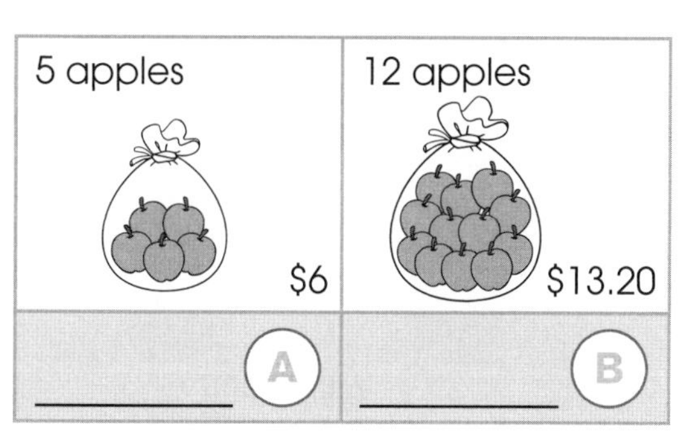

⑯

⑰

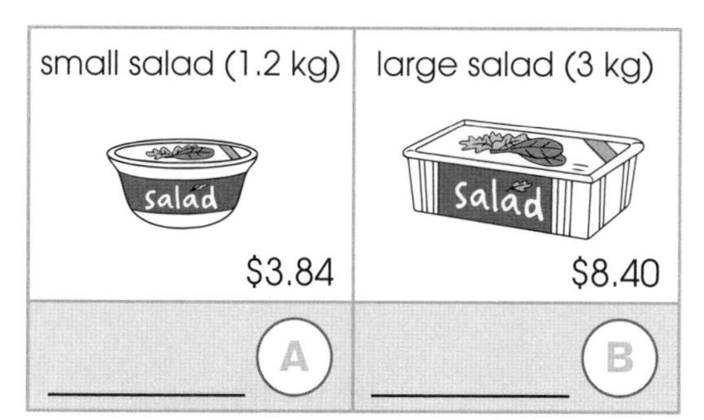

⑱

⑲

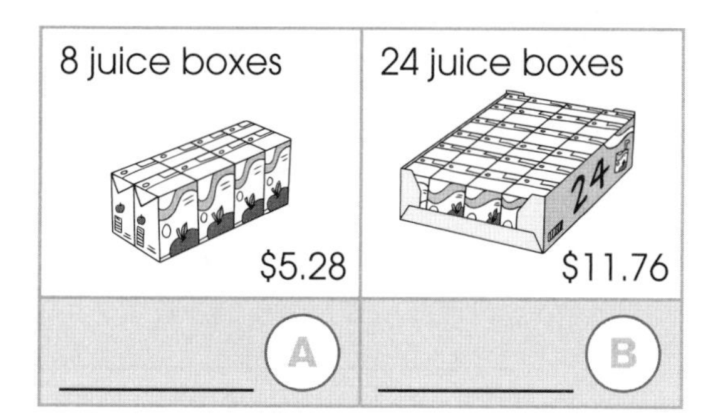

⑳

Find the speeds of the runners.

Speed is a rate that describes how fast someone or something moves.

Speed = Distance ÷ Time

e.g. 120 km in 2 h
Speed = 120 ÷ 2
= 60 (km/h)

㉑ Start — 5100 m — Checkpoint A — 2400 m — Checkpoint B — 3430 m — Finish

Runner	Checkpoint A	Checkpoint B	Finish
Jamie	40 min	25 min	35 min
Louis	48 min	30 min	25 min
Andrea	51 min	20 min	28 min

a. Speed from Start to Checkpoint A:

Jamie: ________ m/min Louis: ________ Andrea: ________

b. Speed from Checkpoint A to Checkpoint B:

Jamie: ________ Louis: ________ Andrea: ________

c. Speed from Checkpoint B to Finish:

Jamie: ________ Louis: ________ Andrea: ________

d. Jamie's speed from Start to Finish:

(______ + ______ + ______) ÷ (______ + ______ + ______) = ________

Find the distance travelled or the time taken.

㉒ **Distance Travelled**

a. 75 km/h for 6.2 h

____

b. 200 m/min for 125 min

____

㉓ **Time Taken**

a. 280 m at 25 m/min

____

b. 765 km at 85 km/h

____

Use the tables to find the amount of calories in each meal.

㉔

Food Item	pizza	burger	fries	salad	soda	fruit cup
Calories/Serving	285	254	365	117	150	124

a. 2 servings of pizza

____

b. 2 servings of burger and 1 serving of fries

____

c. 3 servings of pizza and 2 servings of soda

____

d. 3 servings of salad and 2 servings of soda

____

e. 1 serving of burger, 1 serving of fries, and 2 servings of fruit cup

____

f. 1 serving of pizza, 2 servings of salad, and 2 servings of soda

____

㉕

Breakfast Item	(eggs)	(pancakes)	(bacon)	(orange slice)
Calories/100 g	155	227	541	47

a. 180 g of eggs and 250 g of bacon

____

b. 300 g of pancakes and 50 g of orange slices

____

Find how many calories are in 1 g of each breakfast item first.

e.g. Calories in 1 g of eggs:

$$\frac{155 \text{ calories}}{100 \text{ g}} = 1.55 \text{ calories/g}$$

↑ unit rate

Then multiply the unit rate by the specified amount to find the total calories in the meal.

$$\text{Total calories} = \text{unit rate} \times \text{specified amount}$$

9 Perimeter and Area

- finding perimeter and area

Read This

Square	Rectangle	Triangle
P = 4s	P = 2(l + w)	P = a + b + c
A = s^2	A = lw	A = bh ÷ 2

Example Find the perimeter and area.

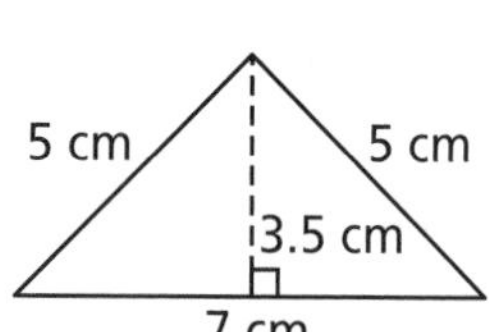

Perimeter: sum of 3 sides
5 + 5 + 7 = 17

Area: base x height ÷ 2
7 x 3.5 ÷ 2 = 12.25

Perimeter: 17 cm   Area: 12.25 cm²

Try It

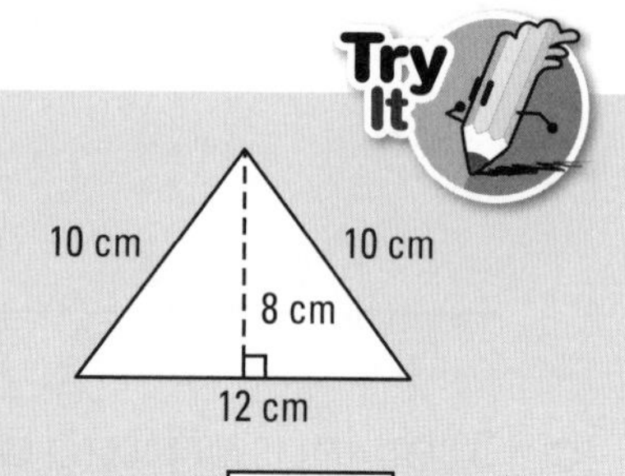

Perimeter: ______ cm

Area:  cm²

Find the perimeter and area of each shape. Show your work.

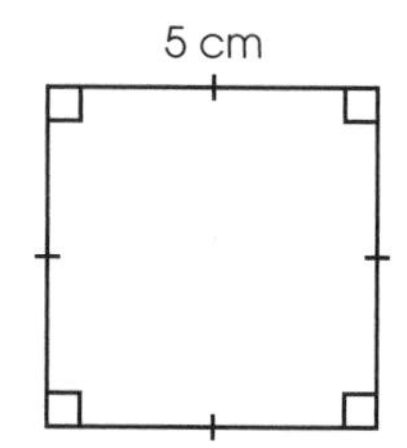

Perimeter:

______ = ______

Area:

______ = ______

② 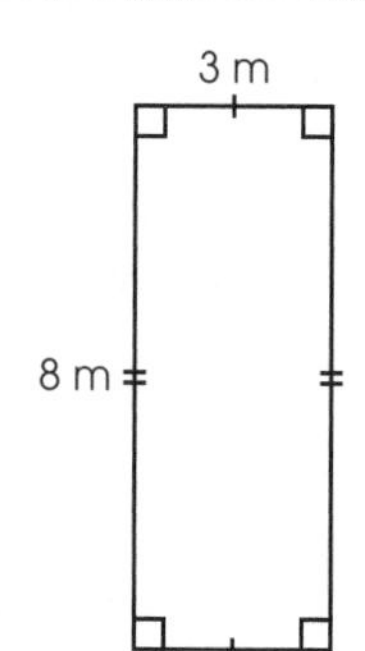

Perimeter:

______ = ______

Area:

______ = ______

③ 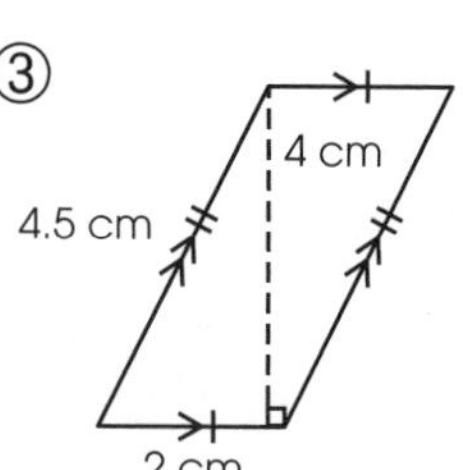

Perimeter:

______ = ______

Area:

______ = ______

④ 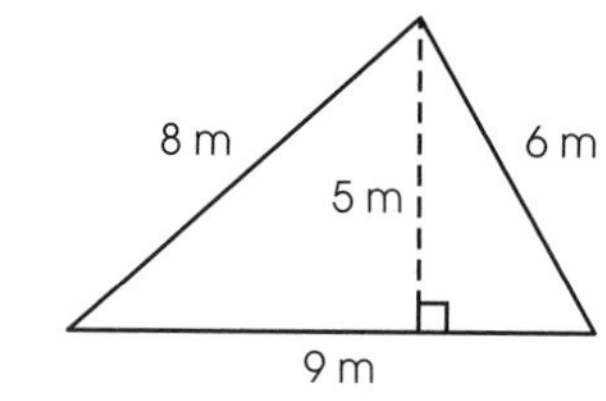

Perimeter:

______ = ______

Area:

______ = ______

⑤ 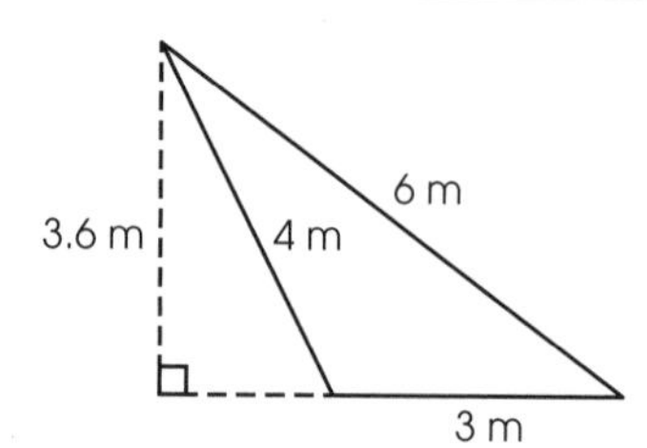

Perimeter:

______ = ______

Area:

______ = ______

Find the area of each trapezoid. Show your work.

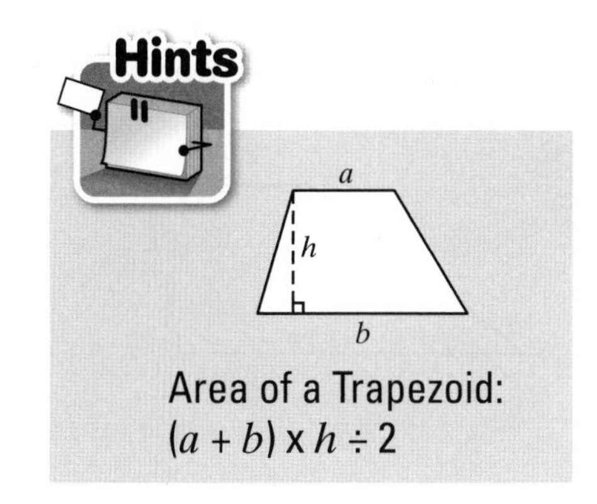

⑥

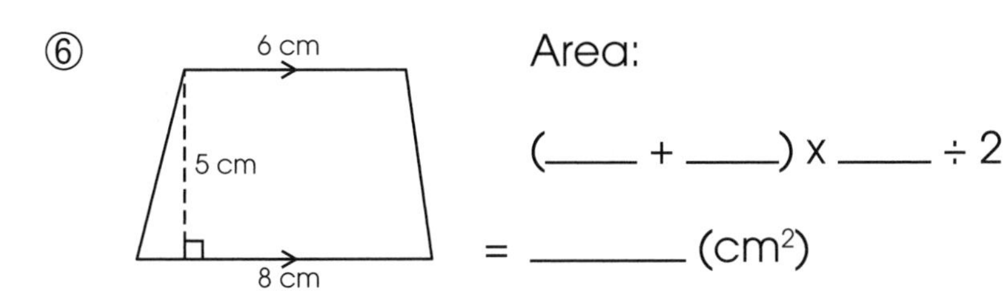

Area:

(____ + ____) x ____ ÷ 2

= ______ (cm^2)

⑦

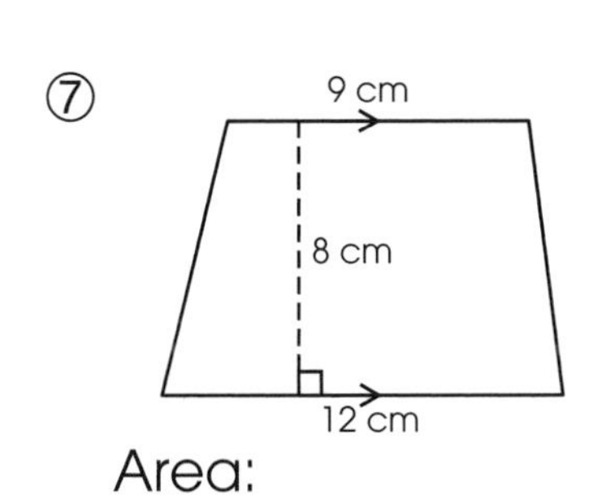

Area:

⑧

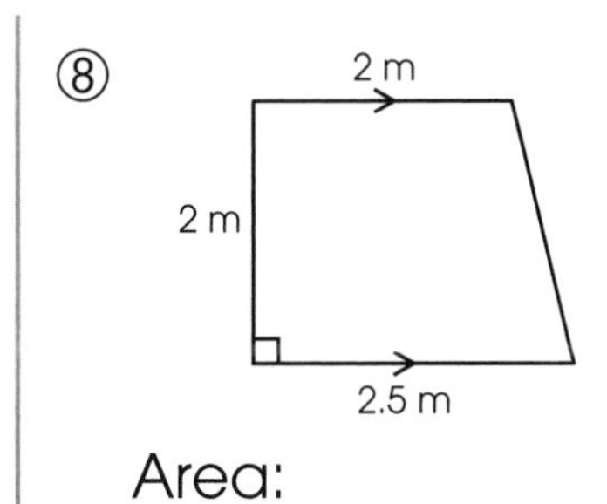

Area:

⑨

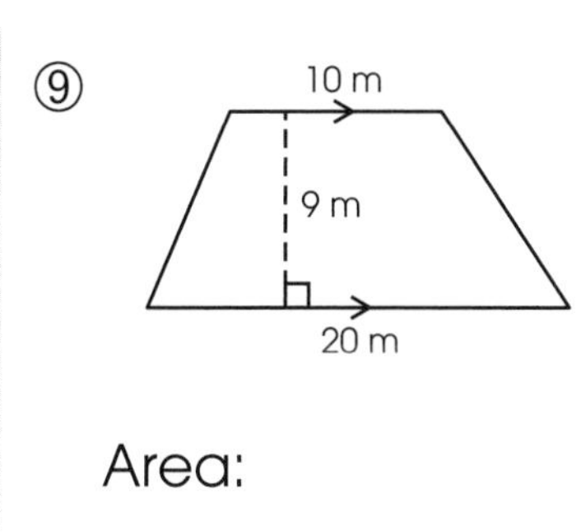

Area:

Find the perimeter and area of each shape.

⑩

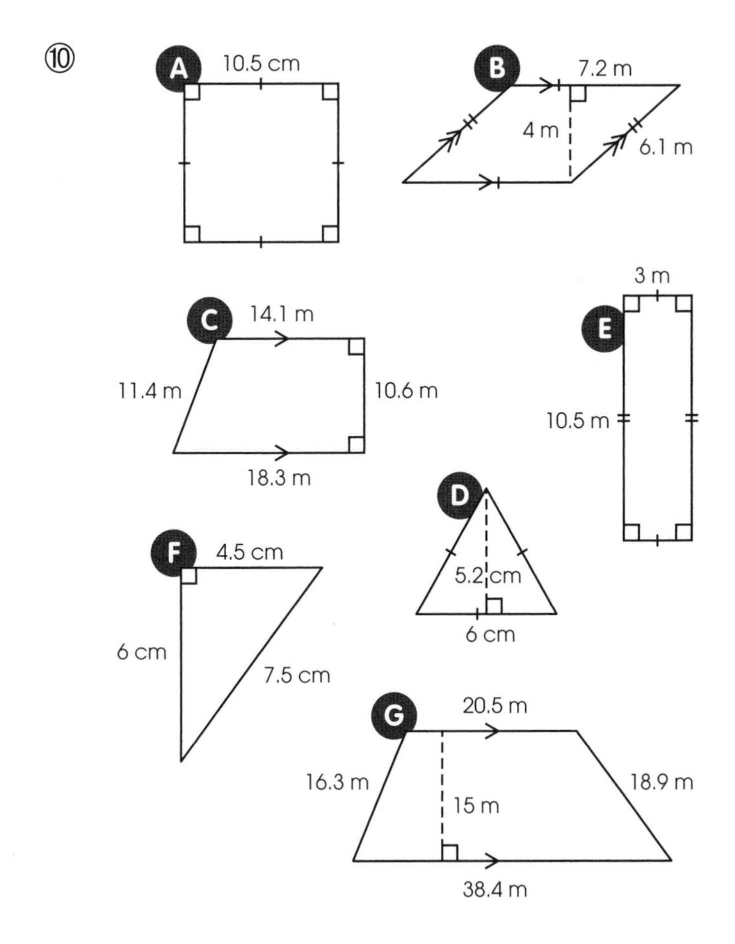

	Perimeter	Area
A	______	______
B	______	______
C	______	______
D	______	______
E	______	______
F	______	______
G	______	______

Find the perimeter and area of each composed shape.

⑪

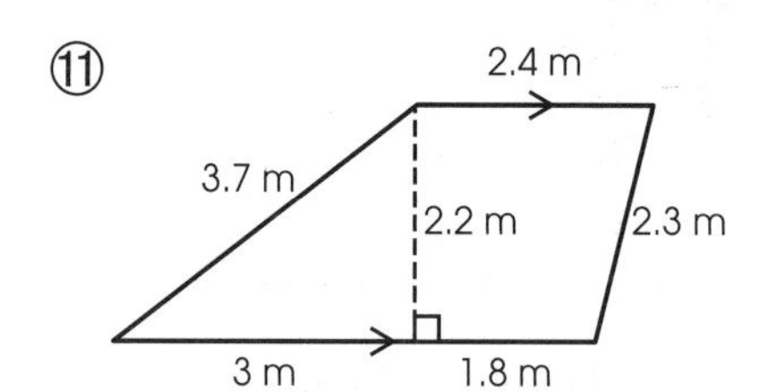

Perimeter: ______

Area: ______

⑫

4 cm
3 cm
4.5 cm
5 cm
8 cm

Perimeter: ______

Area: ______

To find the area of a composed shape, you may use addition or subtraction.

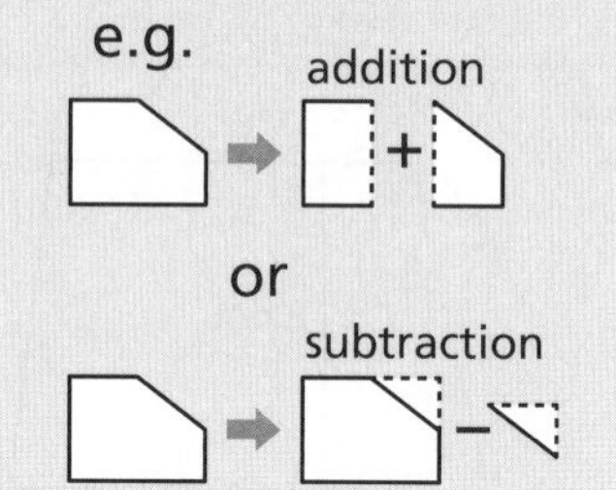

⑬

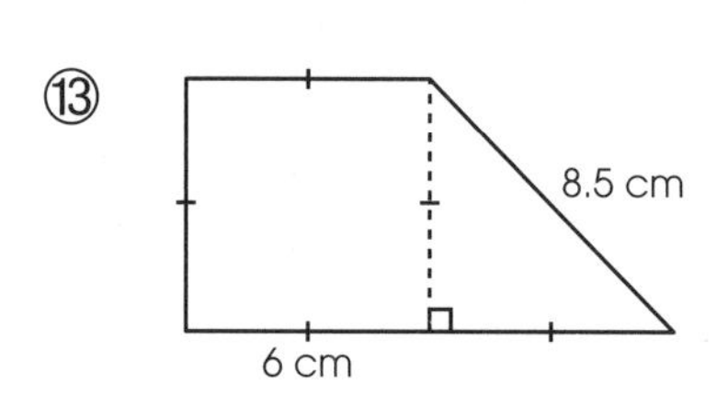

Perimeter: ______

Area: ______

⑭

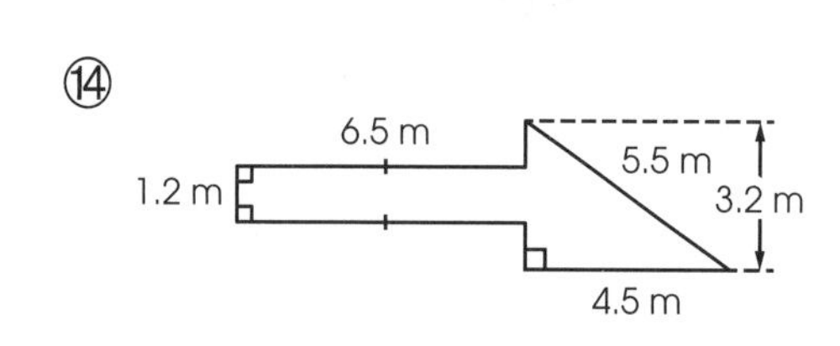

Perimeter: ______

Area: ______

⑮

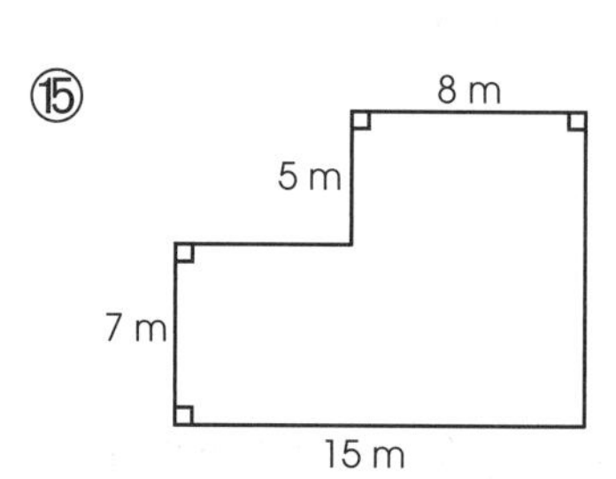

Perimeter: ______

Area: ______

Find the areas of the shaded parts.

⑯

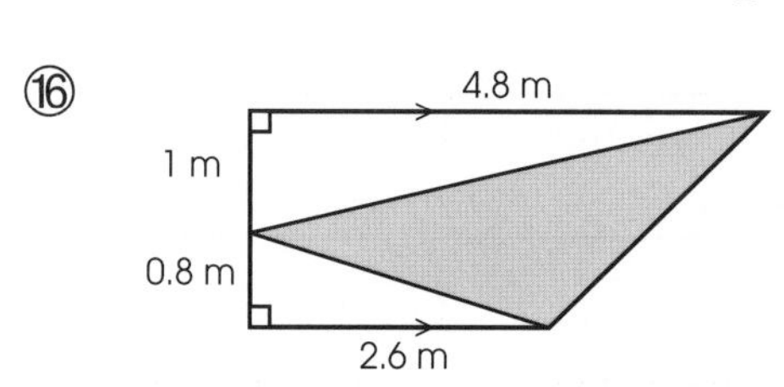

Area: ______

⑰

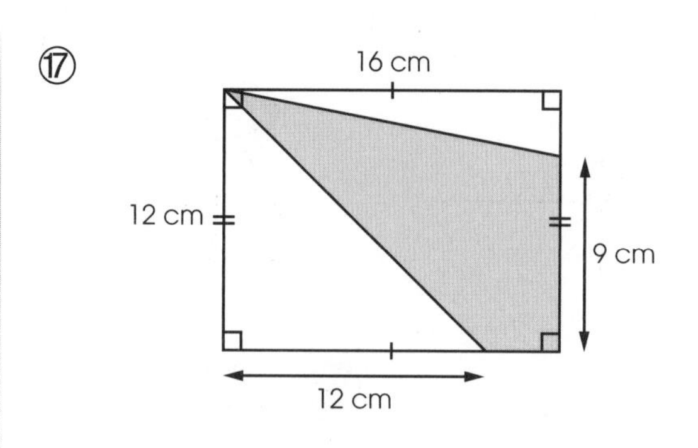

Area: ______

Circle "T" for true and "F" for false.

⑱ If two shapes are congruent,

a. they can have the same area but different perimeters. **T / F**

b. they must have the same area and perimeter. **T / F**

⑲ If two shapes are similar,

a. the one that has a greater area also has a greater perimeter. **T / F**

b. they can have the same area but different perimeters. **T / F**

c. the one that has a greater perimeter also has a smaller area. **T / F**

Hints

Two shapes are congruent if they have the same shape and size.

e.g. 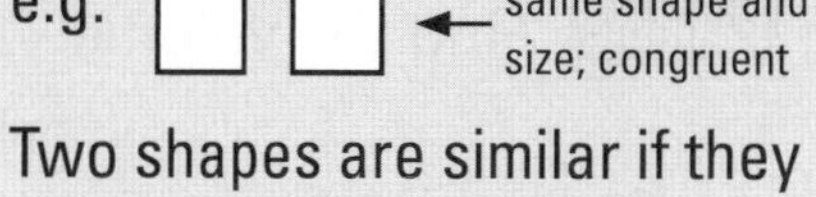

Two shapes are similar if they have the same shape but different sizes.

e.g. 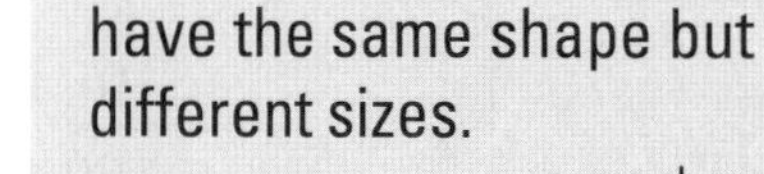

Do the conversions. Then estimate the areas of the shapes in both cm^2 and m^2.

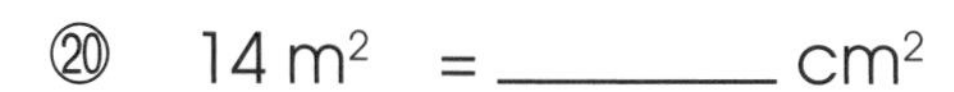

⑳ $14\ m^2$ = ________ cm^2 ㉑ $16.5\ m^2$ = ____________

㉒ $8.45\ m^2$ = ____________ ㉓ $940\ m^2$ = ____________

㉔ $1\ 260\ 000\ cm^2$ = ____________

㉕ $47\ 320\ cm^2$ = ____________

Hints

$1\ m^2 = 10\ 000\ cm^2$

e.g.

$42\ m^2 = 420\ 000\ cm^2$

$130\ 000\ cm^2 = 13\ m^2$

㉖

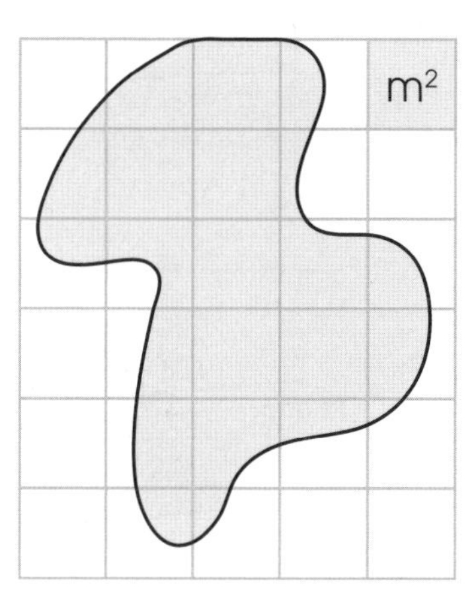

__________ m^2

__________ cm^2

㉗

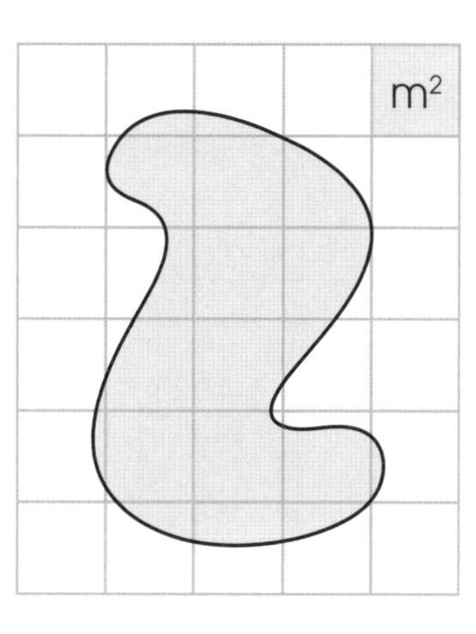

__________ m^2

__________ cm^2

㉘

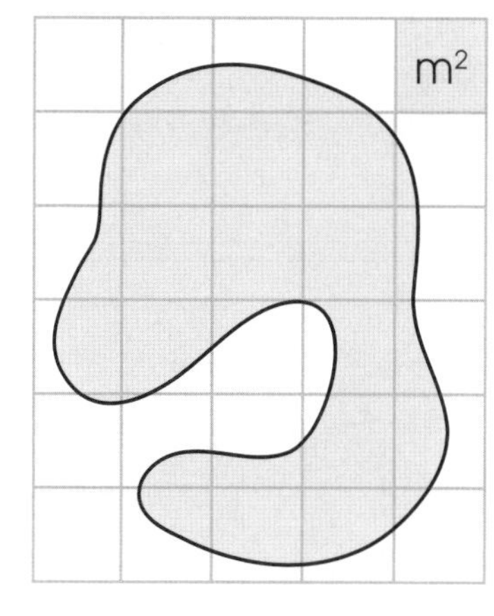

__________ m^2

__________ cm^2

10 Volume

- finding the volume of 3-D shapes

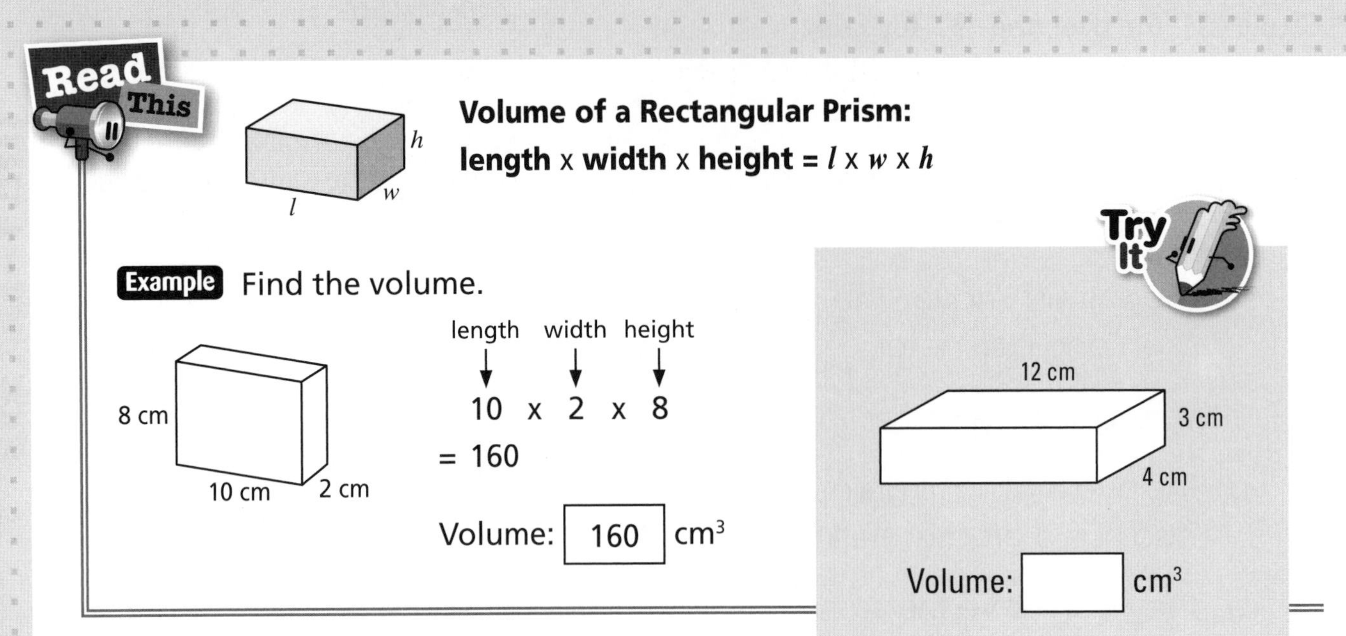

Volume of a Rectangular Prism:
length x **width** x **height** = $l \times w \times h$

Example Find the volume.

length width height

10 x 2 x 8

= 160

Volume: 160 cm^3

Try It

Volume: ______ cm^3

Find the volume of each prism.

①

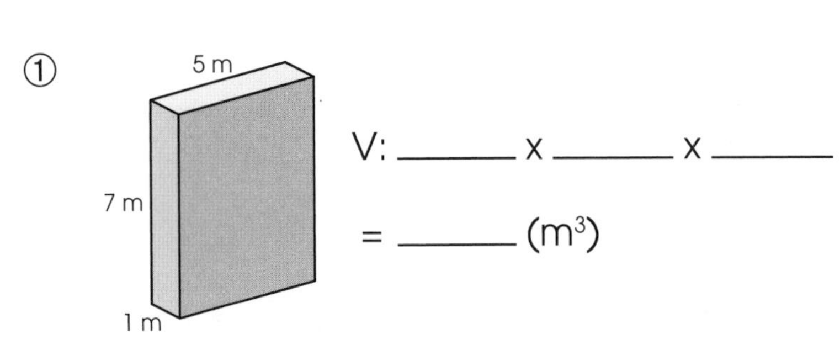

V: ______ x ______ x ______

= ______ (m^3)

Make sure all measurements are in the same unit before finding the volume.

②

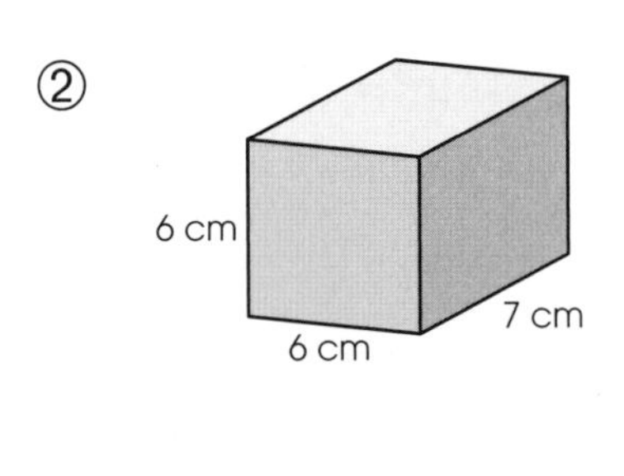

______

③

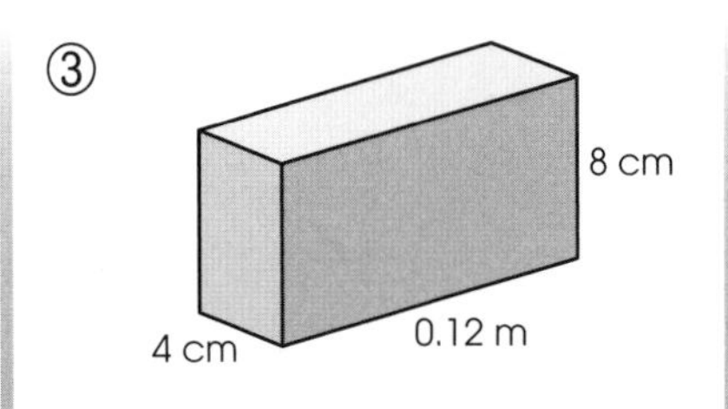

______

④

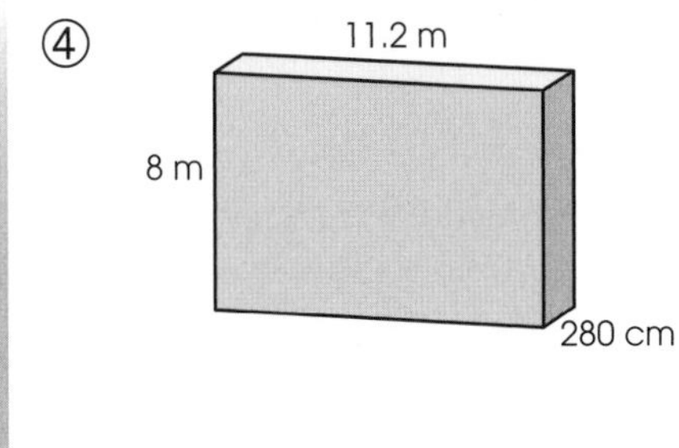

______

⑤

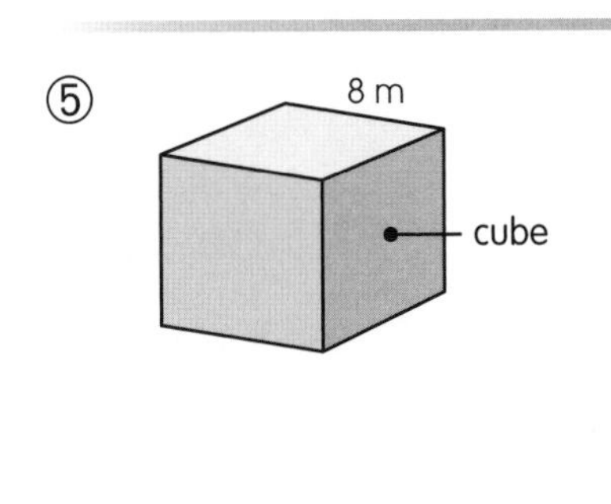

______

⑥

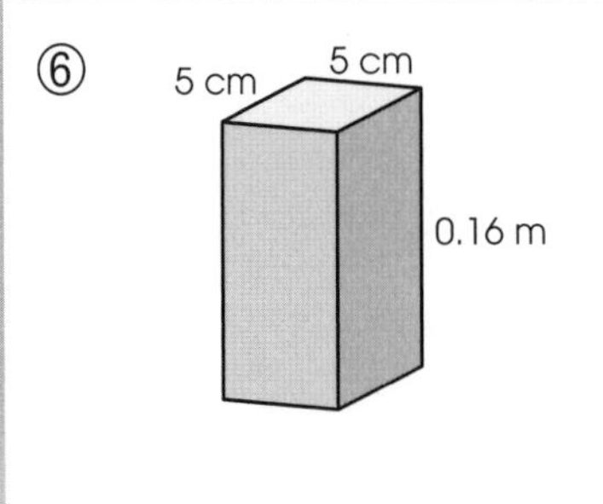

______

⑦

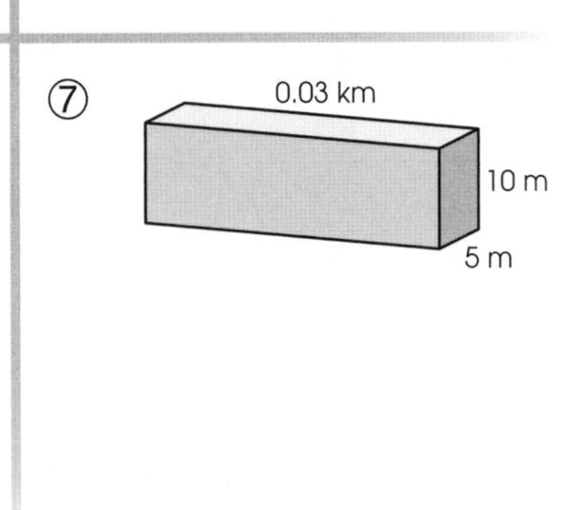

______

The shaded surface of each prism is its base. Find the volume of the prism. Show your work.

⑧

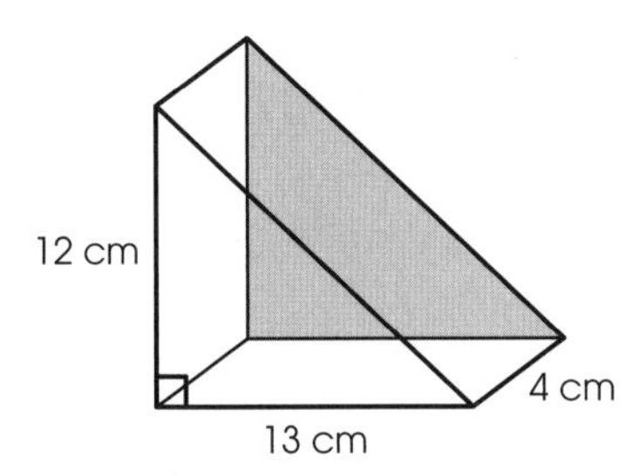

⑨

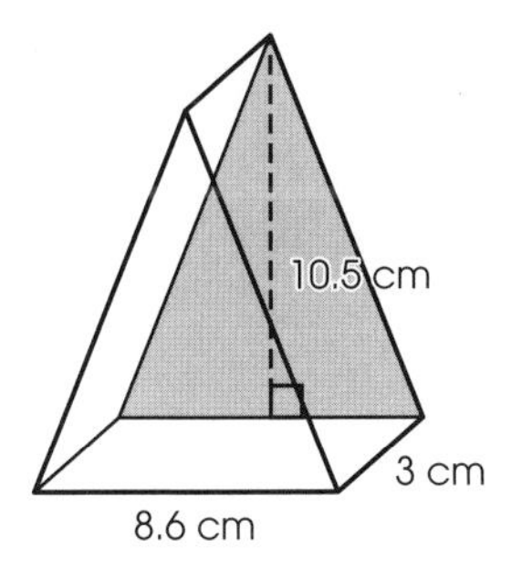

Find the volume of a prism by multiplying the area of its base by its height.

Volume of a Prism:
area of base x height

e.g.

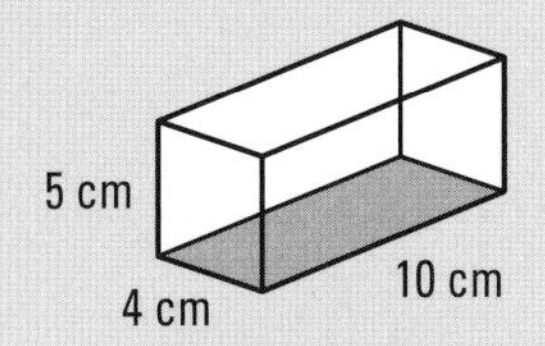

area of base height

V: 10 x 4 x 5

= 200 (cm^3)

⑩

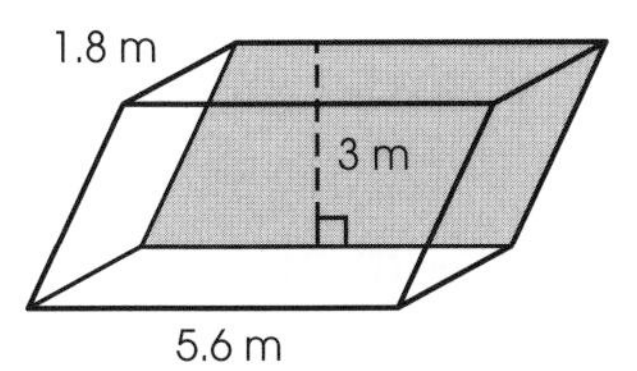

⑪

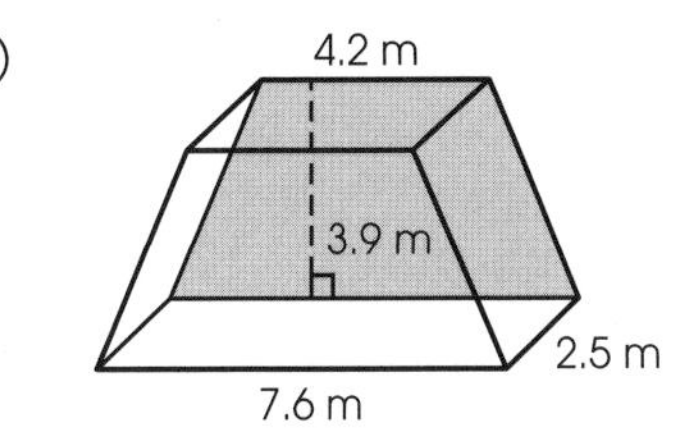

⑫

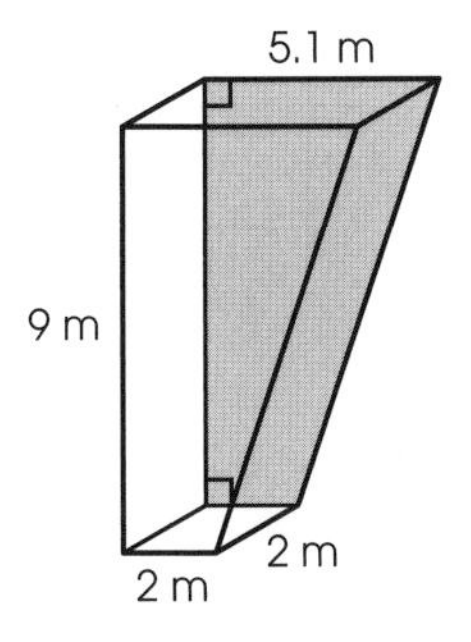

⑬

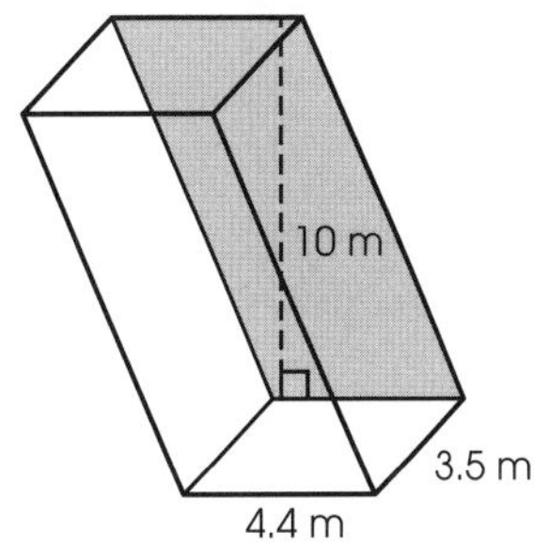

⑭

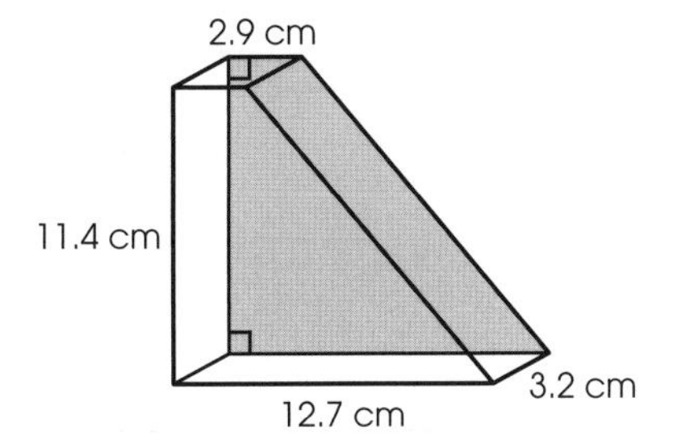

⑮

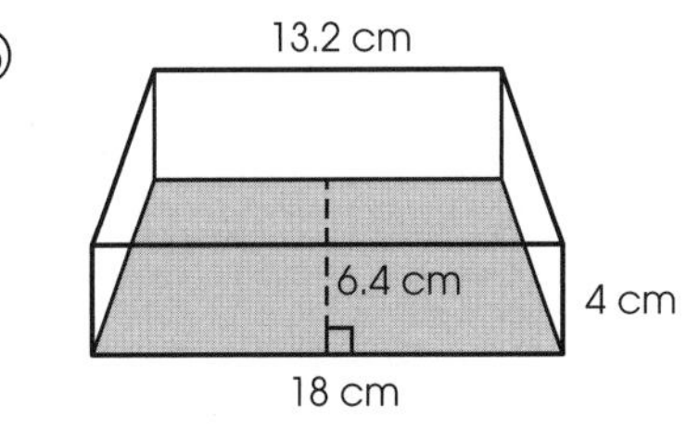

LEVEL 1 – BASIC SKILLS

Find the missing measurement, m, using the given volume and measurements for each prism.

⑯ V: 1024 cm^3

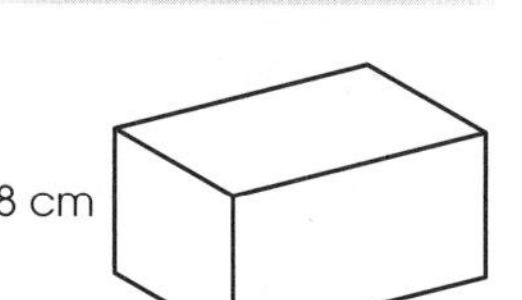

m = ________

⑰ V: 125 m^3

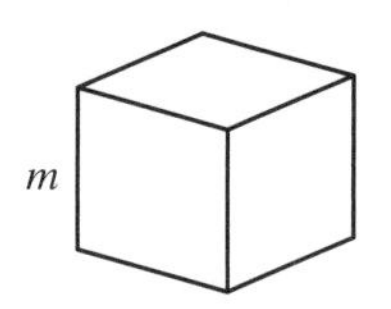

m = ________

⑱ V: 378 cm^3

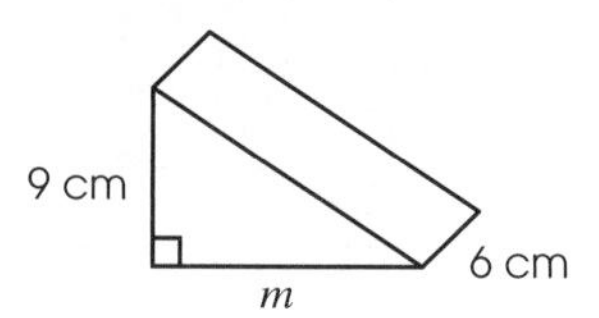

m = ________

⑲ V: 3840 cm^3

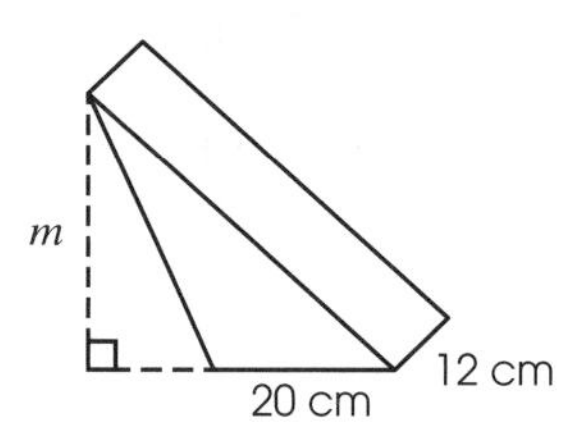

m = ________

⑳ V: 330 m^3

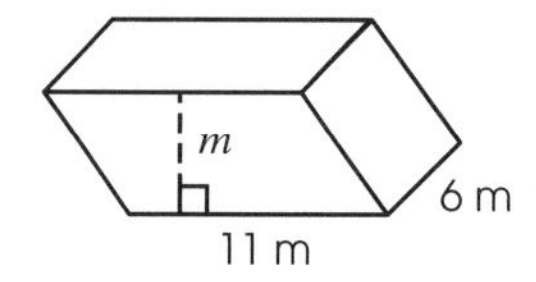

m = ________

㉑ V: 500 cm^3

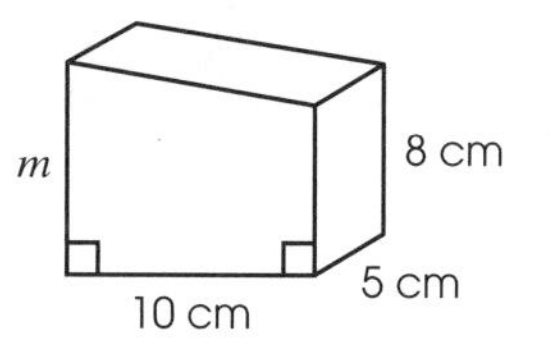

m = ________

Find the volume of each block. Then find the volumes of the stacked blocks.

㉒

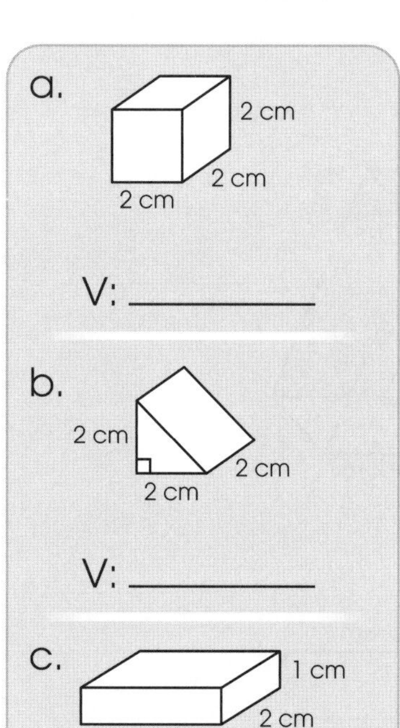

A

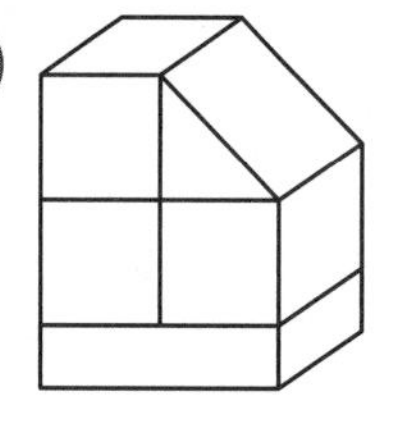

B

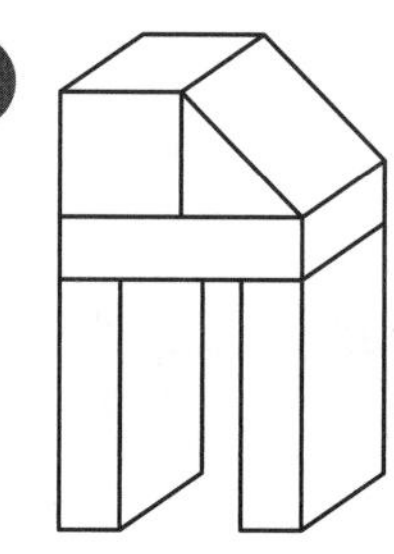

C

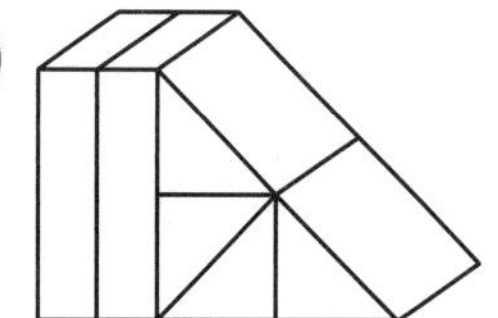

D

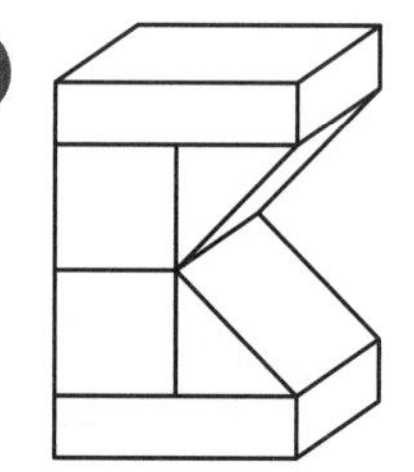

For each solid, find the area of its base and its volume.

Instead of adding the areas of the simpler shapes, you may also find the area by subtraction.

e.g.

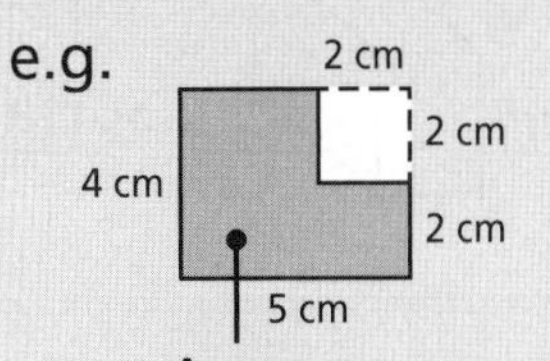

Area:
$5 \times 4 - 2 \times 2$
$= 16\ (cm^2)$

This area can be thought of as a rectangle with a square removed.

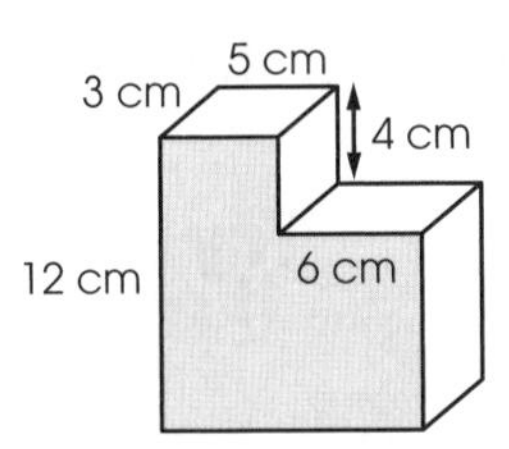

Area of base:

Volume:

㉔

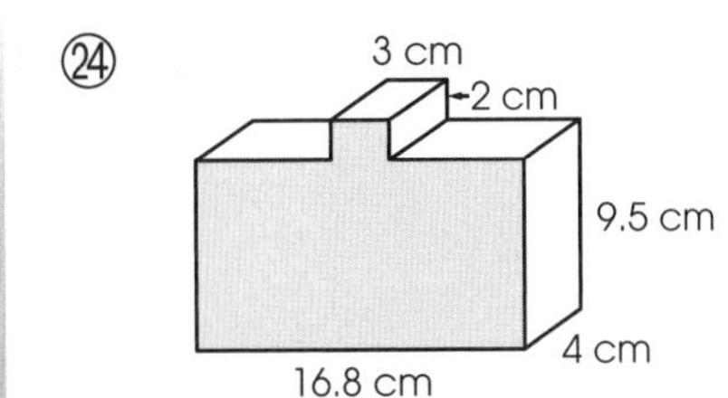

Area of base:

Volume:

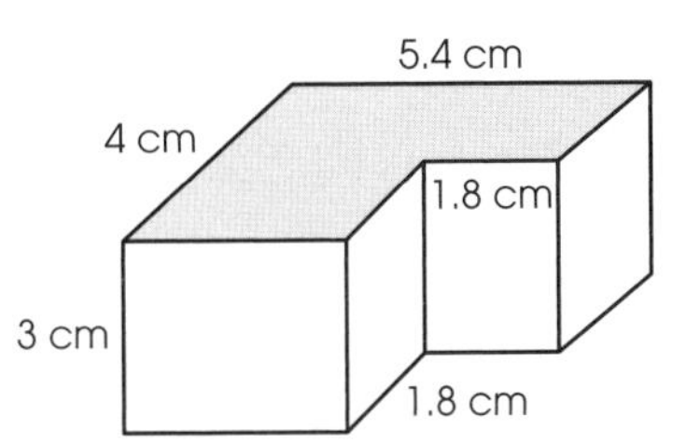

Area of base:

Volume:

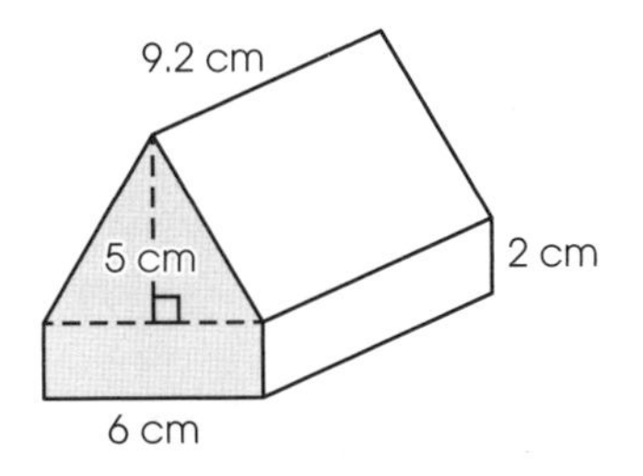

Area of base:

Volume:

㉗

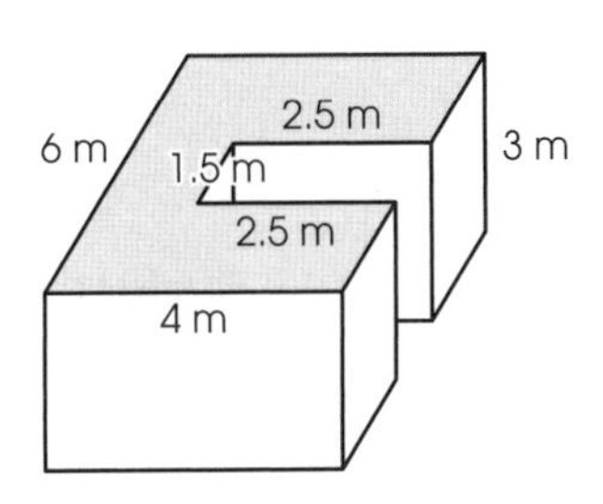

Area of base:

Volume:

11 Surface Area

- finding the surface area of 3-D figures

The surface area of a 3-D figure is the sum of the areas of all the faces of the figure.

Example Find the surface area.

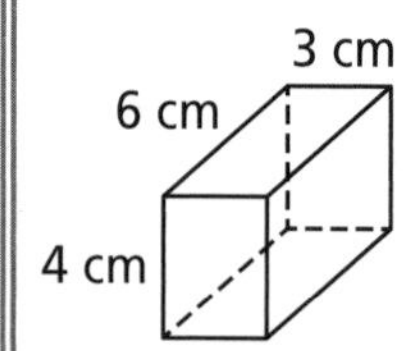

$(6 \times 4) \times 2 + (4 \times 3) \times 2 + (6 \times 3) \times 2$
$= 108$

Surface Area: 108 cm^2

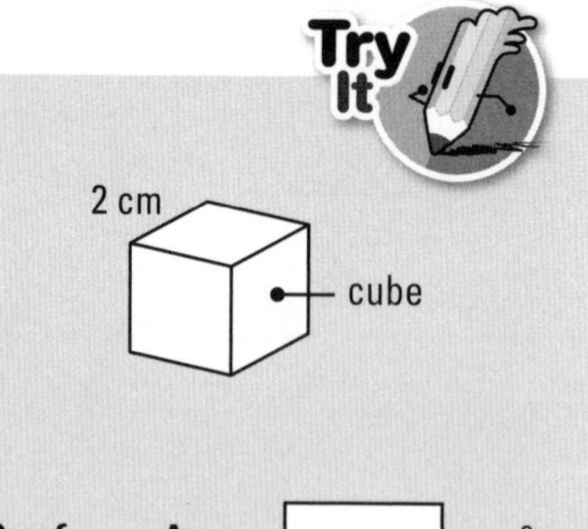

Surface Area: ______ cm^2

Check the correct expression for finding the surface area of each prism. Then find the surface area.

① 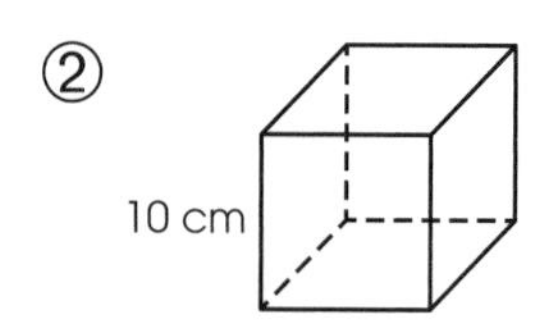

- A $(5 \times 8 \times 4) + (8 \times 5 \times 4) + (4 \times 5 \times 8)$
- B $(5 \times 8) \times 2 + (5 \times 4) \times 2 + (8 \times 4) \times 2$

Surface Area

② 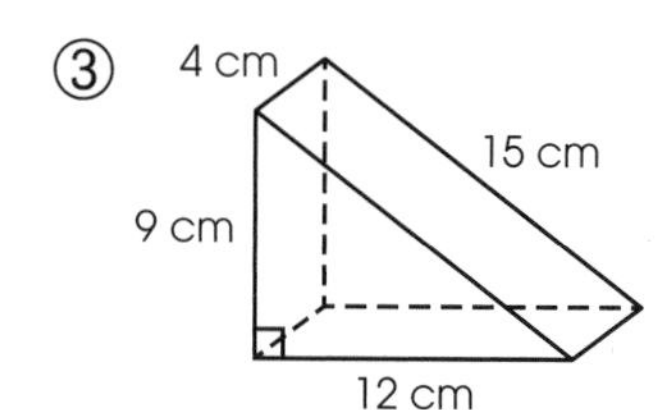

- A $(10 \times 10) \times 6$
- B $(10 + 10) \times 2 + (10 + 10) \times 2 + (10 + 10) \times 2$

Surface Area

③ 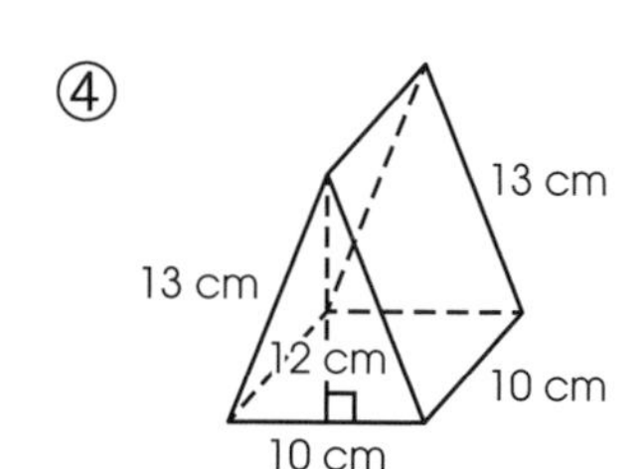

- A $(9 \times 12 \div 2) \times 2 + 9 \times 4 + 12 \times 4 + 15 \times 4$
- B $(9 \times 12) \times 2 + (9 + 12 + 15) \times 4$

Surface Area

④

- A $(10 \times 12 \div 2) \times 2 + (13 \times 10) \times 2 + 10 \times 10$
- B $(12 \times 10) \times 2 + 10 \times 10 + (10 \times 13 \div 2) \times 2$

Surface Area

⑤ 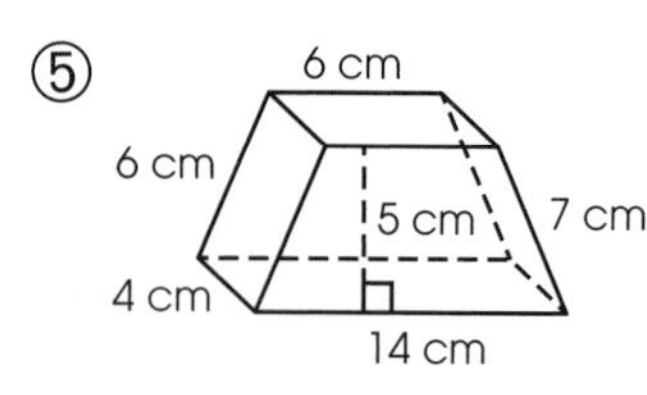

- A $(14 \times 5 \div 2) \times 2 + (6 \times 5 \div 2) \times 2 + (14 + 7 + 6 + 6) \times 4$
- B $((6 + 14) \times 5 \div 2) \times 2 + 14 \times 4 + 7 \times 4 + (6 \times 4) \times 2$

Surface Area

Find the surface areas. Show your work.

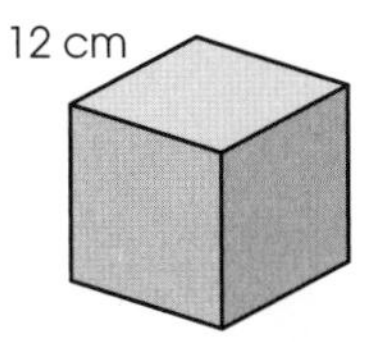

⑦

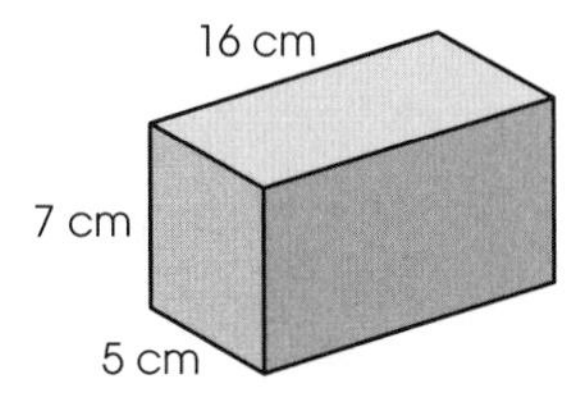

⑧

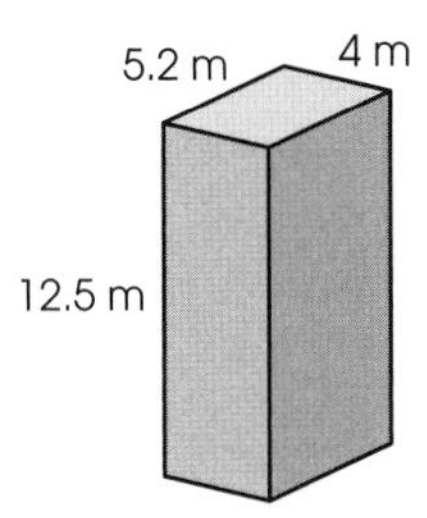

⑨

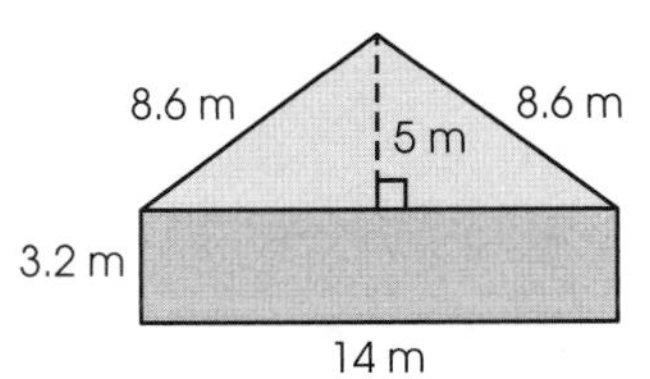

⑩

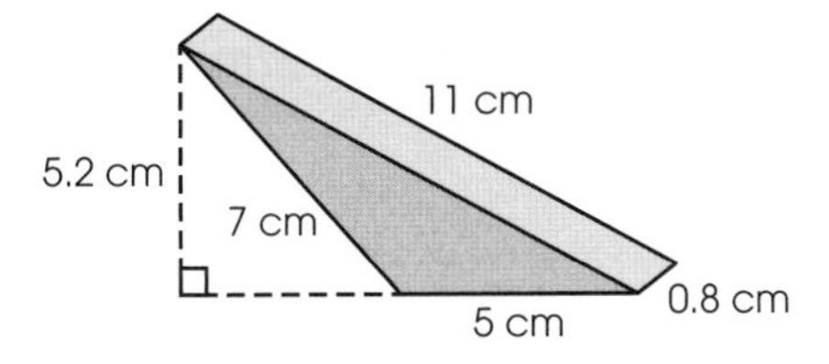

⑪

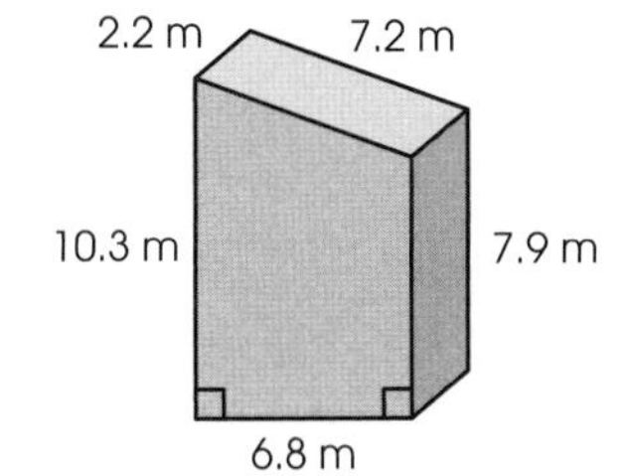

Jamie is combining cubes to create new structures. For each new structure, find the number of cube faces and the surface areas.

⑫ Each cube has a side length of 2 cm.

Area of 1 cube face: ________ cm^2

Tips

Do not count the faces that are joined.

a. No. of Cube Faces: ________

Surface Area: ________

b. No. of Cube Faces: ________

Surface Area: ________

c. No. of Cube Faces: ________

Surface Area: ________

d. No. of Cube Faces: ________

Surface Area: ________

e. No. of Cube Faces: ________

Surface Area: ________

Find the surface areas of the combined shapes using the given measurements.

⑬ a.

b.

Find the surface areas. Show your work.

⑭

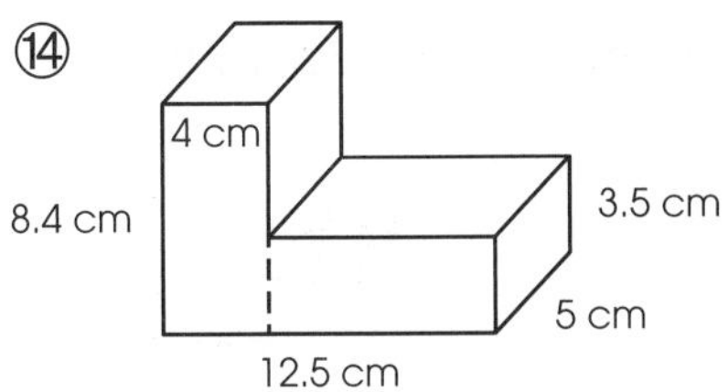

⑮

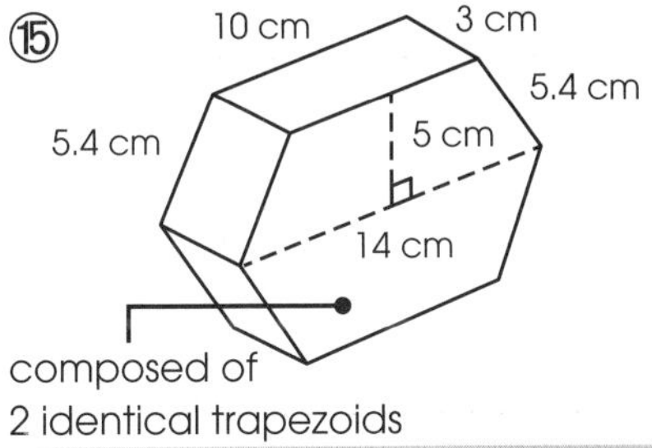

⑯

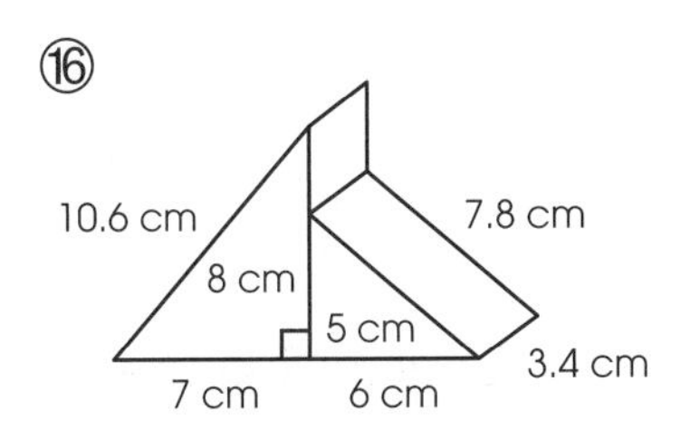

⑰

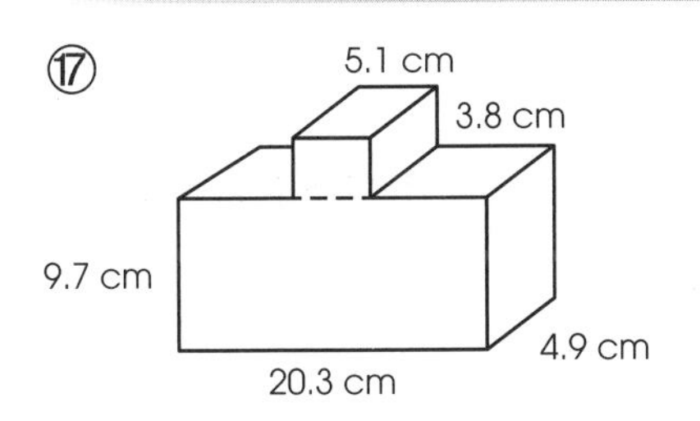

⑱

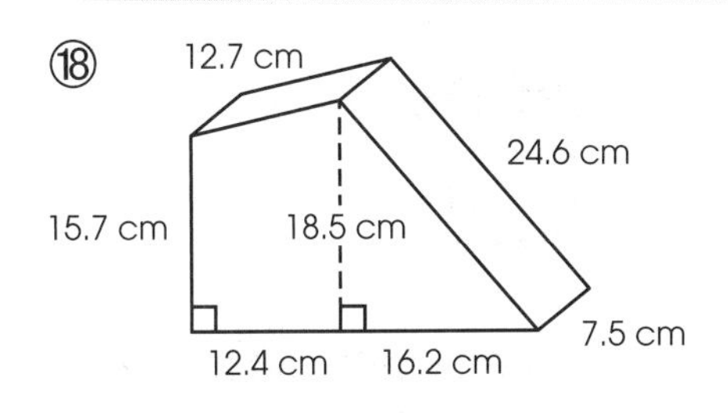

⑲

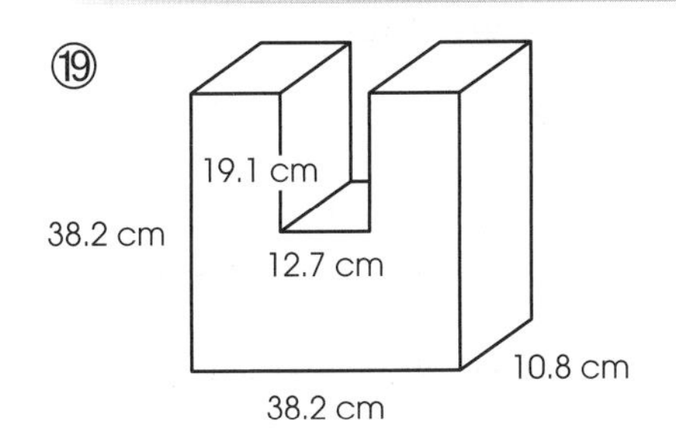

Here is a quick way to find the surface area of a prism:

1. Find the area of the bases.
2. Add the lengths and widths of the rectangular faces.
3. Multiply the sum by the height.
4. Add the areas of the bases and faces.

e.g.

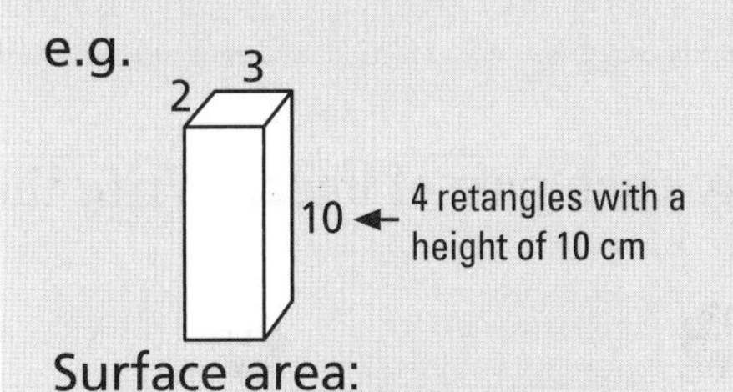

Surface area:

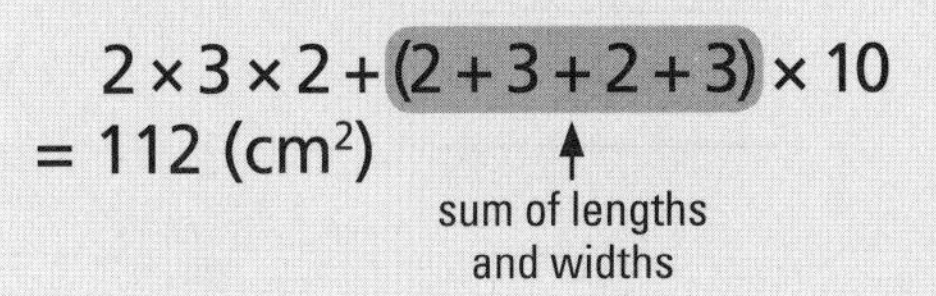

$2 \times 3 \times 2 + (2 + 3 + 2 + 3) \times 10$
$= 112\ (cm^2)$

sum of lengths and widths

12 Lines and Angles

- understanding the relationships between lines and angles

Intersecting lines are lines that cross each other at one point. Parallel lines are lines that never intersect.

Example Name each pair of lines.

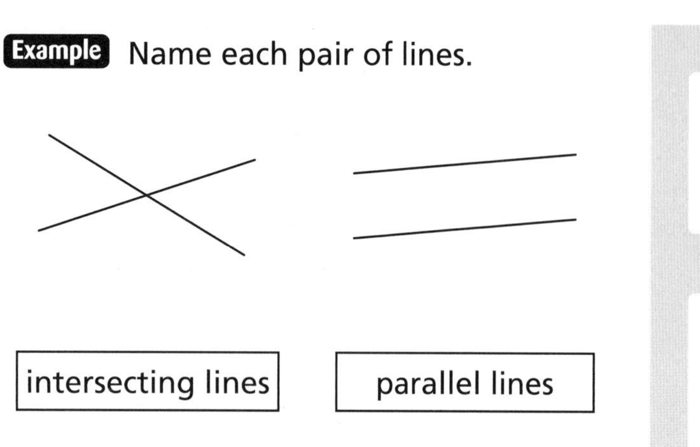

Identify each pair of lines. Write the letters.

①

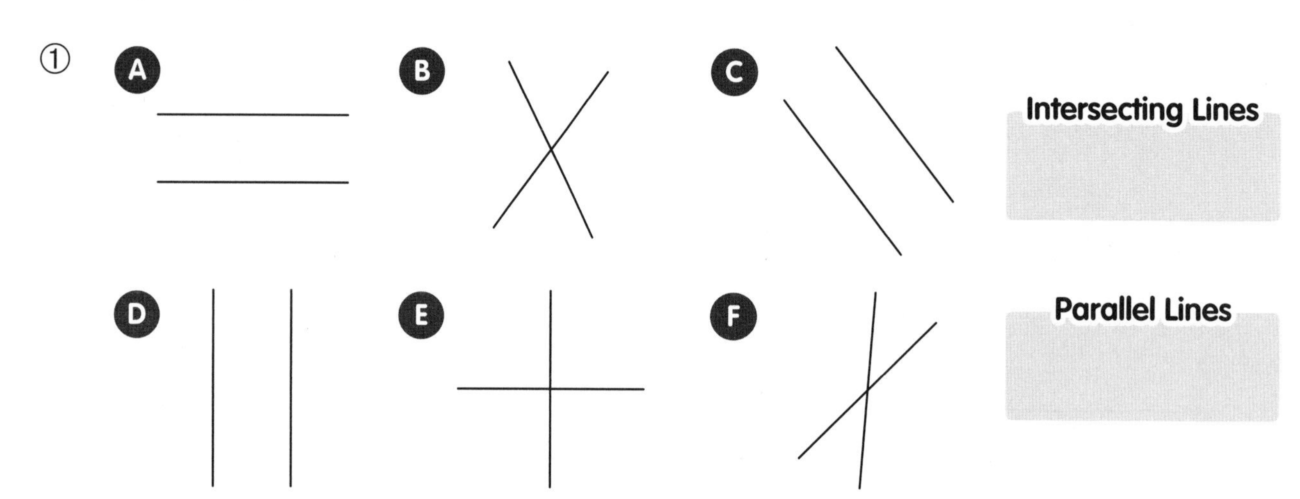

Identify and check the perpendicular lines.

②

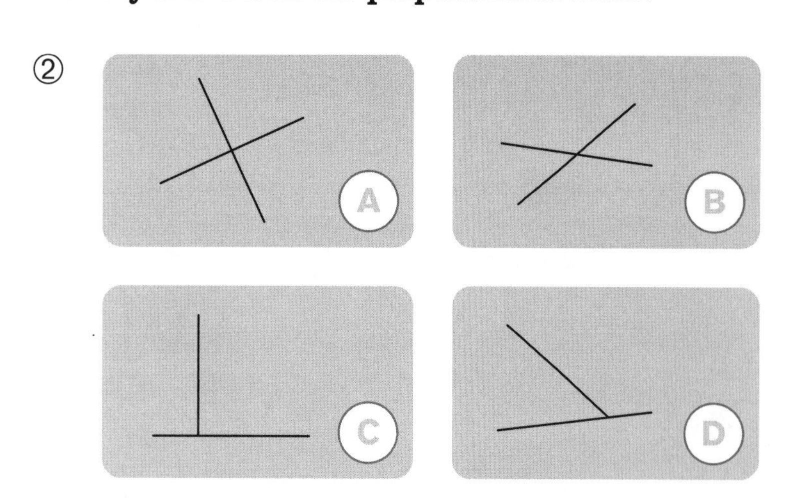

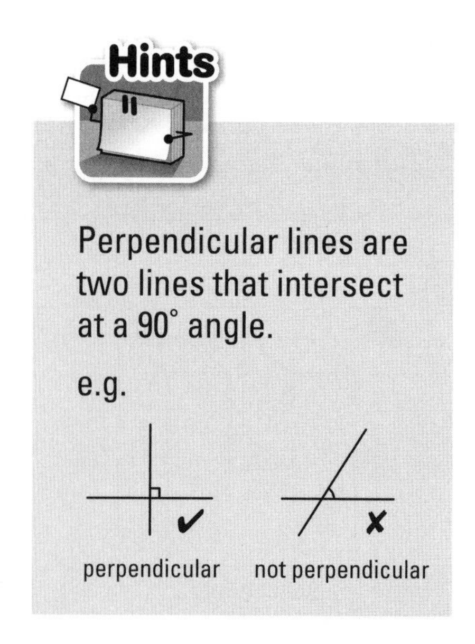

Hints
Perpendicular lines are two lines that intersect at a 90° angle.
e.g.
perpendicular
not perpendicular

SCAN IT!

Determine whether each line in bold is a bisector. If it is, determine whether it is perpendicular. Check the answers.

③

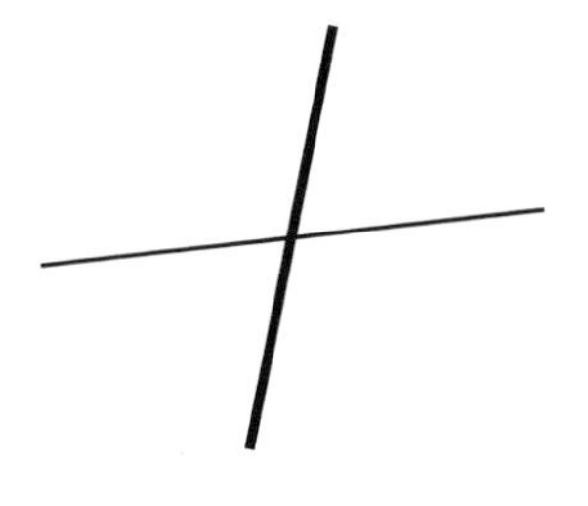

Ⓐ not bisector

Ⓑ bisector
- ◯ perpendicular
- ◯ not perpendicular

④

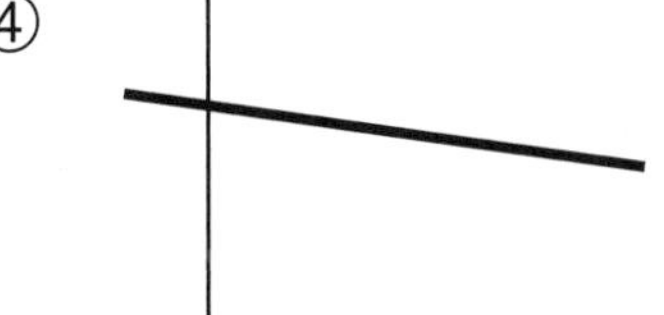

Ⓐ not bisector

Ⓑ bisector
- ◯ perpendicular
- ◯ not perpendicular

⑤

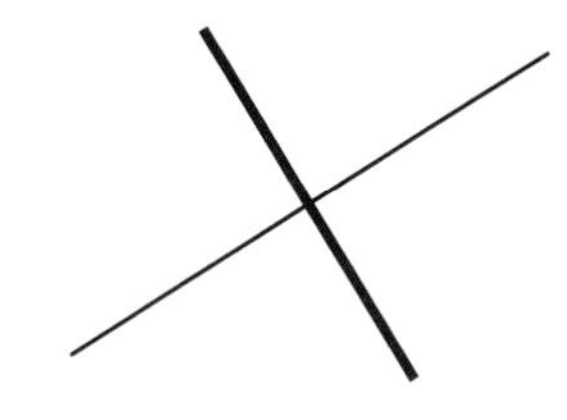

Ⓐ not bisector

Ⓑ bisector

- ◯ perpendicular
- ◯ not perpendicular

⑥

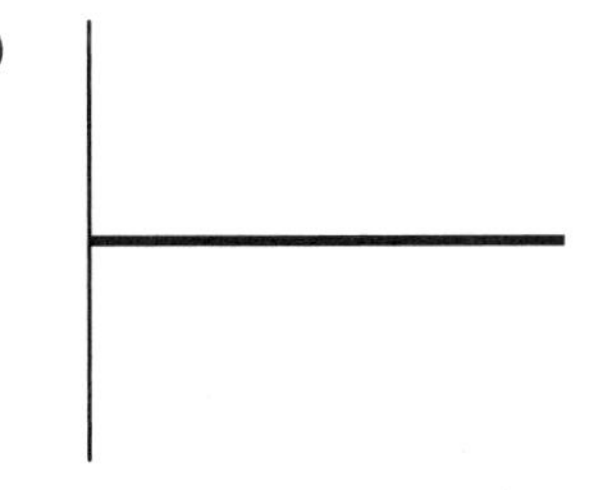

Ⓐ not bisector

Ⓑ bisector

- ◯ perpendicular
- ◯ not perpendicular

A bisector divides a line or an angle into two equal parts.

e.g.

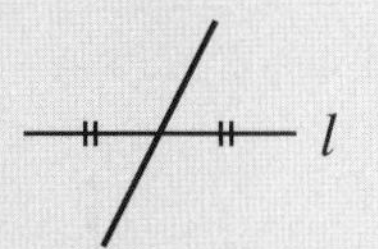

The bisector divides l into 2 equal parts.

A perpendicular bisector bisects a line at a 90° angle.

e.g.

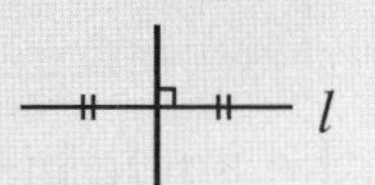

The perpendicular bisector divides l into 2 equal parts at a 90° angle.

Trace the perpendicular bisector for each line.

⑦

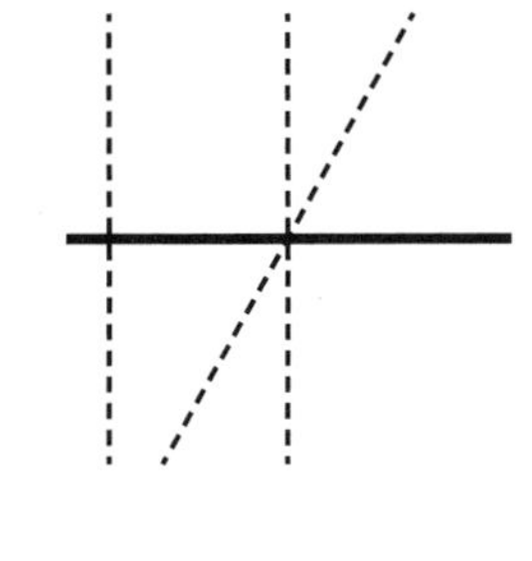

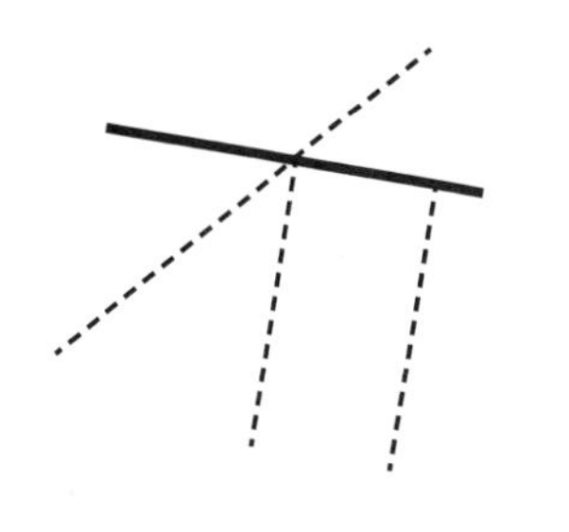

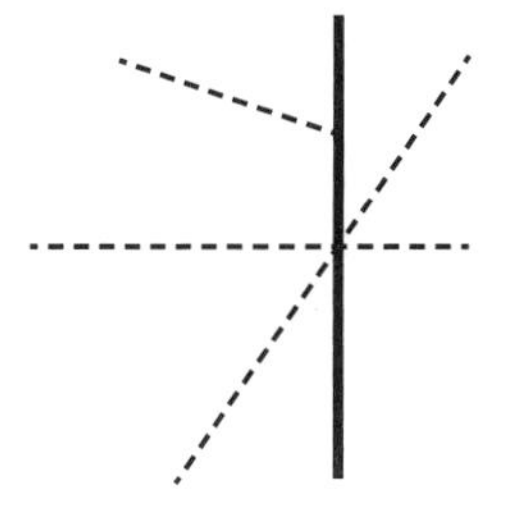

Determine whether each line in bold is an angle bisector. Put a check mark or a cross in the circle.

⑧ angle bisector ◯

⑨ angle bisector ◯

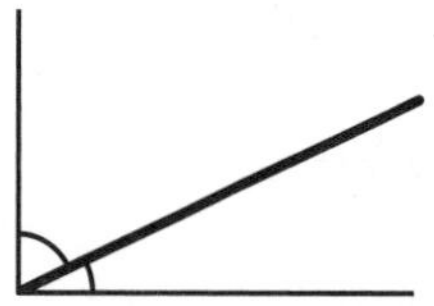

⑩ angle bisector ◯

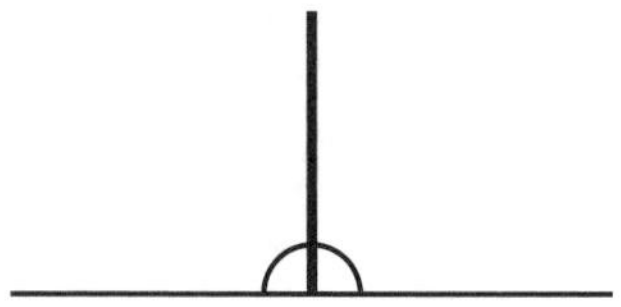

⑪ angle bisector ◯

Tips

The two angles that an angle bisector makes must be the same size.

e.g.

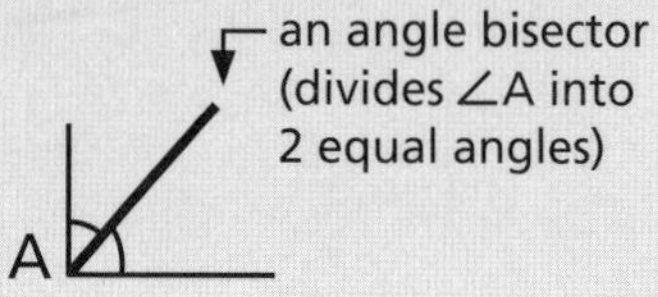

an angle bisector (divides ∠A into 2 equal angles)

A

Trace the angle bisector for each angle. Then measure and record the size of the new angles created by the bisector.

⑫ ________

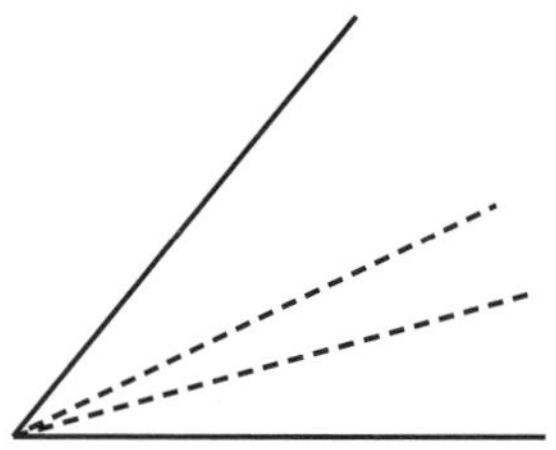

⑬ ________

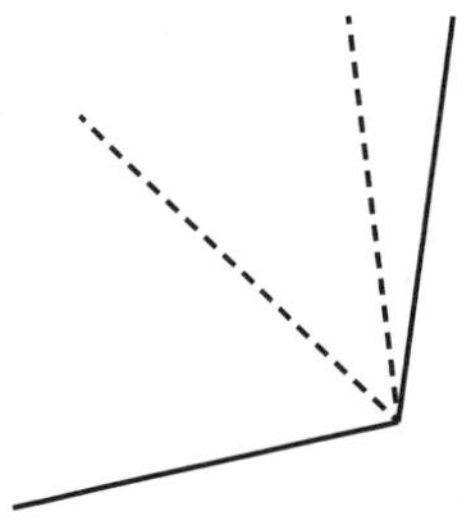

⑭ ________

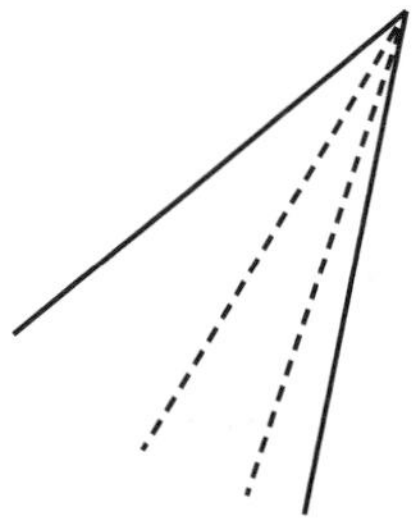

⑮ ________

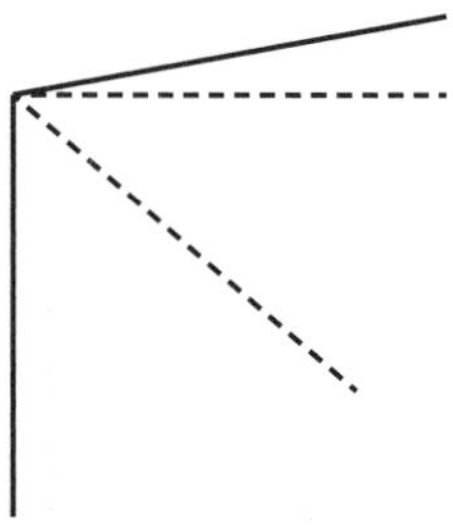

⑯ ________

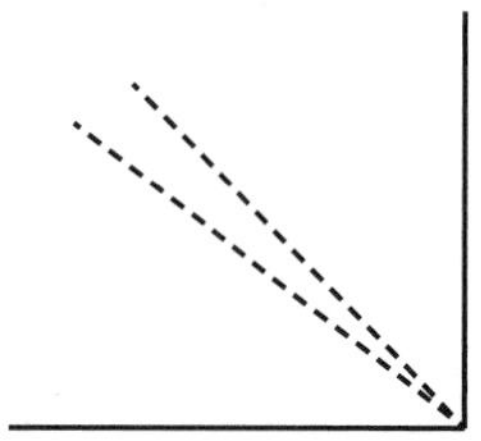

⑰ ________

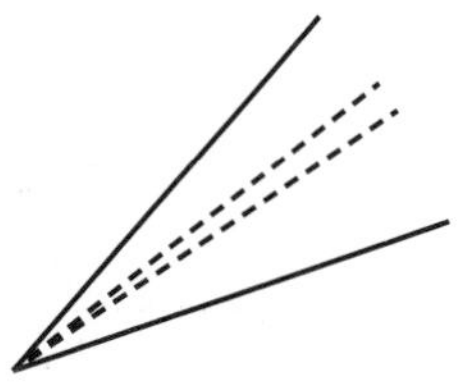

Draw the perpendicular bisector for each line and the angle bisector for each angle.

⑱ **Perpendicular Bisectors**

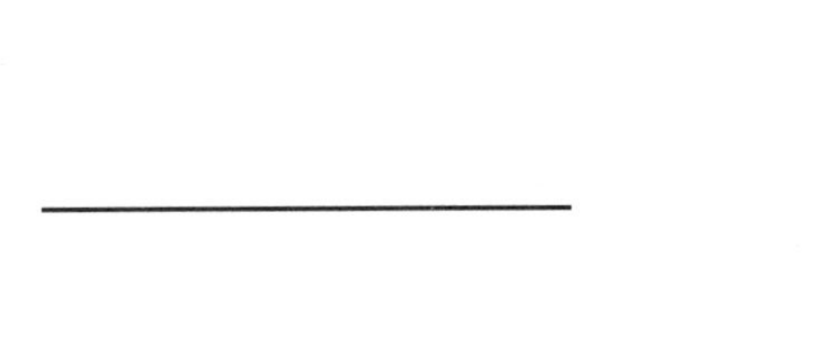

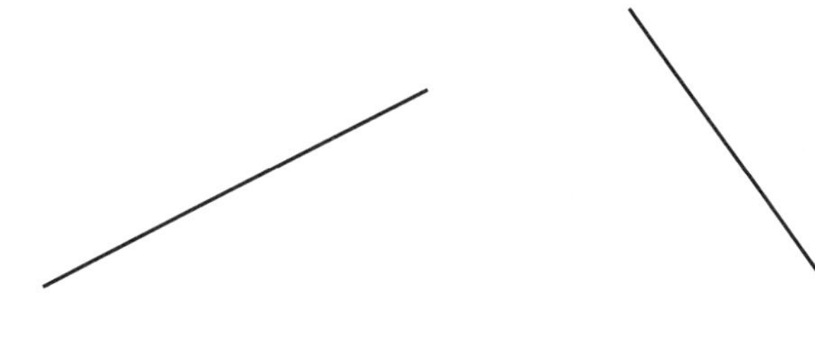

⑲ **Angle Bisectors**

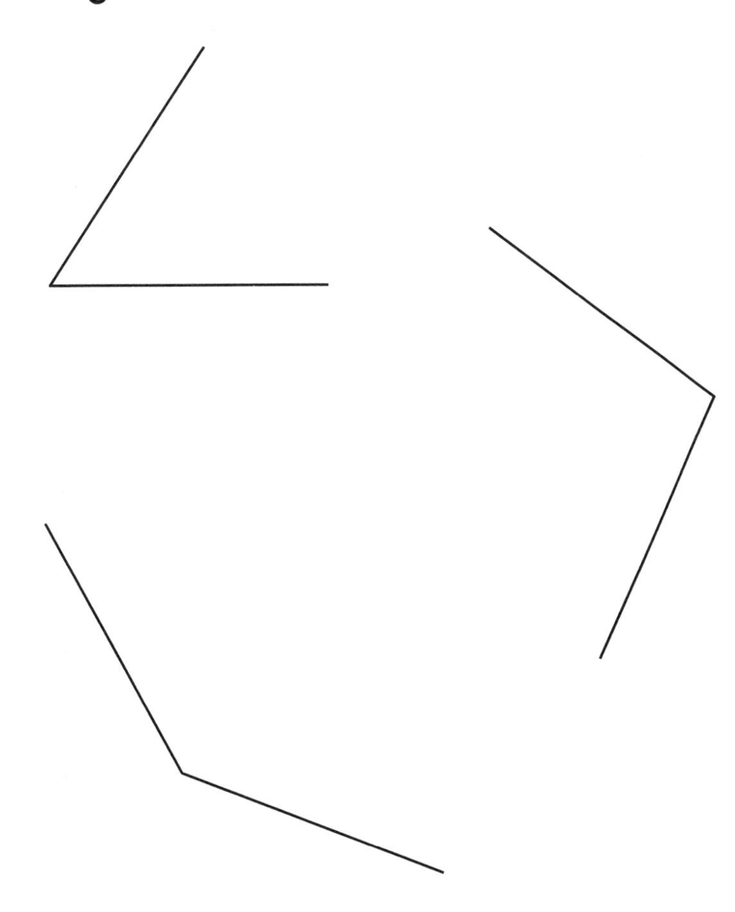

Hints

Follow the steps below to draw the bisectors.

Perpendicular Bisectors

1. Use a compass to draw two arcs from both ends on each side of the line.

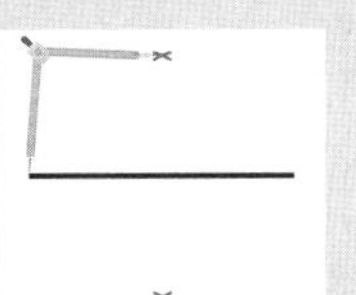

2. Connect the intersections of the arcs.

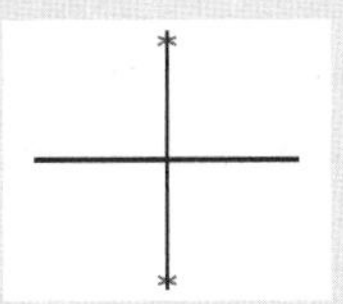

Angle Bisectors

1. Use a compass to draw an arc that intersects the arms of the angle.

2. Use each new intersection to draw another arc.

3. Connect the vertex with the point where the last two arcs met.

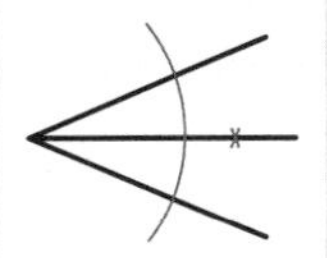

LEVEL 1 – BASIC SKILLS

13 Angles and Shapes

- understanding the relationships between angles and shapes

Triangles can be named by their angles and their sides.

Example Check the correct names of the triangle.

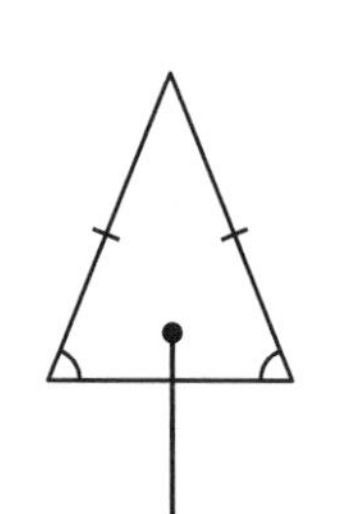

By Angles	By Sides
✔ acute triangle	◯ equilateral triangle
◯ obtuse triangle	✔ isosceles triangle
◯ right triangle	◯ scalene triangle

- all acute angles
- 2 equal sides

Try It

Name the triangle in two ways.

____

____

Name each triangle in two ways. Measure the sides and angles if needed.

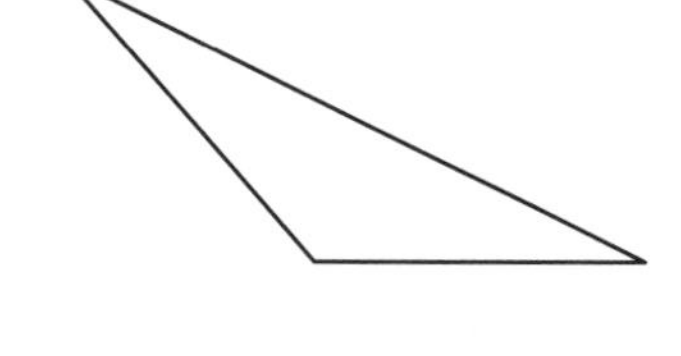

____

____

②

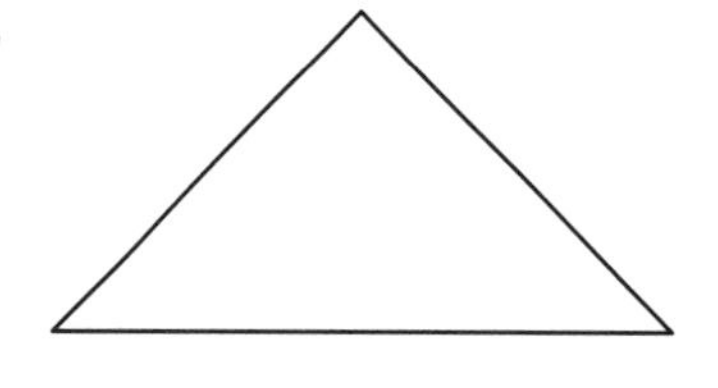

____

____

③

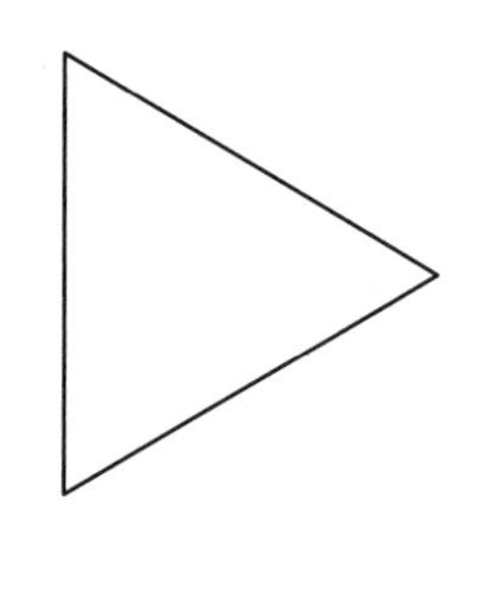

____

____

Circle "T" for the true statements and "F" for the false ones.

④ All isosceles triangles have a pair of equal angles. **T / F**

⑤ The angles in all equilateral triangles are 60°. **T / F**

⑥ A triangle can have both a right angle and an obtuse angle. **T / F**

⑦ Not all quadrilaterals can be cut into two triangles. **T / F**

⑧ Triangles made by cutting a square in half are always right isosceles triangles. **T / F**

⑨ Triangles cut from parallelograms are always obtuse triangles. **T / F**

Find the third angle of each triangle.

⑩ 60° 25° ______

⑪ 36° 115° ______

⑫ 82° 82° ______

⑬ 64° 32° ______

⑭ 120° 20° ______

⑮ 45° 110° ______

⑯ 70° 62° ______

⑰ 23° 86° ______

The sum of all angles in a triangle is 180°.

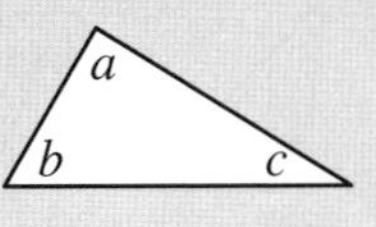

$a + b + c = 180°$

Draw triangles with the given lengths.

⑱ 5 cm, 3 cm, 4 cm

⑲ 2 cm, 3 cm, 4 cm

⑳ 4 cm, 3 cm, 4 cm

Follow these steps to draw a triangle.

5 cm, 2 cm, 4 cm

1. Draw a 5-cm line.

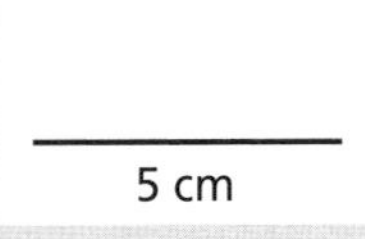

2. Take each end point as a centre. Draw an arc with a radius of 2 cm from one end and one with a radius of 4 cm from the other.

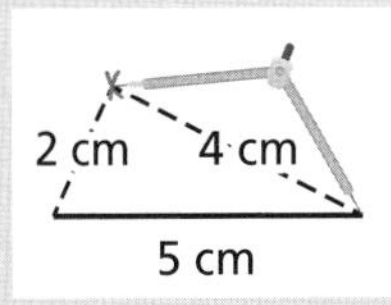

3. Join the end points of the line with the intersection of the arcs.

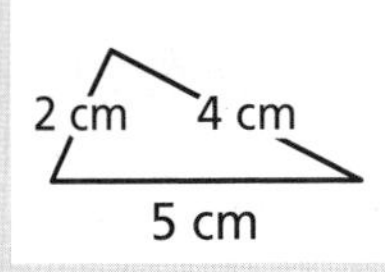

For each group, sketch to determine how many triangles you can draw using the given measurements. Then circle the correct words about unique triangles. (Hint: A triangle is unique if it can be drawn in exactly one way using a set of measurements.)

㉑ **Group A: knowing one side/two sides/three sides**

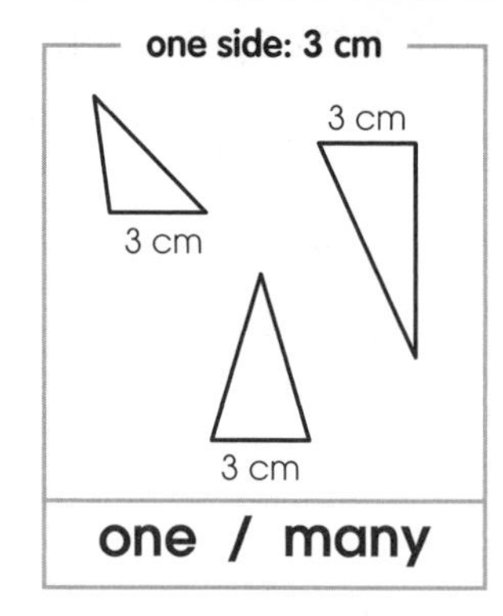

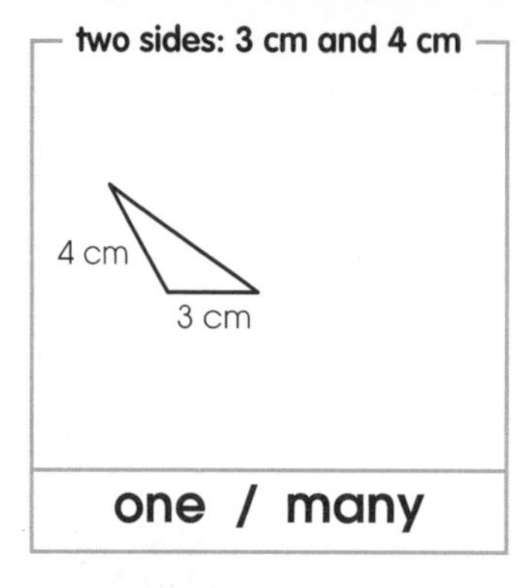

two sides: 3 cm and 4 cm
4 cm
3 cm
one / many

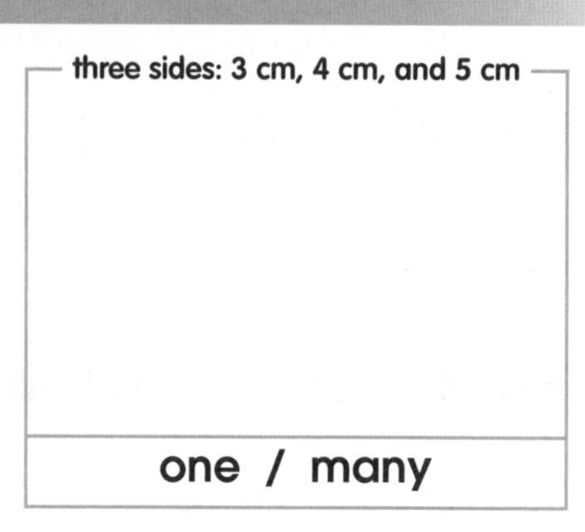

three sides: 3 cm, 4 cm, and 5 cm
one / many

A unique triangle can be drawn if the **lengths / angles** of the three sides are given.

㉒ **Group B: knowing one angle/one angle and one side/one angle and two sides**

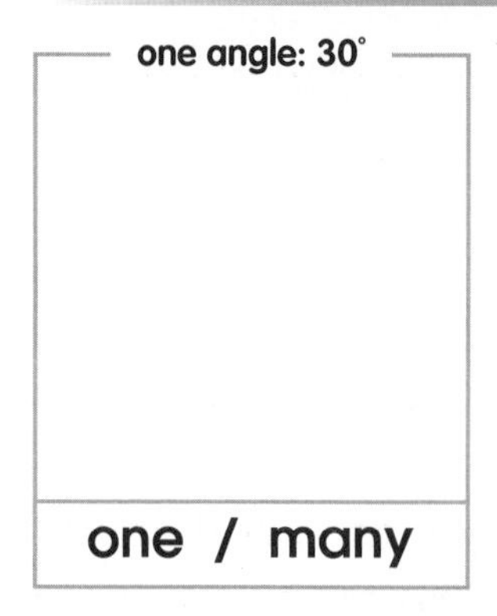

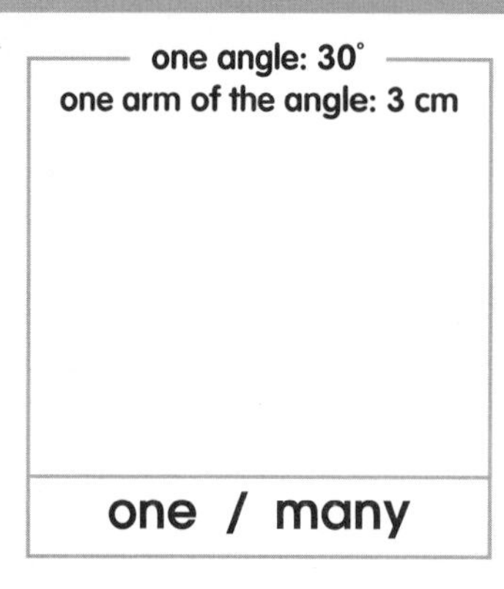

one angle: 30°
one arm of the angle: 3 cm
one / many

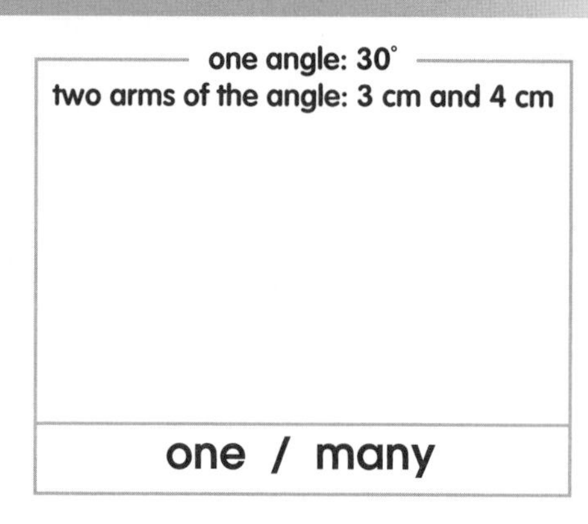

one angle: 30°
two arms of the angle: 3 cm and 4 cm
one / many

A unique triangle can be drawn if one **side / angle** and the **lengths / angles** of the angle's arms are given.

㉓ **Group C: knowing one side/one side and one angle/one side and two angles**

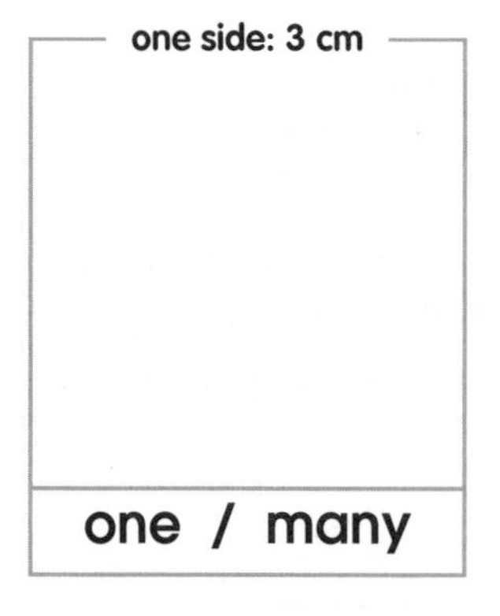

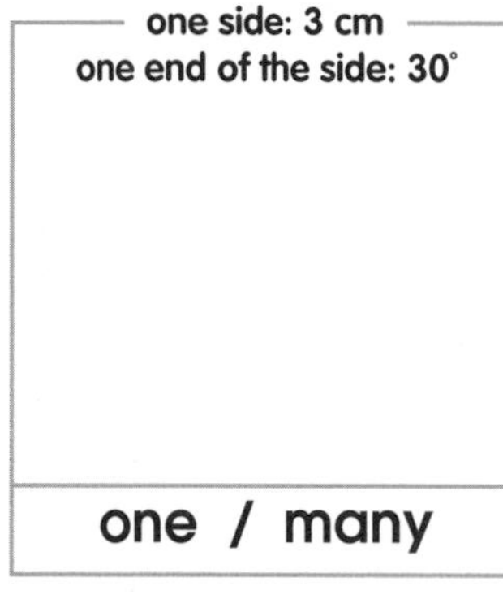

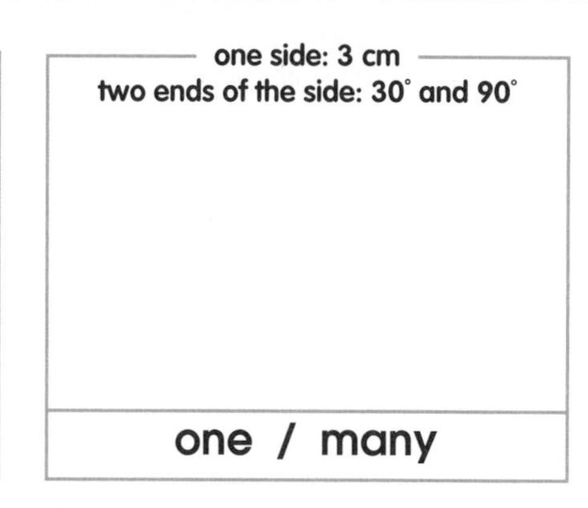

one side: 3 cm
two ends of the side: 30° and 90°
one / many

A unique triangle can be drawn if the **length / angle** of one side and the sizes of the two **lengths / angles** of the side are given.

For each quadrilateral, mark the angles and sides where applicable.

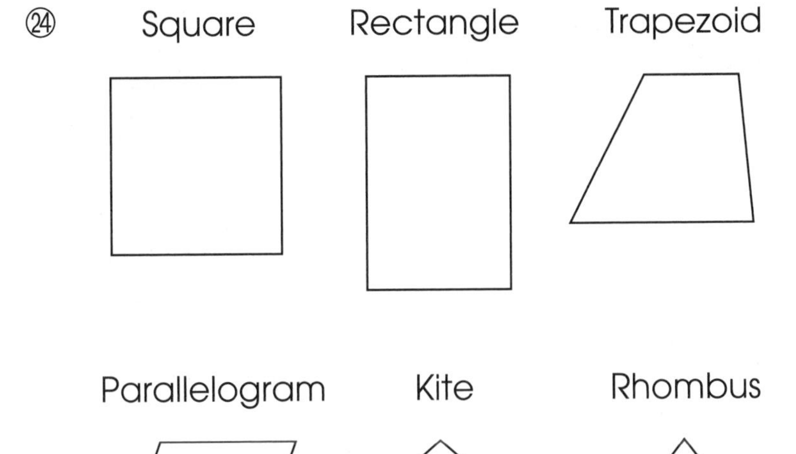

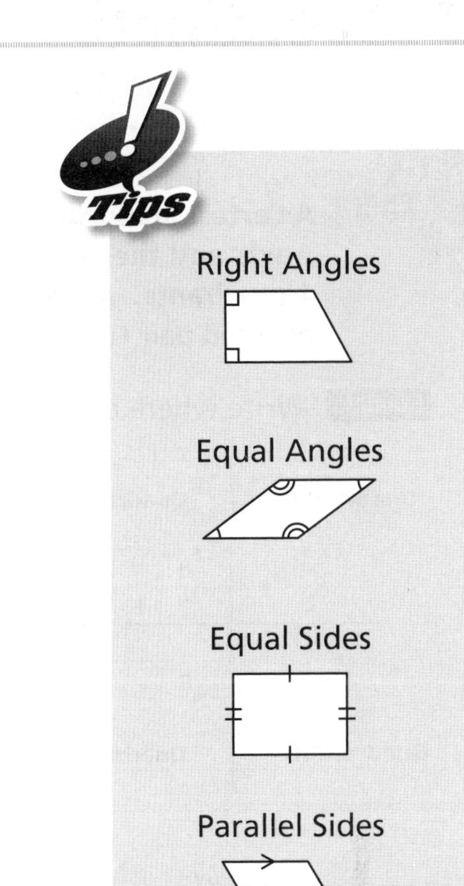

Draw the quadrilaterals. Then mark the angles and sides.

25 **Parallelogram**

- a side length of 3 cm
- an angle of 110°

26 **Rhombus**

- a side length of 3 cm
- an angle of 40°

27 **Trapezoid**

- two right angles and an angle of 100°
- 2 side lengths of 4 cm

14 Coordinates

- understanding coordinates

A Cartesian coordinate plane contains two axes – the *x*-axis and the *y*-axis – which divide the plane into 4 quadrants. A point on the plane is named by its ordered pair (*x*,*y*).

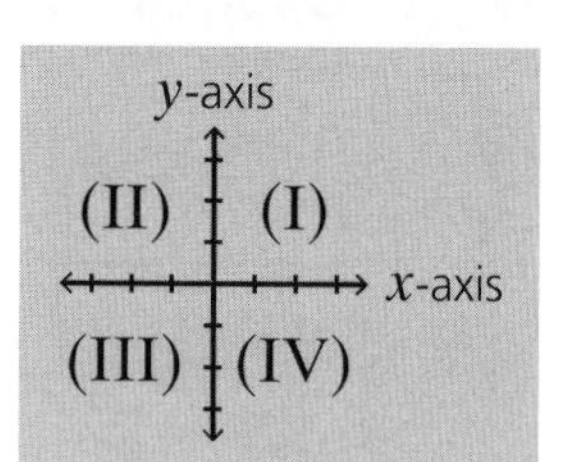

Example Write where each point is.

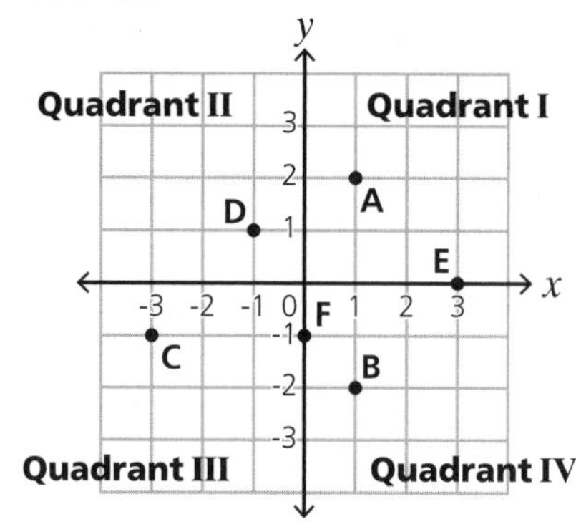

A(1,2): Quadrant I

B(1,-2): Quadrant IV

C(-3,-1): Quadrant III

D(-1,1): Quadrant ☐

E(3,0): ☐ -axis

F(0,-1): ☐ -axis

Locate and write the coordinates of each point. Then plot the others and answer the questions.

①

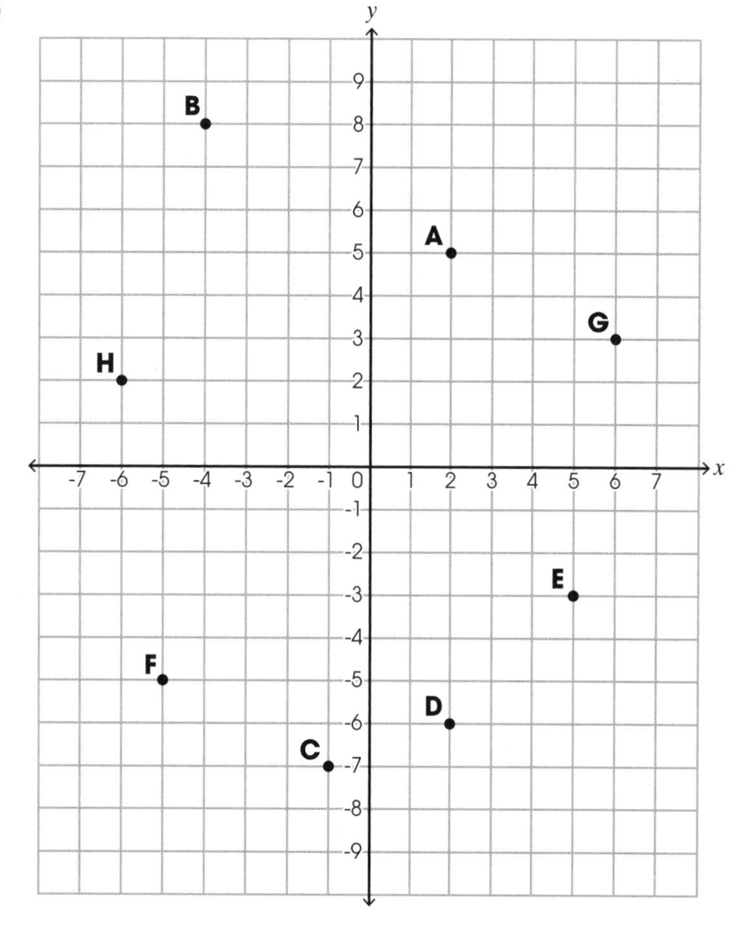

a. Write the coordinates and plot the points.

A(,)	I(0,6)
B(,)	J(3,0)
C(,)	K(-2,-2)
D(,)	L(-3,-7)
E(,)	M(-5,0)
F(,)	N(6,8)
G(,)	O(0,0)
H(,)	P(3,-8)

b. Which points are in

- Quadrant I? ________
- Quadrant IV? ________

Answer the questions about the map.

②

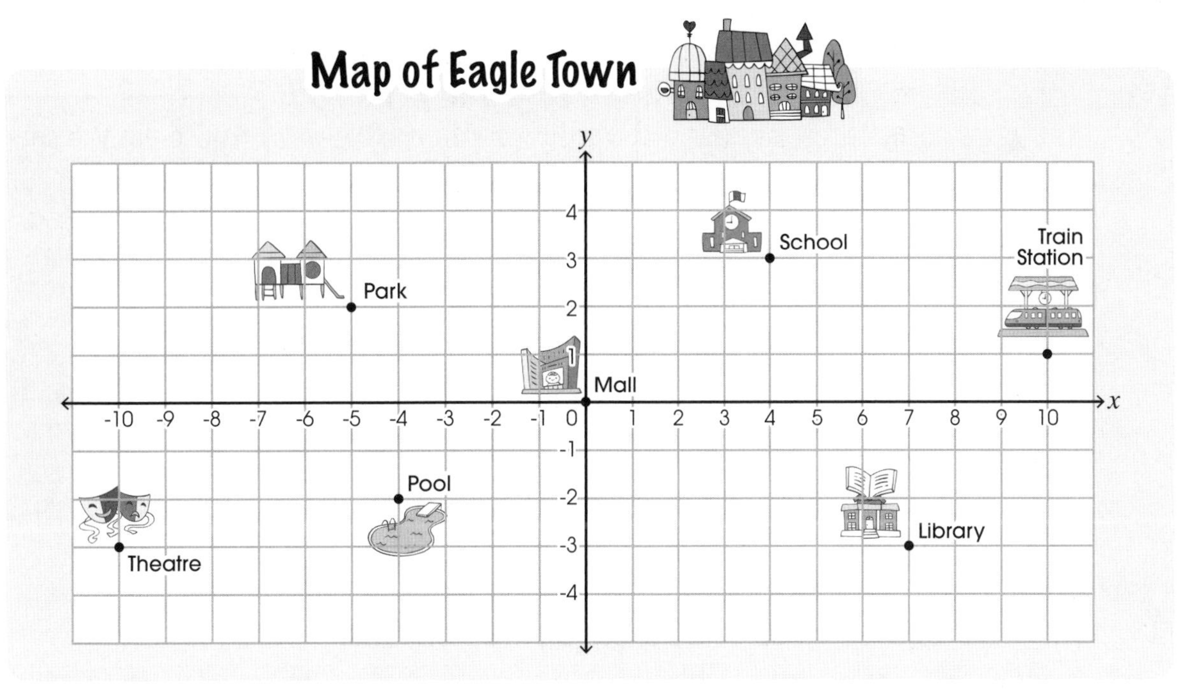

a. Plot the children on the map at the given coordinates.

b. What is located at the origin? __________

c. How many units is the pool from the mall? __________

d. How many units is the train station from the school? __________

e. Which is closer to the pool, the theatre or the park? __________

f. If Dave moves 8 units to the left, what will his new coordinates be? __________

g. If Helen moves 3 units down and 3 units to the left, where will she be? __________

h. How should Elaine move if she wants to reach the school?

____

i. How should Steve move if he wants to reach the theatre?

____

Look at Tony's coordinate plane and answer the questions.

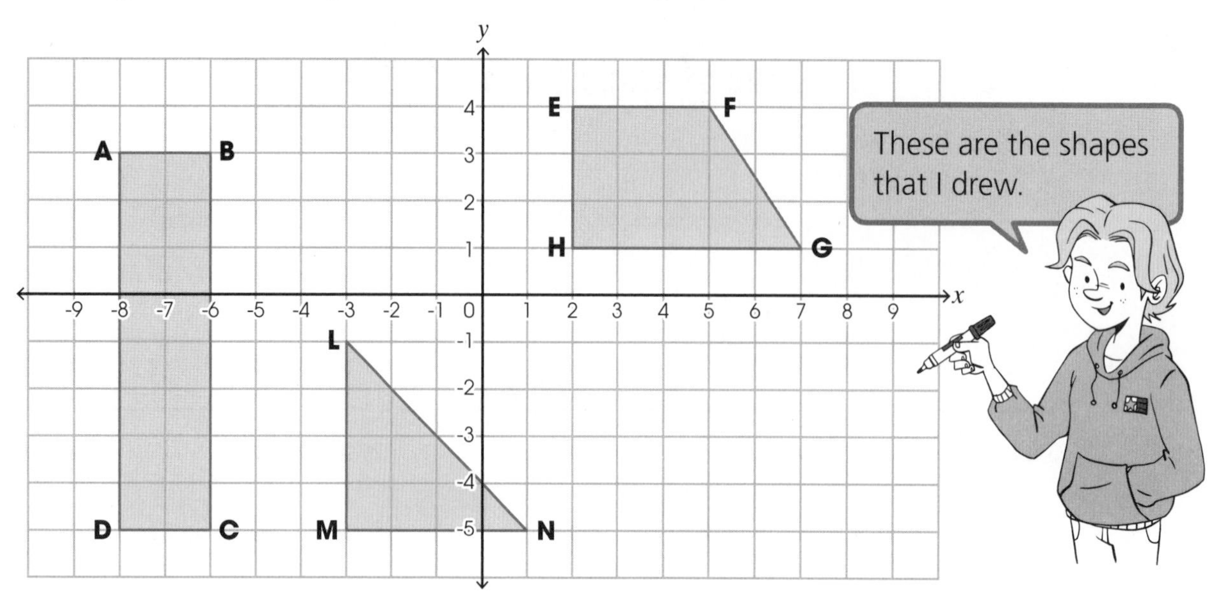

③ Write the coordinates of each shape's vertices.

Rectangle	Trapezoid	Triangle
A(,)	E(,)	L(,)
____________	____________	____________
____________	____________	____________
____________	____________	
How many units are between	How many units are between	How many units are between
• A and B?	• E and F?	• L and M?
_____ units	_____ units	_____ units
• A and D?	• G and H?	• M and N?
_____ units	_____ units	_____ units
What is the area?	What is the area?	What is the area?
_____ square units	_____ square units	_____ square units

④ Which shape is in Quadrant I? ____________

⑤ Which shape has a vertex in Quadrant IV? ____________

Plot each set of points on the coordinate plane and connect the dots in each set in alphabetical order to make the shapes. Name each shape. Then answer the questions.

⑥

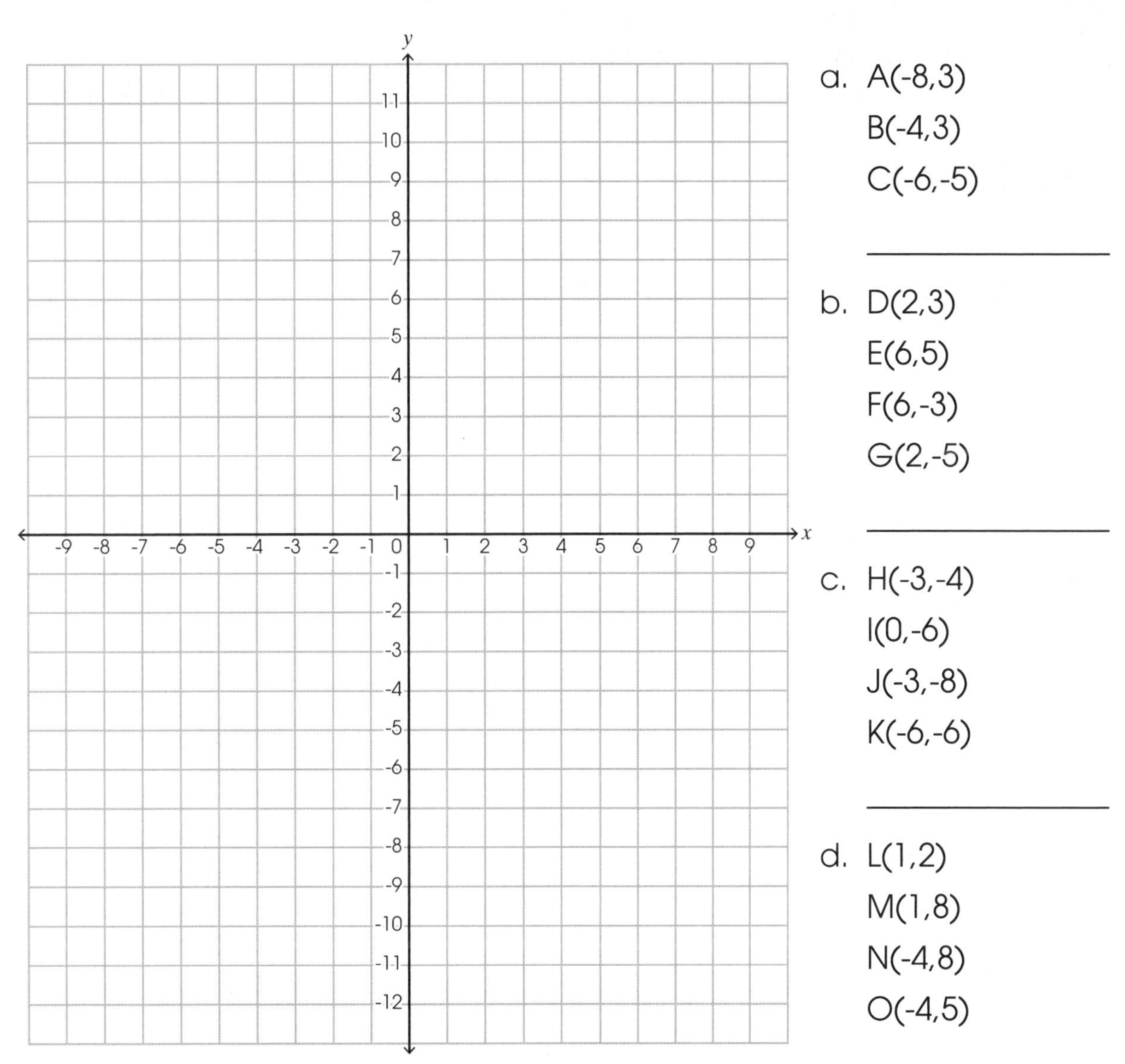

a. A(-8,3)
B(-4,3)
C(-6,-5)

____

b. D(2,3)
E(6,5)
F(6,-3)
G(2,-5)

____

c. H(-3,-4)
I(0,-6)
J(-3,-8)
K(-6,-6)

____

d. L(1,2)
M(1,8)
N(-4,8)
O(-4,5)

____

e. Which quadrant(s) is each shape in?

- Shape ABC: ____________
- Shape DEFG: ____________
- Shape HIJK: ____________
- Shape LMNO: ____________

f. Help the children make the shapes with the given information.

Jenny's Square

- in Quadrants I and IV
- a vertex at (3,-1) and a vertex at (8,-1)

Oliver's Rectangle

- in Quadrant III only
- a vertex at (-3,-10) and a vertex at (-3,-12)
- an area of 10 square units

15 Transformations

- understanding transformations

Translations, rotations, and reflections do not change the size of shapes. However, dilatations do – they either enlarge or reduce.

Example Identify the transformations.

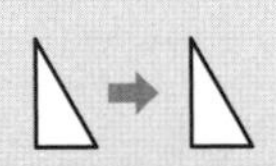

translation

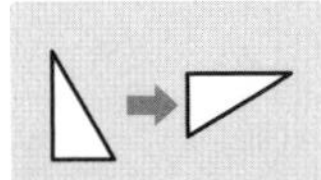

rotation

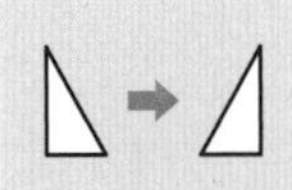

reflection

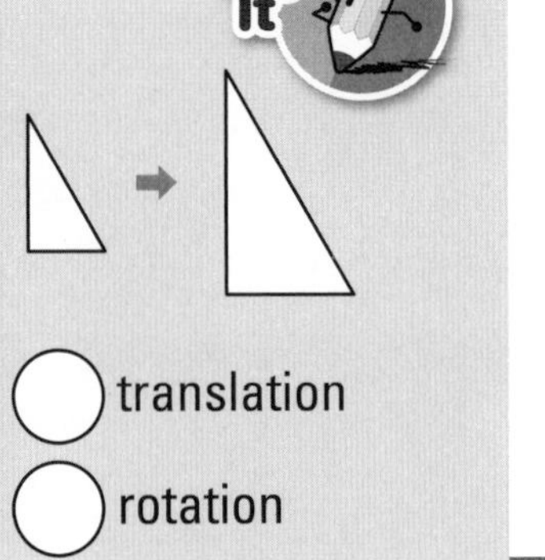

◯ translation
◯ rotation
◯ reflection
◯ dilatation

Identify the transformation for each pair. If it is a dilatation, specify whether it is an enlargement or reduction.

①
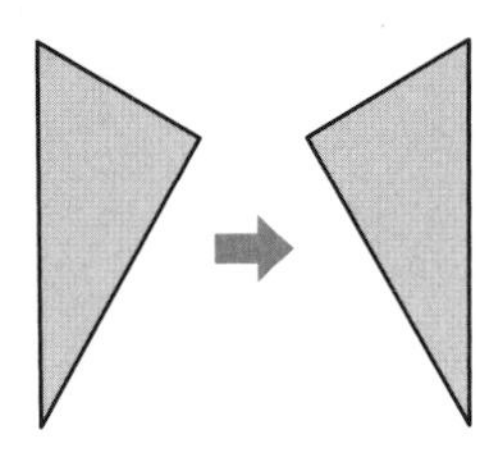

____

②
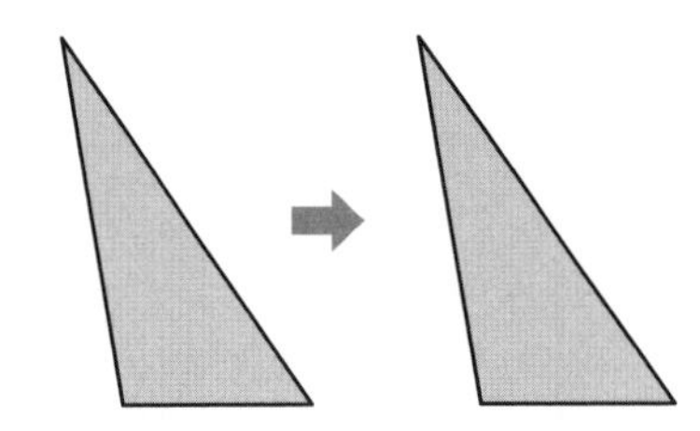

____

A dilatation is a transformation that enlarges or reduces a shape to a similar shape by a scale factor.

③
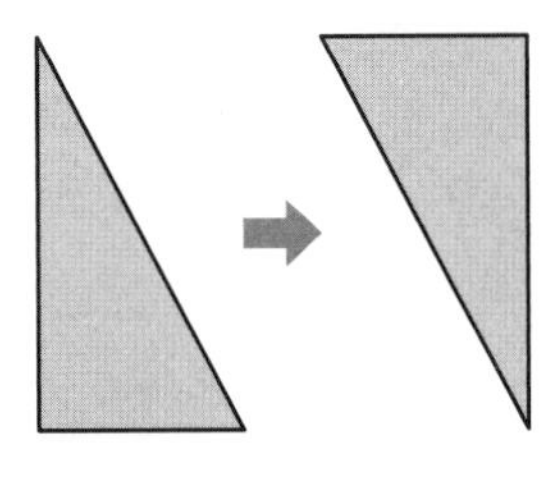

____

④
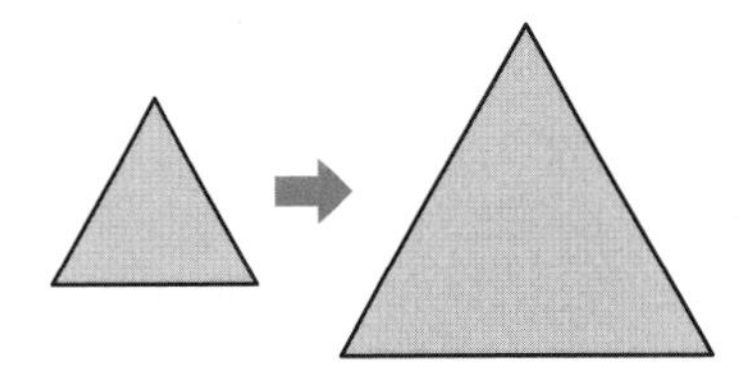

____

⑤

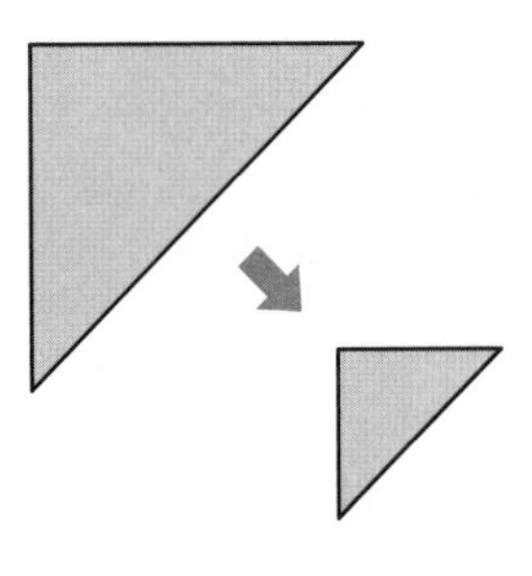

____

⑥
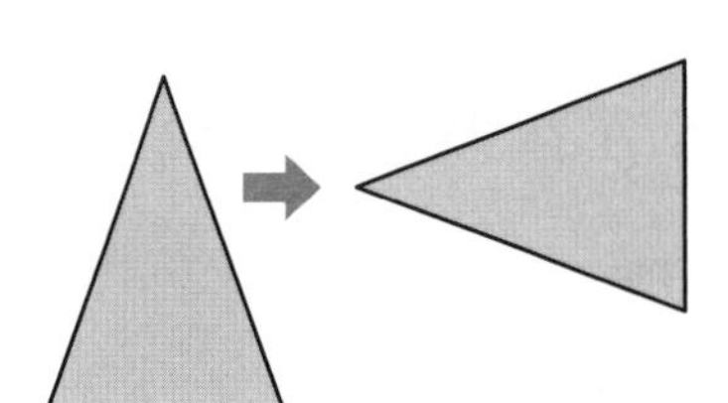

____

⑦
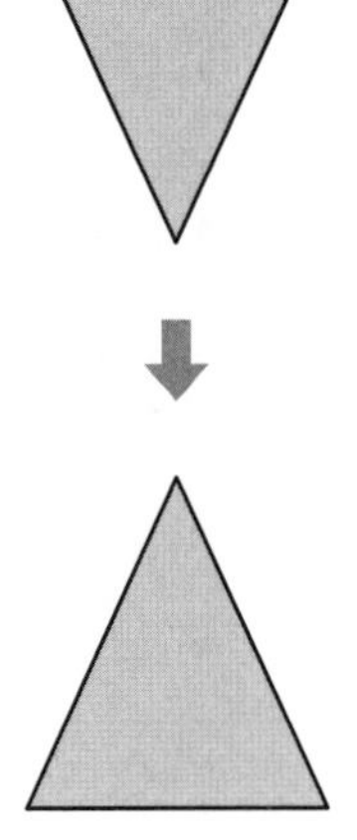

____

Describe the transformations.

⑧

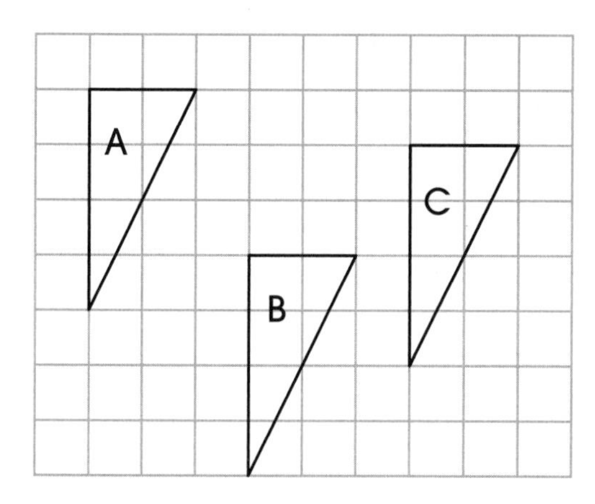

a. A to B:

a translation of ______ unit(s) down and ______ unit(s) to the right

b. B to C:

____

____

⑨

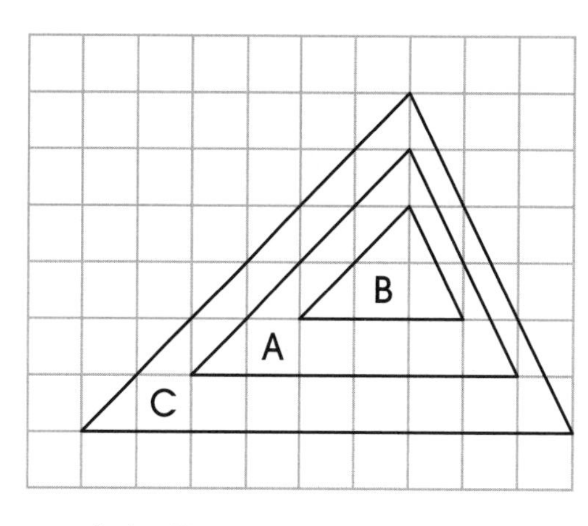

a. A to B:

a reduction by a scale factor of ______
2/3

(Hint: Compare their bases.)

b. B to C:

____

____

⑩

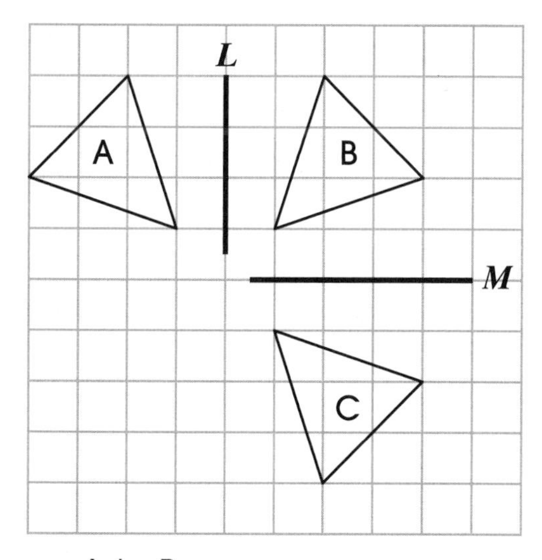

a. A to B:

a reflection in Line ______

b. B to C:

____

____

⑪

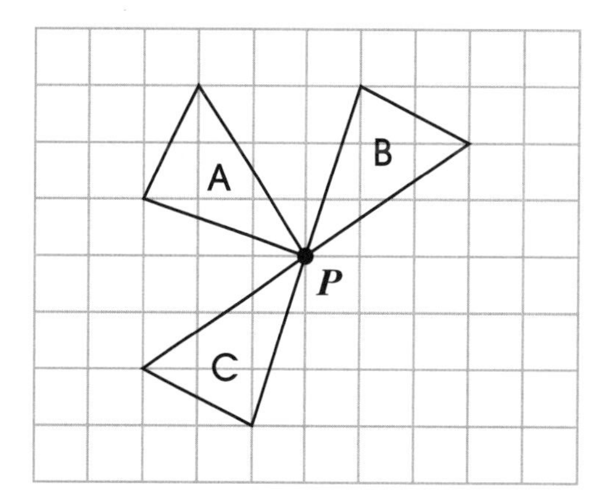

a. A to B:

a $\frac{1}{4}$ clockwise rotation about Point ______

b. B to C:

____

____

Describe the combined transformations of each shaded shape.

⑫

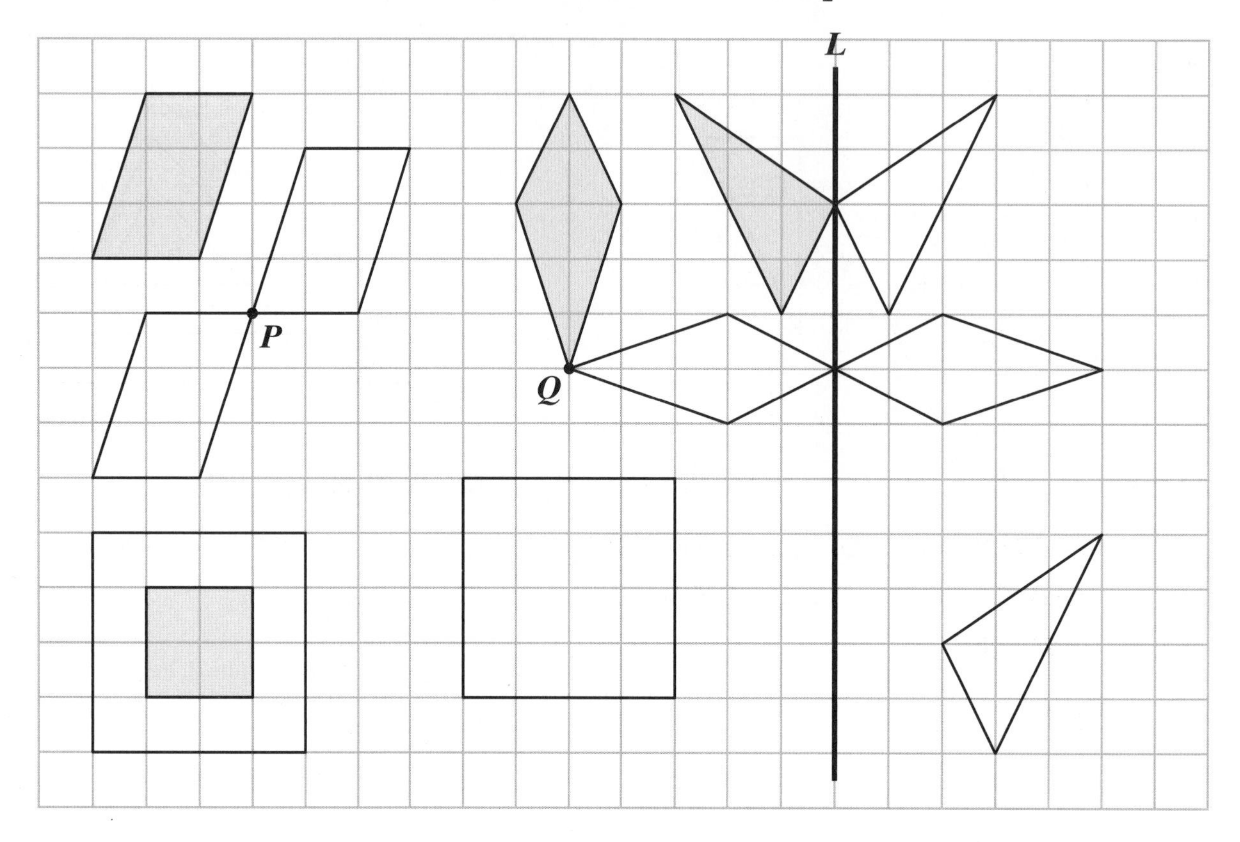

a. Parallelogram

Translate it ______ unit(s) to the right and ______ unit(s) down and then rotate it ______ ° about Point P.

There can be more than one way to describe a combined transformation.

b. Kite

____

____

c. Square

____

____

d. Triangle

____

____

Draw to show the transformations.

⑬ a. **Shape A**

- Reflect A in Line L. Name it B.
- Translate B 4 units down and 5 units to the left. Name it C.

b. **Shape X**

- Translate X 3 units down. Name it Y.
- Rotate Y $\frac{1}{4}$ clockwise about Point P. Name it Z.

c. **Shape I**

- Reflect I in Line M. Name it J.
- Translate J 13 units down and 7 units to the right. Name it K.

d. **Shape U**

- Rotate U $\frac{1}{4}$ clockwise about Point Q. Name it V.
- Translate V 5 units up and 4 units to the right. Name it W.

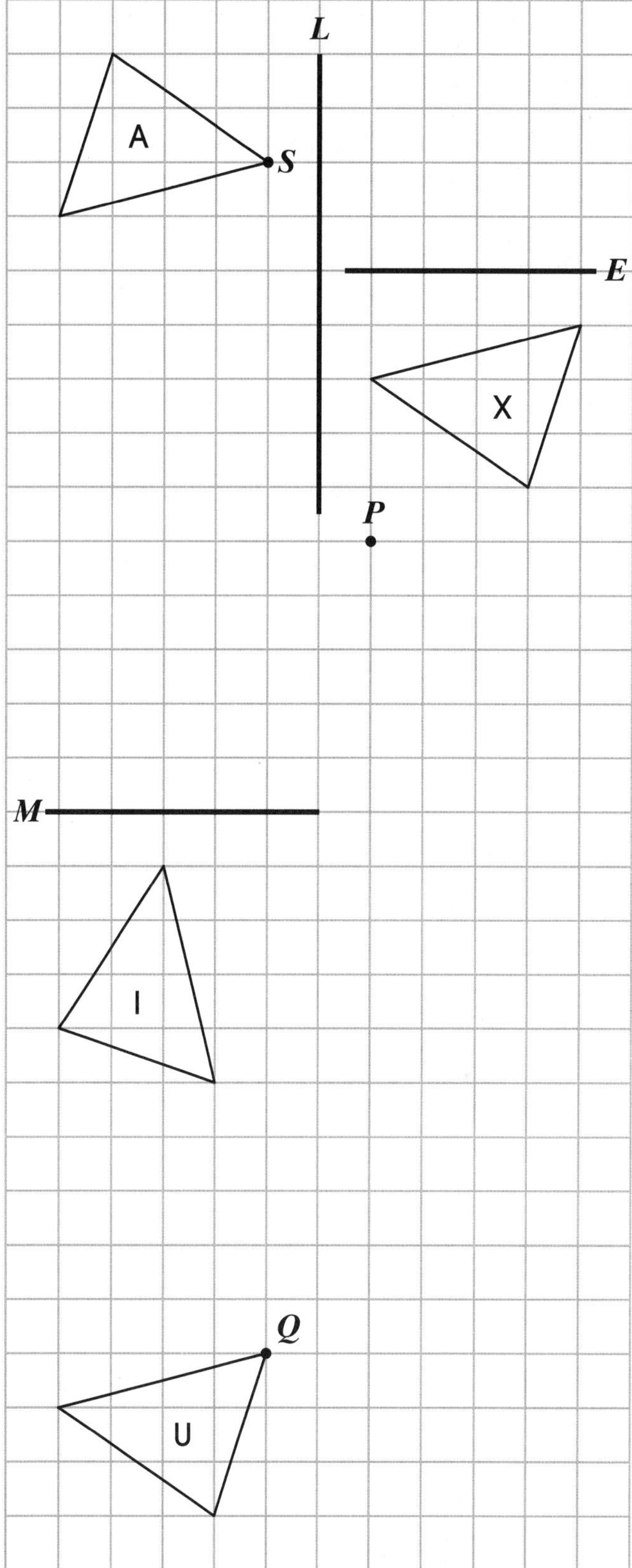

⑭ Describe two ways to transform Shape A into Shape X.

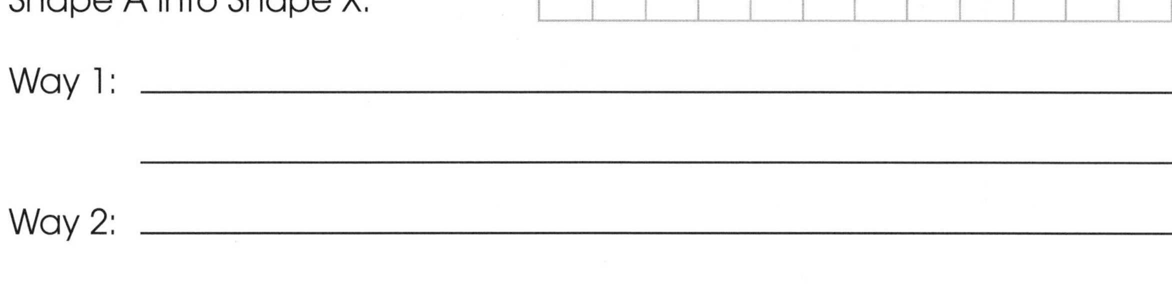

Way 1: ______________________________

____

Way 2: ______________________________

____

16 Algebraic Expressions

- understanding algebraic expressions

An algebraic expression can contain numbers, variables, and operations.

Check the correct algebraic expression.

the sum of 8 and a number

(✔) $8 + x$ (B) $8 - x$

the product of a number and 5

(A) $x \div 5$

(B) $5x$

Check the correct algebraic expressions.

① the difference of 7 and the product of 3 and a number

(A) $7 - 3x$

(B) $7 + 3x$

② subtract 4 from a number divided by 3

(A) $x \div 3 - 4$

(B) $x \div 4 - 3$

③ the sum of 1 and half of a number

(A) $1 + x \div 2$

(B) $1 + 2x$

④ triple the sum of a number and 1

(A) $1 + 3x$

(B) $3(x + 1)$

Match. Write the letters.

⑤ A the sum of 2 and the product of 3 and a number

B half the difference of 3 and a number

C divide a number by 2 and subtract 3

D the difference of 3 and the product of 2 and a number

E twice the sum of a number and 3

F triple the difference of 2 and a number

G divide 3 by a number and add 2

Algebraic Expressions

( ) $2x - 3$

( ) $x \div 2 - 3$

( ) $2 + 3x$

( ) $(3 - x) \div 2$

( ) $2(x + 3)$

( ) $3 \div x + 2$

( ) $3(2 - x)$

Write an algebraic expression for each.

⑥ the product of a number and 4 __________

⑦ a number divided by 8 __________

⑧ half of a number __________

⑨ a number subtracted by 7 and divided by 2 __________

⑩ the sum of 3 and two times a number __________

⑪ divide 4 by a number and subtract 5 __________

⑫ a quarter of a number subtract 2 __________

⑬ triple a number and add 7 __________

Determine which operation (+, –, x, or ÷) is related to each keyword. Then write each algebraic expression in words.

⑭

Keyword	Operation
increase	
product	
sum	
divide	
twice	
difference	
reduce	
half	
decrease	

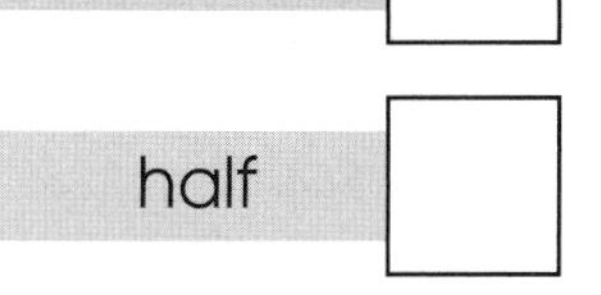

⑮ $5x$ __________

⑯ $9 \div x$ __________

⑰ $x + 1$ __________

⑱ $10 - x$ __________

⑲ $8 + x$ __________

⑳ $x \div 2$ __________

㉑ $4x + 1$ __________

㉒ $6(y - 2)$ __________

㉓ $x \div 4 - 9$ __________

Evaluate each expression using the given value.

㉔ $a = 2$

$3a + 1$

$= 3 \times$ ____ $+ 1$

$=$ ____

㉕ $d = 1$

$(5 - d) \div 2$

㉖ $x = 9$

$x \div 3 - 2$

㉗ $k = 2$

$10 - 4k$

Hints

To evaluate an algebraic expression, substitute the value for the variable.

e.g. Evaluate $x + 2$ where $x = 1$.

$x + 2$

$= 1 + 2$ ← Substitute 1 for x.

$= 3$

㉘ $y = 5$

$8 \times (2 + y)$

㉙ $s = 4$

$5s \div 2$

㉚ $w = 6$

$30 \div (w \div 2)$

For each algebraic expression, evaluate using the given values.

㉛ **$4x - 3$**

a. $x = 1$

b. $x = 2$

c. $x = 3$

㉜ **$(y + 1) \div 2$**

a. $y = 3$

b. $y = 5$

c. $y = 7$

Answer the questions about each scenario.

33

Depending on how many farmers there are, we can plant $3x + 5$ trees each day.

a. What does x represent?

(A) height of trees

(B) no. of farmers

b. How many trees can be planted if

- $x = 4$?

_______ trees

- $x = 7$?

____

I can bake $8(x - 1)$ cookies. The more time I have, the more cookies I can bake.

a. What does x represent?

(A) amount of sugar

(B) no. of hours

b. How many cookies can she bake if

- $x = 3$?

____

- $x = 5$?

____

I weigh more as I get older. I weigh $0.5(x - 1)$ kg now.

a. What does x represent?

(A) no. of weeks

(B) length of tail

b. How much does the dog weigh if

- $x = 5$?

____

- $x = 9$?

____

LEVEL 2

FURTHER YOUR UNDERSTANDING

1 Multiples and Factors

- understanding multiples and factors

Multiples and factors are closely related. For example, a number is always a multiple of each of its factors.

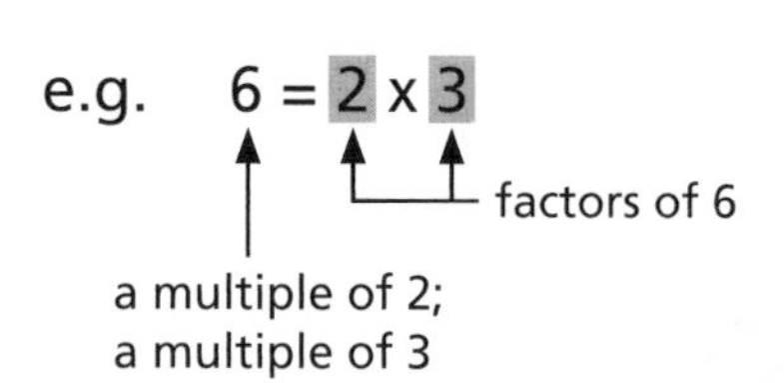

Example Find the common factors of 6 and 10.

Factors of 6: 1, 2, 3, 6

Factors of 10: 1, 2, 5, 10

Common factors of 6 and 10: 1, 2

Try It

List the first 10 multiples of 6 and 10.
Write their common multiples.

6: ____________________

10: ____________________

Common multiples of 6 and 10: ________

List the first 5 multiples of the numbers in each pair and write the common multiples. Then list the factors and write the common factors.

① **4 and 5**

Multiples

4: ____________________

5: ____________________

Common multiples: ________

Factors

4: ____________________

5: ____________________

Common factors: ________

② **8 and 12**

Multiples

8: ____________________

12: ____________________

Common multiples: ________

Factors

8: ____________________

12: ____________________

Common factors: ________

③ **7 and 14**

Multiples

7: ____________________

14: ____________________

Common multiples: ________

Factors

7: ____________________

14: ____________________

Common factors: ________

④ **9 and 15**

Multiples

9: ____________________

15: ____________________

Common multiples: ________

Factors

9: ____________________

15: ____________________

Common factors: ________

Look at each pair of numbers. Match to tell whether the given numbers are common factors of the pair, common multiples, or neither.

⑤ **11 and 13**

1 •
143 •
132 •

• common factor
• common multiple
• neither

⑥ **24 and 28**

168 •
4 •
84 •

• common factor
• common multiple
• neither

⑦ **10 and 25**

100 •
5 •
2 •

• common factor
• common multiple
• neither

⑧ **12 and 20**

4 •
120 •
5 •

• common factor
• common multiple
• neither

⑨ **16 and 18**

72 •
2 •
144 •

• common factor
• common multiple
• neither

⑩ **15 and 21**

5 •
105 •
3 •

• common factor
• common multiple
• neither

If the number in bold is the least common multiple or the greatest common factor, check the box; otherwise, put a cross and write the correct number beside it. Then answer the questions.

⑪ **Least Common Multiple**

a. 10 and 15: **150** ☐ ______
b. 16 and 24: **48** ☐ ______
c. 12 and 15: **60** ☐ ______
d. 14 and 20: **70** ☐ ______
e. 9 and 25: **450** ☐ ______
f. 6 and 16: **48** ☐ ______

⑫ **Greatest Common Factor**

a. 12 and 16: **2** ☐ ______
b. 5 and 7: **1** ☐ ______
c. 8 and 12: **4** ☐ ______
d. 6 and 10: **1** ☐ ______
e. 15 and 18: **3** ☐ ______
f. 10 and 16: **4** ☐ ______

⑬ What is the least common factor of all numbers? ______

⑭ Is there a greatest common multiple? ______

Find the factors and multiples of the given numbers. Then fill in the blanks to show which ones they have in common.

⑮

16 30

45

36 40

a. Common factors of ______ and ______ : 1, 2, 3, 6

b. Common factors of ______ and ______ : 1, 2, 4, 8

c. Common multiples of ______ and ______ : 90, 180

d. Common multiples of ______ and ______ : 120, 240

⑯

60 39

52

45 30

a. Common factors of ______ and ______ : 1

b. Common factors of ______ and ______ : 1, 13

c. Common multiples of ______ and ______ : 60, 120

d. Common multiples of ______ and ______ : 390, 780

Solve each problem by finding the multiples or factors of the numbers. Fill in the blanks.

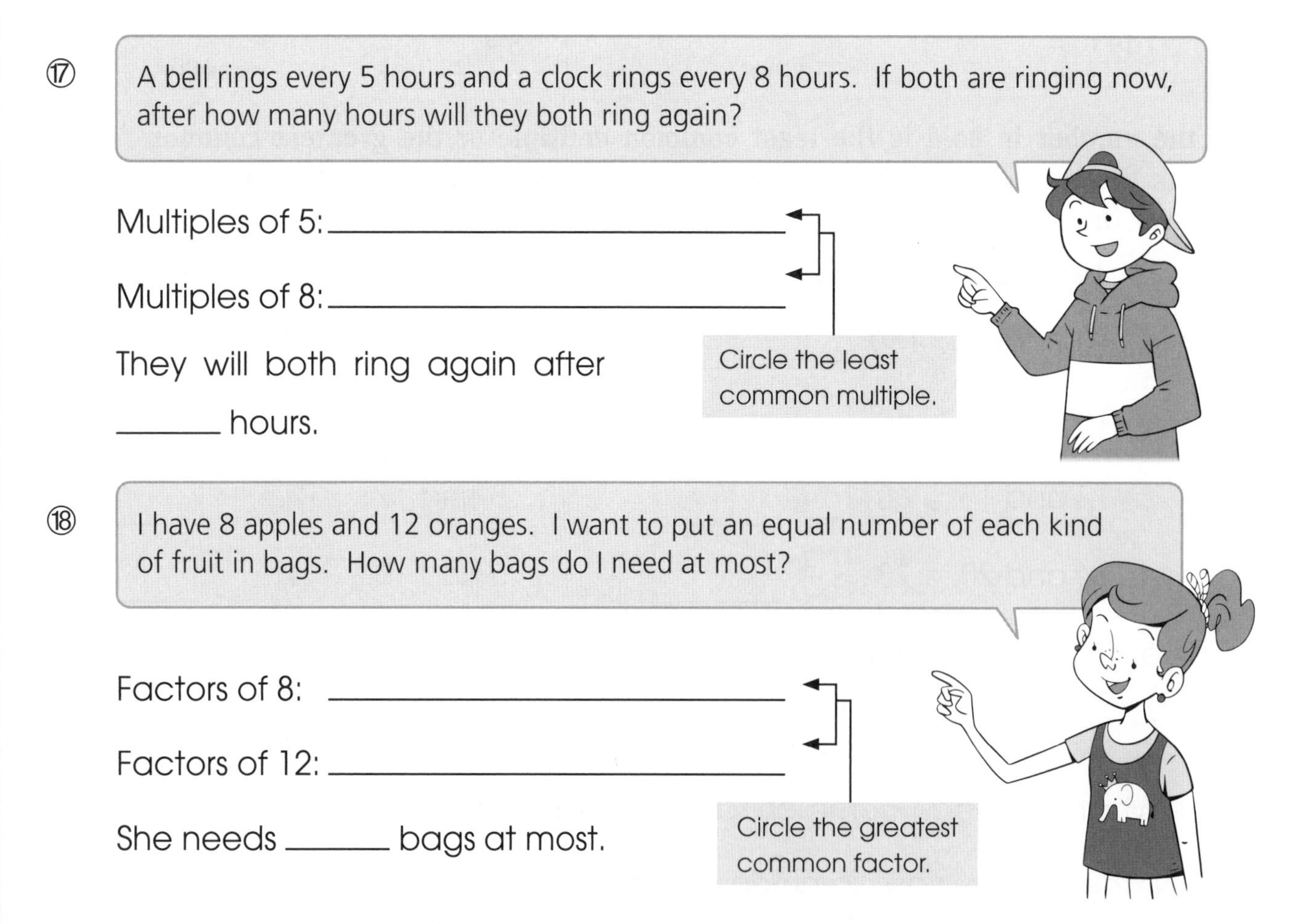

⑰ A bell rings every 5 hours and a clock rings every 8 hours. If both are ringing now, after how many hours will they both ring again?

Multiples of 5: ______________________

Multiples of 8: ______________________

They will both ring again after ______ hours.

Circle the least common multiple.

⑱ I have 8 apples and 12 oranges. I want to put an equal number of each kind of fruit in bags. How many bags do I need at most?

Factors of 8: ______________________

Factors of 12: ______________________

She needs ______ bags at most.

Circle the greatest common factor.

For each problem, check whether it involves multiples or factors. Then solve it. Show your work.

⑲ Alice visits her grandma every 2 weeks and her aunt every 3 weeks. If she visited her grandma and aunt today, when will she visit both of them in the same week again?

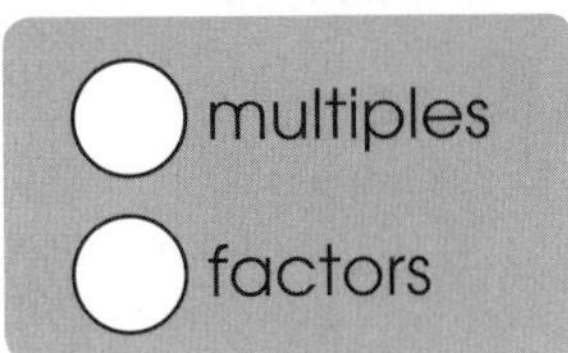

She will visit them again after ______ weeks.

⑳ Ms. Via is putting 14 boys and 21 girls into boys' teams and girls' teams. How many players will be on each team if all the teams are the same size?

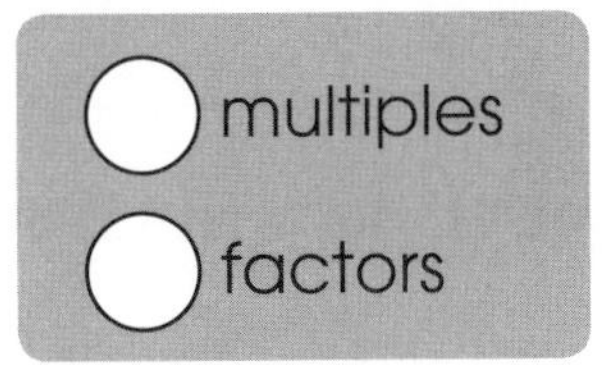

There will be ______ players on each team.

㉑ Agnes and Kyle played a carnival game. Every time they played, Agnes scored 10 points and Kyle scored 12. They played a different number of times, but ended with the same total score. At least how many points did each child score in total?

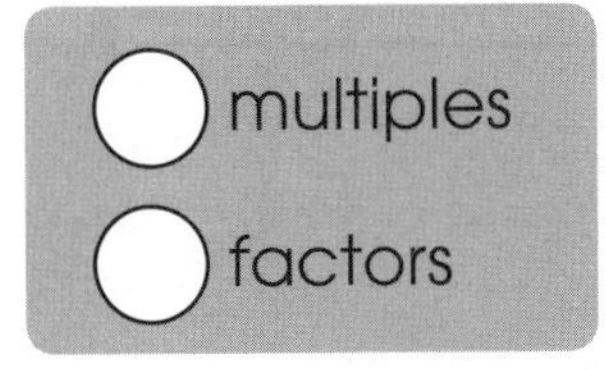

Each child scored at least ______ points in total.

㉒ A candy store was doing a giveaway. Every 5th customer in line got a lollipop and every 8th customer got a candy bar. Jason was the first customer to get both a lollipop and a candy bar. What was his number in line?

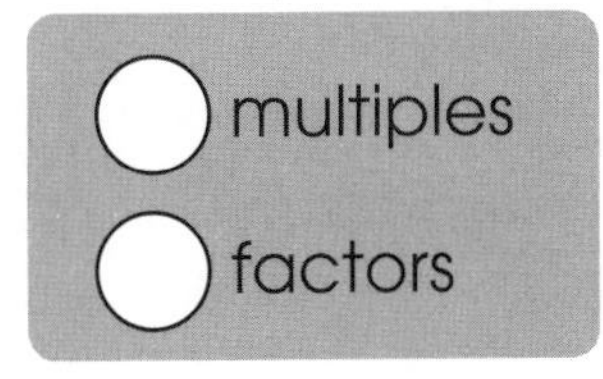

His number in line was ______ .

2 Exponents

- using exponents to represent numbers

Exponents show repeated multiplication of a number called the base.

e.g. 5 x 5 x 5 = 5^3 (power; 3 ← exponent; 5 ← base; 5 x 5 x 5 ← a repeated multiplication)

Example Check the expression(s) that can be represented as a power.

A) 3 + 3 + 3 + 3 + 3 ← not a repeated multiplication

✔ B) 4 x 4 x 4 x 4 x 4 ← 4^5

C) 3 x 3 x 3 + 3 x 3 x 3 ← not a repeated multiplication

A) 1 x 1 x 1 x 1 B) 1 + 1 + 1 + 1

C) 2 x 2 D) 2 x 2 + 2 x 2

Determine whether each expression can be written as a power. If so, write it on the line; otherwise, put a cross.

① A 5 x 5 x 5 ______

B 2 x 2 x 2 x 3 ______

C 3 + 3 ______

D 10 x 10 x 10 ______

E 9 + 9 + 9 + 9 ______

F 11 x 11 x 11 x 11 ______

G 200 x 200 ______

H 4 + 4 + 4 + 4 + 4 ______

I 8 x 8 x 8 x 8 ______

J 5 x 5 x 5 x 5 x 5 ______

K 10 + 10 + 10 + 10 ______

L 2 + 2 + 2 + 2 ______

M 6 x 6 x 6 ______

N 15 x 15 x 15 x 15 x 15 ______

Fill in the missing base or exponent.

② 25 = $5^{\square}$

③ 64 = $\square^2$

④ 36 = $\square^2$

⑤ 81 = $\square^2$

⑥ 27 = $3^{\square}$

⑦ 100 = $10^{\square}$

⑧ 625 = $5^{\square}$

⑨ 49 = $7^{\square}$

⑩ 16 = $2^{\square}$

Write each expression as a power.

An expression can be written as a power as long as each number has the same base.

e.g. $4^2 \times 4$ (same base)

$= \underline{4 \times 4} \times 4$

$= 4^3$

⑪ $5^2 \times 5 =$ ______

⑫ $2^3 \times 2 =$ ______

⑬ $4^3 \times 4 =$ ______

⑭ $3 \times 3^4 =$ ______

⑮ $9 \times 9^2 \times 9 =$ ______

⑯ $8^2 \times 8 \times 8^2 =$ ______

⑰ $6 \times 6^3 \times 6^2 =$ ______

⑱ $2^4 \times 2 \times 2^3 =$ ______

⑲ $3^2 \times 3 \times 3^4 =$ ______

⑳ $5^1 \times 5^2 \times 5^3 =$ ______

㉑ $8^3 \times 8^0 \times 8^2 =$ ______

㉒ $4^4 \times 4^3 \times 4^2 =$ ______

㉓ $10 \times 10 \times 10^2 \times 10 \times 10^3 \times 10^2 \times 10 \times 10 =$ ______

㉔ $25 \times 25^2 \times 25^3 \times 25^4 \times 25 \times 25 \times 25 \times 25 \times 25 =$ ______

㉕ $18 \times 18^2 \times 18 \times 18 \times 18^3 \times 18^0 \times 18^0 \times 18^0 \times 18^0 =$ ______

㉖ $100 \times 100 \times 100^2 \times 100 \times 100 \times 100^3 \times 100 \times 100 =$ ______

Check the correct products of powers. Then write the products of powers.

㉗ $2 \times 2 \times 2 \times 3 \times 3$

- (A) $2^3 \times 3^2$
- (B) $2^3 + 3^2$

㉘ $4 \times 4 \times 4 \times 4 \times 5 \times 5$

- (A) $5^4 \times 4^5$
- (B) $4^4 \times 5^2$

㉙ $9 \times 9 \times 2 \times 9 \times 9 \times 9$

- (A) 2×9^5
- (B) $2 + 9^5$

㉚ $8 \times 8 \times 8 \times 7 \times 7 \times 7$

- (A) $3^7 \times 3^8$
- (B) $7^3 \times 8^3$

㉛ $3 \times 3 \times 3 \times 5 \times 5 \times 5 \times 5$ ______

㉜ $4 \times 4 \times 4 \times 7 \times 7$ ______

㉝ $7 \times 7 \times 7 \times 2 \times 3 \times 3$ ______

㉞ $4 \times 4 \times 4 \times 5 \times 5 \times 6$ ______

㉟ $3 \times 2 \times 3 \times 7 \times 3 \times 2$ ______

㊱ $8 \times 8 \times 6 \times 6 \times 8 \times 9$ ______

㊲ $5 \times 5 \times 5 \times 5 \times 6 \times 7 \times 7$ ______

㊳ $2 \times 7 \times 5 \times 9 \times 9 \times 9 \times 7 \times 5$ ______

㊴ $3 \times 3 \times 5 \times 5 \times 5 \times 2$ ______

Evaluate each power. Then write each expression as a power of 10.

40. **Powers of 10**

10^0 ____________

10^1 ____________

10^2 ____________

10^3 ____________

10^4 ____________

10^5 ____________

10^6 ____________

10^7 ____________

10^8 ____________

10^9 ____________

10^{10} ____________

Hints

10, 100, 1000, etc. can be expressed as powers of 10. The exponents tell how many zeros come after the 1.

e.g. $10^3 = 1000$ (3 zeros)

41. 10 x 10 x 10 ________
42. 10 x 10 x 10 x 10 x 10 ________
43. 1000 ________
44. 10 000 000 ________
45. ten thousand ________
46. one million ________
47. one hundred thousand ________

Circle the correct values. Then write the values.

48. 5×10^2 = **50 / 500**
49. 2×10^3 = **2000 / 300**
50. 9×10^4 = **90 000 / 40 000**
51. 7×10^2 = **700 / 2700**
52. 9×10^0 = **0 / 9**
53. 3×10^2 = **30 / 300**
54. 5×10^4 = **50 000 / 10 000**
55. 12×10^3 = **123 / 12 000**
56. 8×10^3 = ____________
57. 2×10^5 = ____________
58. 4×10^3 = ____________
59. 6×10^6 = ____________
60. 8×10^1 = ____________
61. 11×10^4 = ____________
62. 7×10^0 = ____________
63. 15×10^7 = ____________

Match the numbers with the products of powers.

64.

20 000 •	• 4×10^4
5000 •	• 5×10^5
40 000 •	• 2×10^3
400 •	• 2×10^4
2000 •	• 4×10^2
500 000 •	• 5×10^3

65.

30 000 •	• 4×10^4
400 000 •	• 3×10^4
40 000 •	• 4×10^5
300 000 •	• 3×10^3
4000 •	• 4×10^3
3000 •	• 3×10^5

Write each number as a product of powers.

66. 900 = ____ x ____
67. 8000 = __________
68. 200 000 = __________
69. 7000 = __________
70. 500 = __________
71. 60 000 = __________
72. 50 000 = __________
73. 6000 = __________
74. 70 000 = __________
75. 60 = __________
76. 9000 = __________
77. 900 000 = __________

Put the products of powers in each set in order from smallest to greatest (1 to 3).

78.
☐ 2×10^3
☐ 5×10^2
☐ 3×10^3

79.
☐ 6×10^5
☐ 5×10^6
☐ 7×10^3

80.
☐ 4×10^3
☐ 2×10^4
☐ 5×10^3

81.
☐ 3×10^2
☐ 9×10^5
☐ 4×10^2

82.
☐ 2×10^5
☐ 8×10^5
☐ 5×10^2

83.
☐ 3×10^3
☐ 2×10^2
☐ 6×10^5

3 Squares and Square Roots

- using squares and square roots

Read This

Perfect squares are the squares of whole numbers.

(whole number)^2 = perfect square

The square root of a number is its squared factor.

square root sign

$\sqrt{\text{a number}}$ = squared factor

e.g. $\sqrt{49} = 7$ ← squared factor

49: perfect square

- 49 is the square of 7.
- 7 is the square root of 49.

The square root of a number that is not a perfect square can be estimated by finding the two closest perfect squares.

Example Estimate $\sqrt{42}$.

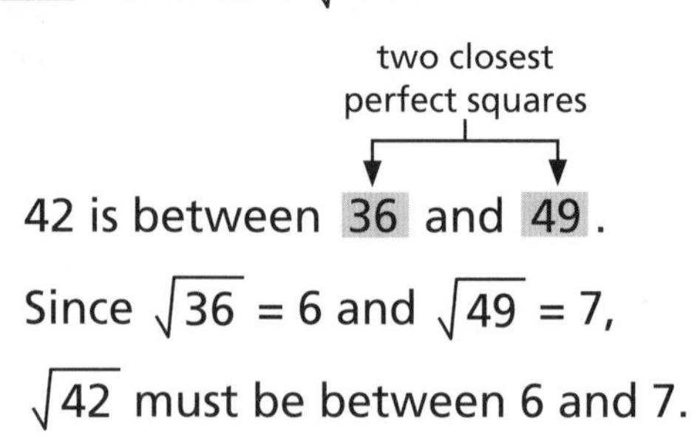

two closest perfect squares

42 is between 36 and 49.

Since $\sqrt{36} = 6$ and $\sqrt{49} = 7$,

$\sqrt{42}$ must be between 6 and 7.

Estimate: 6.5

Circle the closer estimates.

$\sqrt{90}$ ← 90 is between 81 and 100.

Estimate: **8.5 / 9.5**

$\sqrt{26}$ ← 26 is between 25 and 36.

Estimate: **5.1 / 6.9**

Fill in the blanks and circle the closer estimates.

① $\sqrt{75}$ between 8 and ____

Estimate: **7 / 8.7**

② $\sqrt{85}$ between ____ and ____

Estimate: **9.2 / 10**

③ $\sqrt{115}$ between ____ and ____

Estimate: **10.7 / 12.1**

④ $\sqrt{27}$ between ____ and ____

Estimate: **5.2 / 6.2**

⑤ $\sqrt{150}$ between ____ and ____

Estimate: **11.8 / 12.2**

⑥ $\sqrt{95}$ between ____ and ____

Estimate: **9.7 / 10.7**

⑦ $\sqrt{15}$ between ____ and ____

Estimate: **3.9 / 4.5**

⑧ $\sqrt{200}$ between ____ and ____

Estimate: **13.3 / 14.1**

Put “<”, “>”, or “=” in the circles.

⑨ $\sqrt{50}$ ◯ 7 ⑩ 11 ◯ $\sqrt{111}$ ⑪ $\sqrt{169}$ ◯ 13

⑫ $\sqrt{170}$ ◯ 14 ⑬ 24^2 ◯ 576 ⑭ 20 ◯ $\sqrt{399}$

⑮ 23 ◯ $\sqrt{529}$ ⑯ 15^2 ◯ 200 ⑰ $\sqrt{625}$ ◯ 25

⑱ 16^2 ◯ 250 ⑲ 18 ◯ $\sqrt{324}$ ⑳ 14 ◯ $\sqrt{300}$

Put each group of numbers in order from smallest to greatest.

㉑ $\sqrt{42}$ 6 $\sqrt{50}$ 7

____

㉒ 5 $\sqrt{30}$ $\sqrt{40}$ 2^2

____

㉓ $\sqrt{60}$ 3^2 $\sqrt{50}$ 8

____

㉔ $\sqrt{10}$ 2^2 $\sqrt{15}$ 4^2

____

Simplify. Show your work.

㉕ $\sqrt{4^2}$

$= \sqrt{\square \times \square}$

$= \square$

㉖ $(\sqrt{5})^2$

$= \sqrt{\square} \times \sqrt{\square}$

$= \square$

㉗ $\sqrt{9^2}$

㉘ $(\sqrt{11})^2$

㉙ $(\sqrt{7})^2$

㉚ $\sqrt{6^2}$

Square and square root are opposite operations. They cancel each other out when they are applied together.

e.g. $\sqrt{3} \times \sqrt{3} = 3$
So, $(\sqrt{3})^2 = 3$.

$\sqrt{3 \times 3} = 3$
So, $\sqrt{3^2} = 3$.

Match. Then fill in the blanks.

31.

square root of 4^2 •	• $\sqrt{2}^2$ = ______
square root of 4 •	• $\sqrt{4^2}$ = ______
square of $\sqrt{4}$ •	• 2^2 = ______
square of $\sqrt{2}$ •	• $\sqrt{4}$ = ______
square of 2 •	• $\sqrt{4}^2$ = ______

32.

square of $\sqrt{64}$ •	• $\sqrt{64}$ = ______
square root of 64 •	• $\sqrt{8^2}$ = ______
square of 8 •	• $\sqrt{64}^2$ = ______
square root of 8^2 •	• 8^2 = ______
square of $\sqrt{8}$ •	• $\sqrt{8}^2$ = ______

Evaluate. Show your work.

33. $\sqrt{5} \times \sqrt{20}$

$= \sqrt{}$

$=$

34. $\sqrt{3} \times \sqrt{27}$

35. $\sqrt{2} \times \sqrt{8}$

36. $\sqrt{2} \times \sqrt{2} \times \sqrt{4}$

37. $\sqrt{6 + 10}$

38. $\sqrt{4} \times \sqrt{2} \times \sqrt{18}$

39. $\sqrt{50 + 50}$

40. $\sqrt{25} + \sqrt{4} + \sqrt{9}$

Square roots can be multiplied.

e.g. $\sqrt{2} \times \sqrt{8} = \sqrt{2 \times 8}$

$= \sqrt{16}$

$= 4$

Remember to simplify the expressions before multiplying the square roots.

e.g. $\sqrt{8 + 12} \times \sqrt{5}$

$= \sqrt{20} \times \sqrt{5}$

$= \sqrt{100}$

$= 10$

Find the answers.

41. $\sqrt{5+11}$

42. $\sqrt{12+13} \times \sqrt{9}$

43. $\sqrt{32+32}$

44. $\sqrt{20+30+50}$

45. $3+\sqrt{30-5}$

46. $\sqrt{4+12}+\sqrt{16}$

47. $\sqrt{15-3} \times \sqrt{3}$

48. $\sqrt{12+13} \times \sqrt{5-1}$

49. $\sqrt{28-10} \times \sqrt{2}$

50. $\sqrt{2+2}+\sqrt{5 \times 5}$

Addition/Subtraction under Square Roots

Add/subtract numbers under the same square root to evaluate.

e.g. $\sqrt{9+16}$

$= \sqrt{25}$ ← added first

$= \underline{5}$

Never add or subtract numbers that are not under the same square root.

e.g. $\sqrt{9}+\sqrt{16}=\sqrt{25}$ ✗

You should evaluate the square roots individually first. Then add.

e.g. $\sqrt{9}+\sqrt{16}=3+4$

$= \underline{7}$

Circle "T" for the true statements and "F" for the false ones.

51. The sum of $\sqrt{8}$ and $\sqrt{12}$ is $\sqrt{20}$. T / F

52. The square root of a number squared is the number itself. T / F

53. Square roots can be expressed as exponents. Therefore, according to the order of operations, square roots and exponents must be evaluated before multiplication. T / F

54. The product of 2 square roots cannot be a whole number. T / F

55. The square of an odd number is always odd. T / F

56. The square root of an even number is always even. T / F

57. A perfect square has a whole number square root. T / F

4 Integers

- adding and subtracting integers

Adding a negative integer is the same as regular subtraction.

e.g. 5 + (-1)
= 5 – 1
= 4

Subtracting a negative integer is the same as regular addition.

e.g. 5 – (-1)
= 5 + 1
= 6

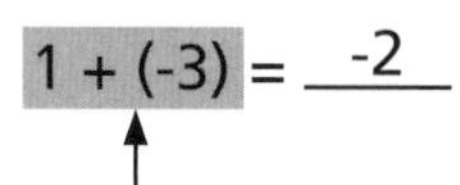

Do the addition on the number lines.

1 + 3 = 4

Move 3 steps to the right.

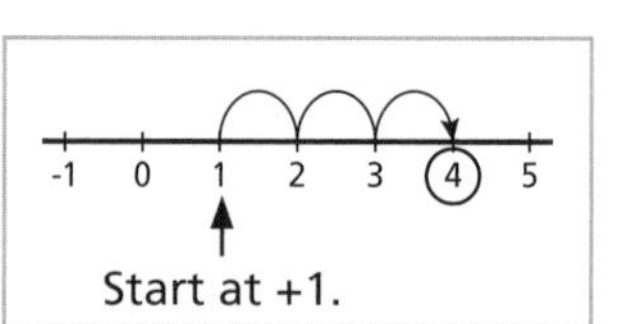

1 + (-3) = -2

Move -3 to the right, which means move 3 steps to the left.

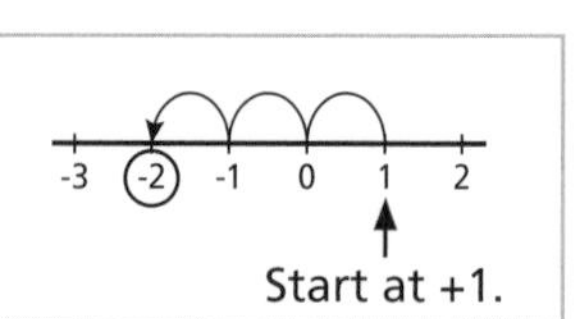

3 + 2 = ______

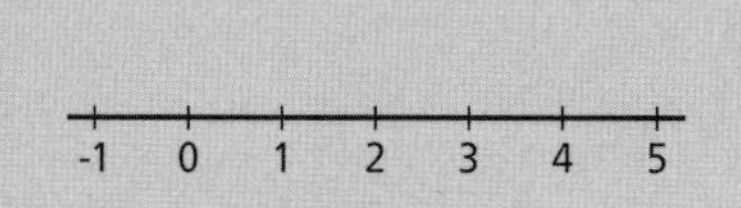

3 + (-2) = ______

-1 0 1 2 3 4 5

Do the addition on the number lines. Then circle the correct answers.

① -3 + 2 = ☐

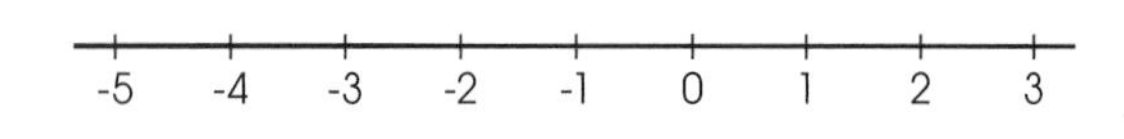

② -1 + (-4) = ☐

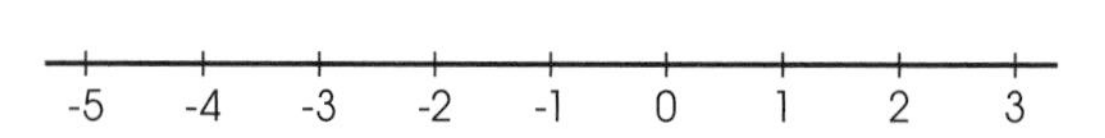

③ 2 + (-4) = ☐

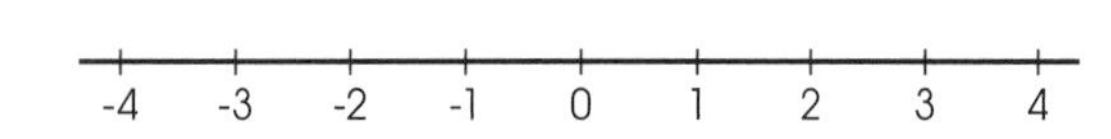

④ -5 + (-1) = ☐

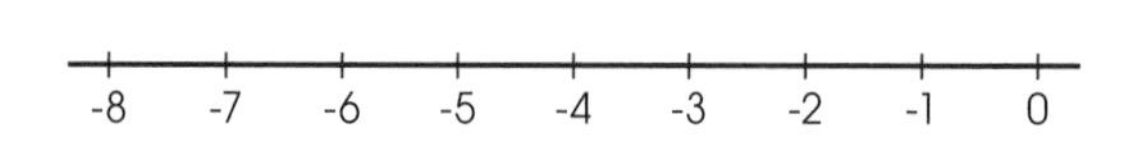

⑤ 5 + (-1) = ☐

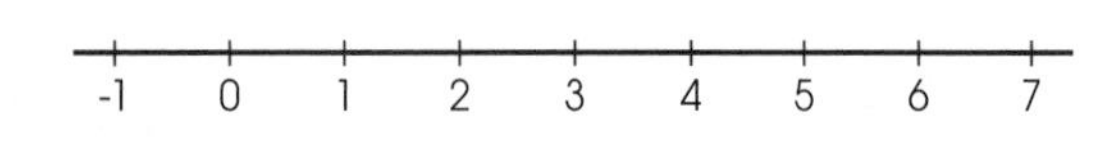

⑥ -2 + (-1) = ☐

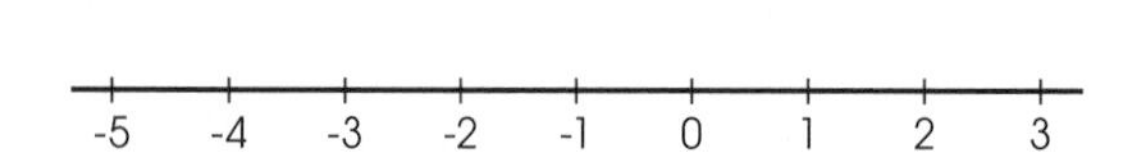

Adding Integers

When adding a positive integer, count **forward / backward** .

When adding a negative integer, count **forward / backward** .

Do the subtraction on the number lines. Then circle the correct answers.

⑦ 3 – 5 = ______

-4 -3 -2 -1 0 1 2 3 4 5 6

⑧ 1 – (-2) = ______

-3 -2 -1 0 1 2 3 4 5 6 7

⑨ -2 – 3 = ______

-8 -7 -6 -5 -4 -3 -2 -1 0 1 2

⑩ -4 – (-1) = ______

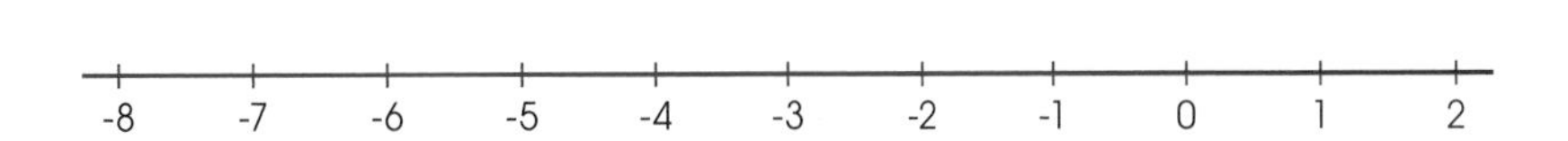

Subtracting Integers

When subtracting a positive integer, count **forward / backward** .

When subtracting a negative integer, count **forward / backward** .

Do the addition and subtraction on the number lines.

⑪ -5 – (-4) = ______

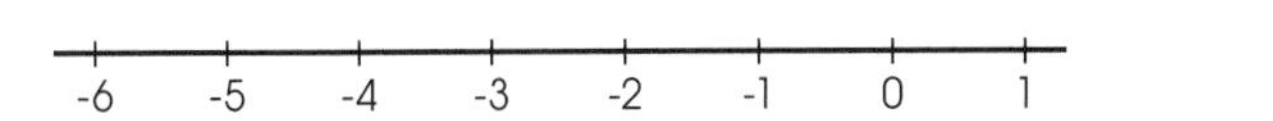

⑫ 3 + (-7) = ______

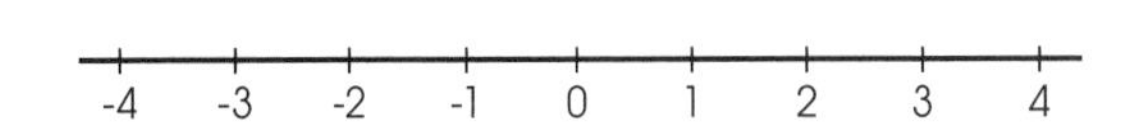

⑬ -2 + (-3) = ______

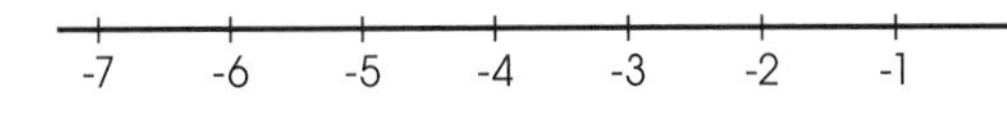

⑭ 4 – (-2) = ______

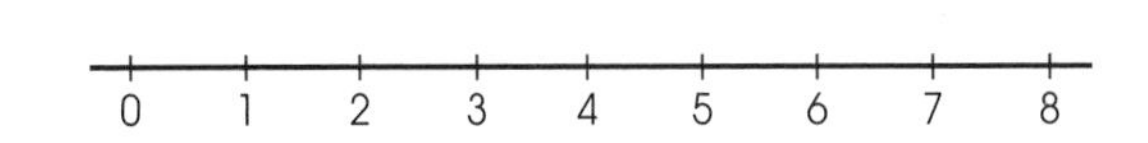

⑮ 8 – (-2) = ______

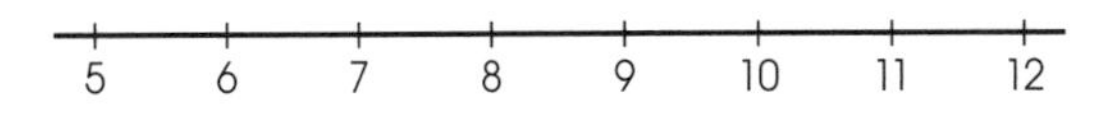

⑯ -6 – 4 = ______

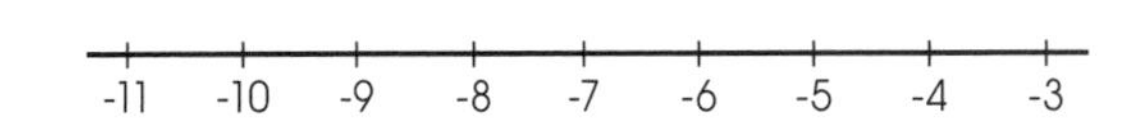

⑰ -4 + 1 = ______

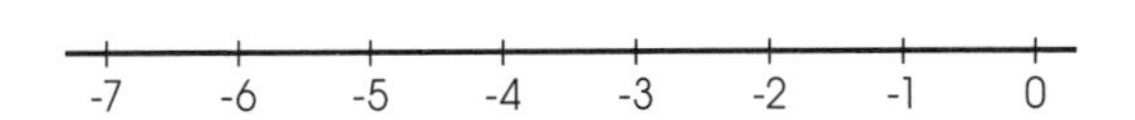

⑱ 7 + (-5) = ______

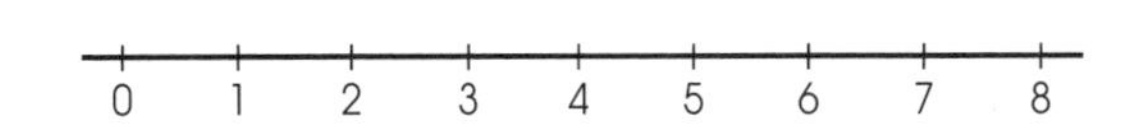

⑲ 5 + (-2) = ______

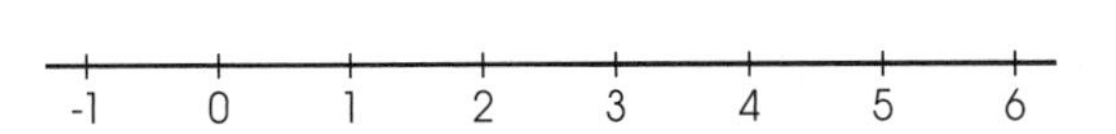

⑳ -3 – (-3) = ______

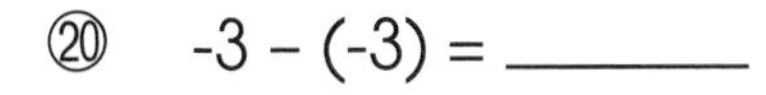

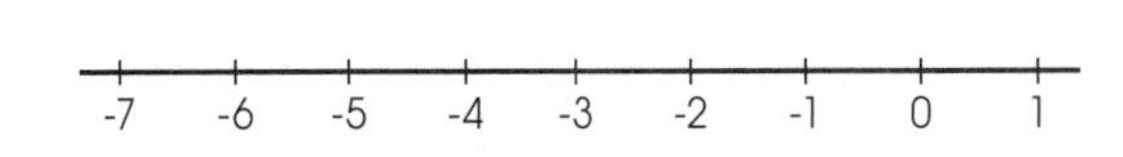

Find the answers. Show your work.

㉑ $2 + (-4)$

$= 2 - 4$

$=$ ______

㉒ $8 - (-1)$

㉓ $3 - (-2)$

㉔ $-1 + 4$

㉕ $-5 + 2$

㉖ $4 - (-2)$

㉗ $-2 - (-6)$

㉘ $-7 + 5$

㉙ $5 + (-5)$

㉚ $-6 - (-3)$

㉛ $-9 + 3$

Hints

Remember these rules when adding and subtracting integers:

- Two like signs yield an addition.

$+ + \Rightarrow +$	$- - \Rightarrow +$
$2 + (+1)$	$2 - (-1)$
$= 2 + 1$	$= 2 + 1$
$= 3$	$= 3$

- Two unlike signs yield a subtraction.

$+ - \Rightarrow -$	$- + \Rightarrow -$
$2 + (-1)$	$2 - (+1)$
$= 2 - 1$	$= 2 - 1$
$= 1$	$= 1$

Match the equivalent expressions by writing the correct ones in the boxes.

㉜ $-2 - 1$ $-1 - (-2)$ $-1 + (-2)$ $-2 + 1$

$-1 + 2$

$-2 - (-1)$

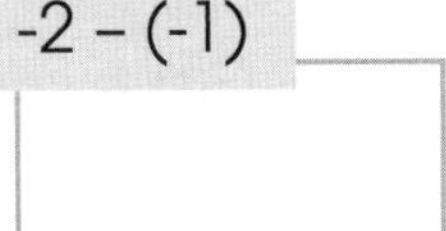

$-2 - (+1)$

$-1 - 2$

㉝ $-3 + (-4)$ $-3 - (-4)$ $-4 + 3$ $-4 - 3$

$-4 + (-3)$

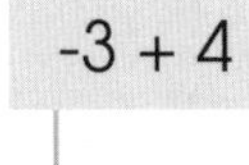

$-3 - 4$

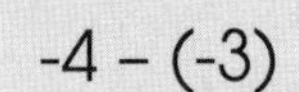

Find the answers using the number lines.

㉞ -2 + (-3) – 1 = ☐

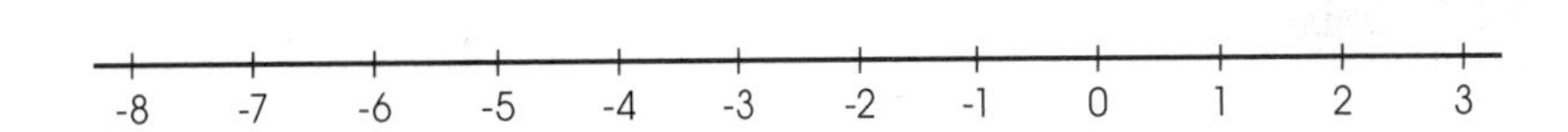

㉟ 5 + 2 + (-3) = ☐

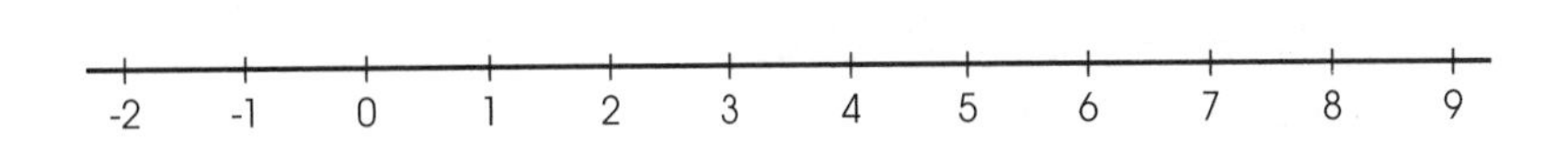

㊱ -10 + 3 – (-6) = ☐

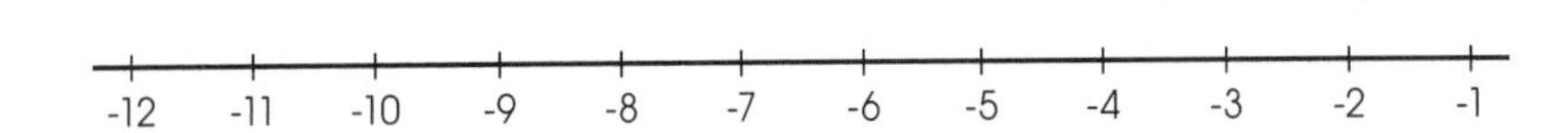

㊲ 3 – (-4) + (-1) = ☐

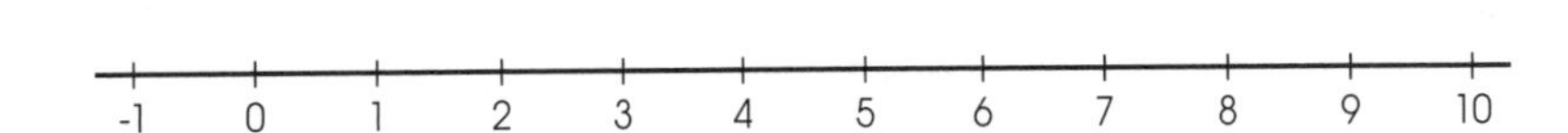

㊳ -7 + 3 + (-5) = ☐

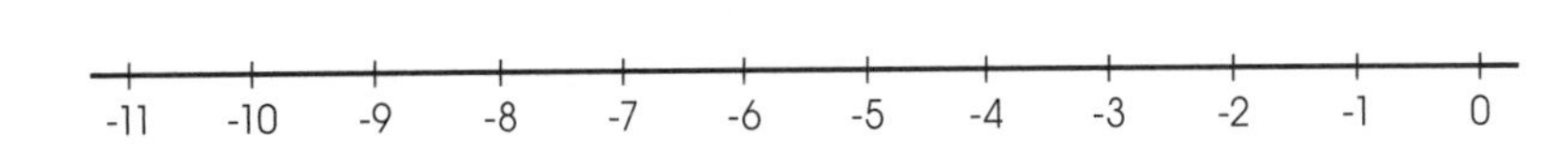

㊴ 6 – 4 – (-5) = ☐

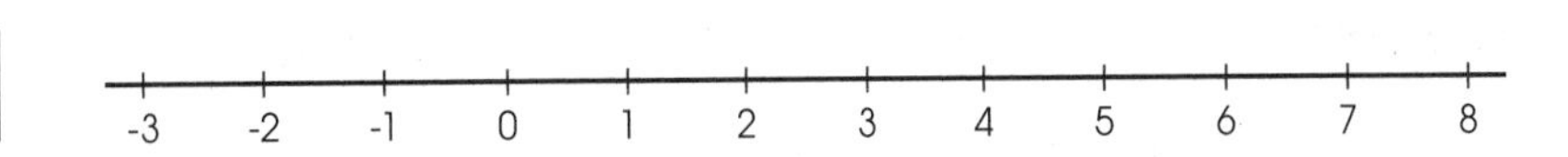

Find the answers. Show your work.

㊵ -3 + (-4) – (-5)

= -3 – ______ + ______

= ______

㊶ -1 + 6 – (-8)

㊷ 8 – 12 + (-3)

㊸ (-7) – (-8) – (-9)

㊹ (-10) + (-6) – 4

㊺ 4 – (-8) + (-3)

㊻ (-3) – (-2) + 6

㊼ 9 – (-4) – (-1)

㊽ (-2) + (-3) + (-4)

㊾ 2 – (-2) + (-5)

㊿ (-15) – 4 – (-2)

(51) 8 + (-1) – (-6)

5 Fractions, Decimals, and Percents

- relating fractions, decimals, and percents

Read This

Parts of a whole can be represented by a fraction, a decimal, or a percent. Each of these forms can be converted to the other two forms.

Example Circle the ones that represent the shaded area.

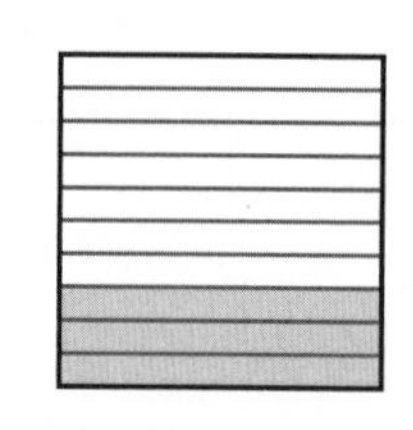

Fraction	Decimal	Percent
$\frac{3}{10}$ (circled)	0.3 (circled)	3%
$\frac{3}{100}$	0.03	30% (circled)

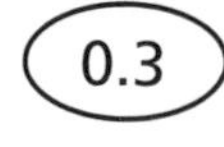

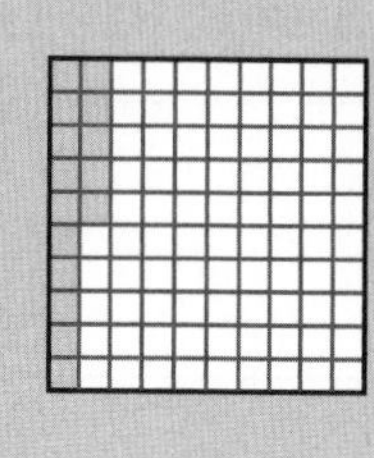

Fraction	Decimal	Percent
$\frac{15}{100}$	1.5	15%
$\frac{15}{10}$	0.15	150%

Convert the fractions to decimals and percents.

① **Fraction ➡ Decimal**

Divide the numerator by the denominator.

e.g. $\frac{1}{2} = 0.5 \leftarrow 1 \div 2$

a. $\frac{3}{5}$ = ______ b. $\frac{1}{4}$ = ______

c. $\frac{2}{5}$ = ______ d. $\frac{3}{4}$ = ______

e. $\frac{3}{8}$ = ______ f. $\frac{4}{5}$ = ______

g. $\frac{7}{10}$ = ______ h. $\frac{3}{2}$ = ______

i. $\frac{6}{5}$ = ______ j. $\frac{5}{8}$ = ______

k. $\frac{2}{4}$ = ______ l. $\frac{3}{10}$ = ______

m. $\frac{7}{8}$ = ______ n. $\frac{7}{5}$ = ______

② **Fraction ➡ Percent**

Multiply by 100%.

e.g. $\frac{1}{2} = 50\% \leftarrow \frac{1}{2} \times 100\%$

a. $\frac{1}{5}$ = ______ b. $\frac{1}{4}$ = ______

c. $\frac{9}{10}$ = ______ d. $\frac{3}{5}$ = ______

e. $\frac{4}{5}$ = ______ f. $\frac{7}{4}$ = ______

g. $\frac{3}{10}$ = ______ h. $\frac{2}{5}$ = ______

i. $\frac{5}{4}$ = ______ j. $\frac{3}{20}$ = ______

k. $\frac{6}{5}$ = ______ l. $\frac{6}{10}$ = ______

m. $\frac{8}{10}$ = ______ n. $\frac{3}{4}$ = ______

Convert the decimals to fractions and percents.

③ **Decimal ➡ Fraction**

Write the decimal as a numerator and 1 as a denominator. Multiply (by 10, 100, 1000, etc.) to make the numerator a whole number. Then simplify.

e.g. $0.25 = \frac{1}{4} \leftarrow 0.25 = \frac{0.25}{1} = \frac{25}{100}$

a. 0.03 = ______ b. 0.5 = ______

c. 0.28 = ______ d. 0.85 = ______

e. 1.2 = ______ f. 0.7 = ______

g. 2.45 = ______ h. 1.05 = ______

④ **Decimal ➡ Percent**

Move the decimal point 2 places to the right. Then add "%".

e.g. 0.25 = 25% ← 0.25

a. 0.35 = ______ b. 0.18 = ______

c. 0.4 = ______ d. 0.95 = ______

e. 2.1 = ______ f. 0.17 = ______

g. 3.5 = ______ h. 1.03 = ______

Convert each decimal to a fraction and a percent. Then match it with the correct diagram. Write the letters.

⑤

Decimal	Fraction	Percent	Diagram
0.55	______	______	◯
0.75	______	______	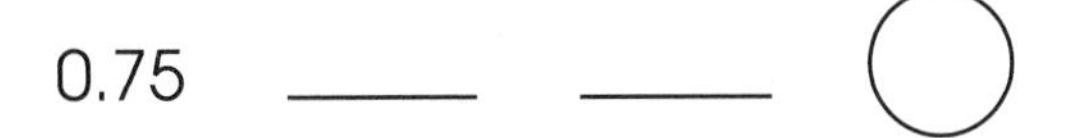
0.4	______	______	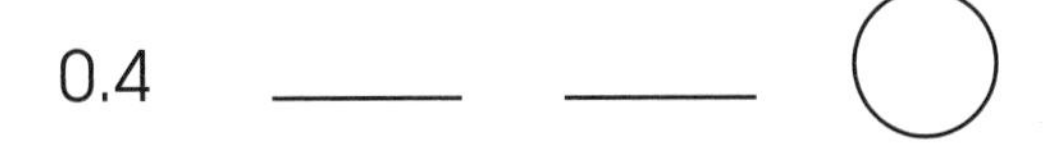
0.6	______	______	◯
0.05	______	______	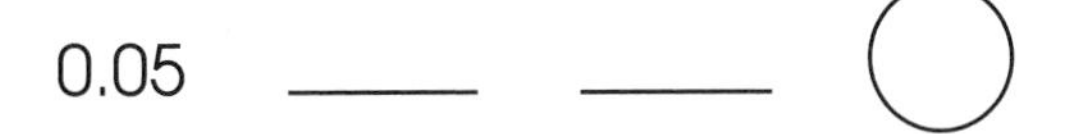
0.625	______	______	◯
0.375	______	______	◯

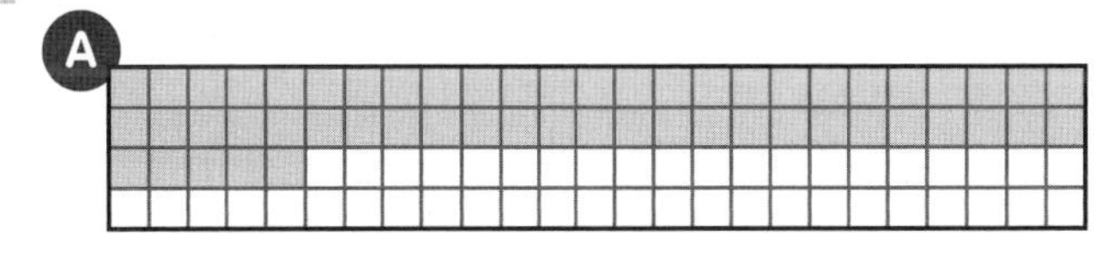

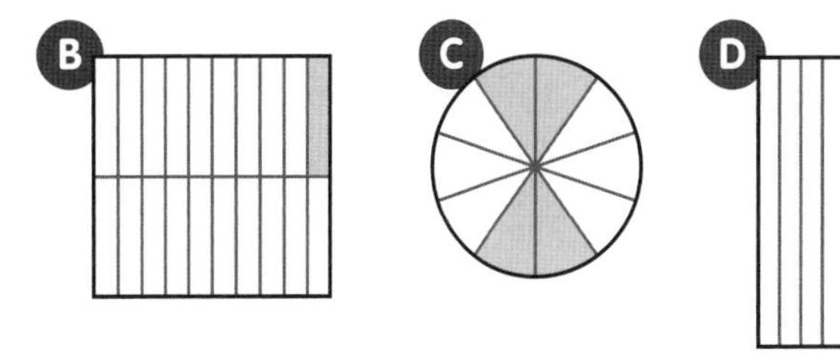

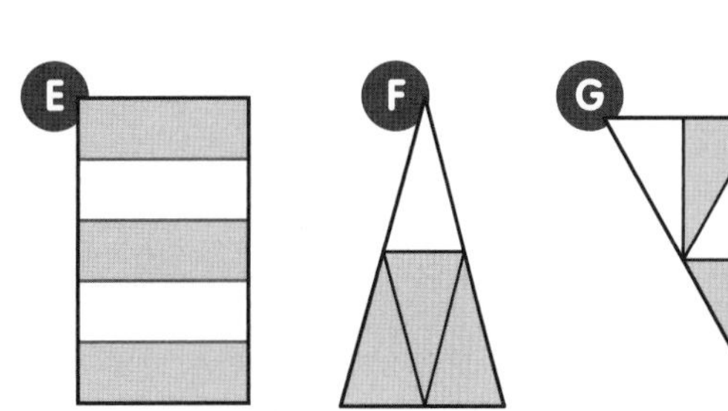

Convert the percents to fractions and decimals.

⑥ **Percent ➡ Fraction**

Remove "%" and add a denominator of 100. Then simplify.

e.g. 40% = $\frac{2}{5}$ ← $\frac{40}{100}$

a. 50% = ______ b. 20% = ______

c. 35% = ______ d. 42% = ______

e. 78% = ______ f. 64% = ______

g. 82% = ______ h. 28% = ______

⑦ **Percent ➡ Decimal**

Remove "%" and move the decimal point 2 places to the left.

e.g. 40% = 0.4 ← .40.

a. 10% = ______ b. 70% = ______

c. 14% = ______ d. 29% = ______

e. 59% = ______ f. 6% = ______

g. 8% = ______ h. 24% = ______

Write each percent as a fraction and a decimal. Then draw a diagram to illustrate it.

⑧ 30%

______

______

⑨ 25%

______

______

⑩ 60%

______

______

⑪ 70%

______

______

⑫ 35%

______

______

⑬ 85%

______

______

Write each number in the other two forms.

⑭ $\frac{9}{25}$ ______ ______

⑮ 24% ______ ______

⑯ 0.18 ______ ______

⑰ 0.15 ______ ______

⑱ $\frac{16}{25}$ ______ ______

⑲ $\frac{7}{20}$ ______ ______

⑳ $1\frac{1}{2}$ ______ ______

㉑ 2.5 ______ ______

㉒ $1\frac{4}{5}$ ______ ______

Put each set of numbers in order from smallest to greatest.

㉓ 0.53 $\frac{8}{25}$ 85%

______ < ______ < ______

㉔ $\frac{29}{40}$ 0.73 70.5%

____

㉕ $1\frac{1}{2}$ 123% 1.15

____

㉖ 0.65 $\frac{14}{20}$ 60%

____

㉗ 110% $1\frac{2}{5}$ 1.5

____

㉘ 2.5 200% $2\frac{1}{4}$

____

For each scenario, put a check mark if the quantity in bold is in the most appropriate form; otherwise, put a cross and write the quantity in the most appropriate form.

㉙ ◯ Anna ate **0.25** of a pie.

㉚ ◯ **0.1** of the students like Math.

㉛ ◯ Alan weighs **47.5** kg.

㉜ ◯ The grasshopper jumped **25%** m.

㉝ ◯ Elvia grew $\mathbf{\frac{1}{2}}$ cm.

㉞ ◯ Jude spent $**3$\mathbf{\frac{3}{4}}$** on a snack.

㉟ ◯ **25%** of the households have trees planted in their backyard.

㊱ ◯ Bernice weighs **7.25** kg more than her sister.

㊲ ◯ The class average on a test is **0.85**.

6 Ratios and Rates

- understanding ratios and rates

Ratios compare quantities of like things, while rates compare two quantities of different units. Always reduce a ratio to its simplest form.

Example Write the ratio of □ to △ in 3 ways.

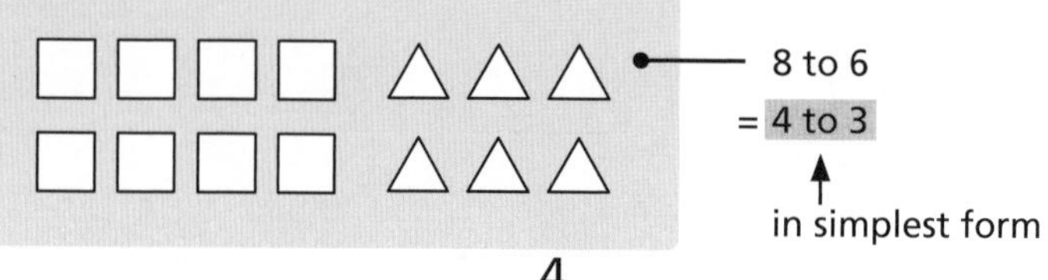

8 to 6
= 4 to 3
in simplest form

4 to 3 (way 1) 4:3 (way 2) $\frac{4}{3}$ (way 3)

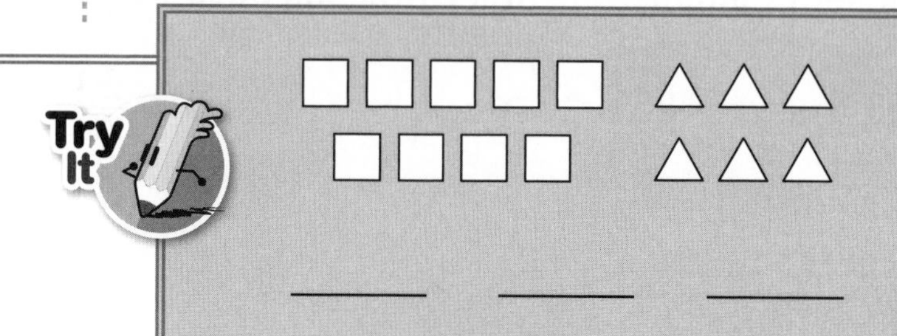

Write each ratio in 3 ways.

①

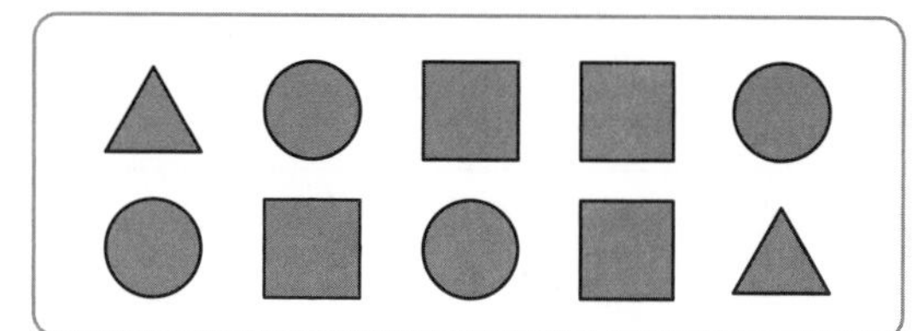

:all ______ ______ ______

②

:______ ______ ______

:all ______ ______ ______

:all ______ ______ ______

③

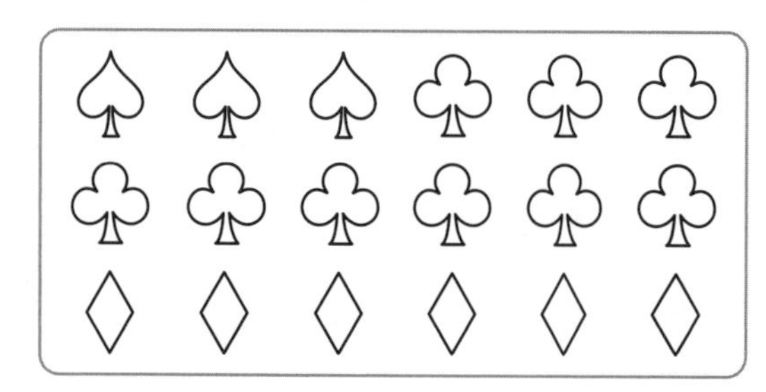

④

A A A A B B B B
B B C C C C C C
C C C C C C D D

A:C ______ ______ ______

D:B ______ ______ ______

C:all ______ ______ ______

Check the ratios that are equivalent to each given one.

⑤ **4:6**

(A) 2:3
(B) 8 to 12
(C) 6:4
(D) 3 to 2

⑥ **5 to 3**

(A) 15 to 9
(B) 6:10
(C) 10:6
(D) 9 to 15

Hints

To find an equivalent ratio, multiply or divide both terms by the same whole number.

e.g. $2:5 = 6:15$ (× 3 on both terms)

2:5 and 6:15 are equivalent ratios.

⑦ $\frac{1}{5}$

(A) 10 to 2
(B) 4 to 20
(C) 5:1
(D) 2:10

⑧ **3:9**

(A) $\frac{3}{1}$
(B) 1:3
(C) 6:18
(D) 9 to 3

⑨ $\frac{5}{2}$

(A) 8 to 20
(B) 10 to 4
(C) 5:2
(D) 6 to 15

Draw to match the ratios.

⑩ □ to △ = 2:3

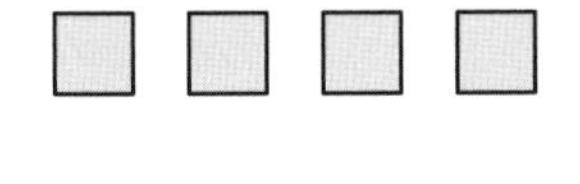

⑪ ○ : □ = 4:1

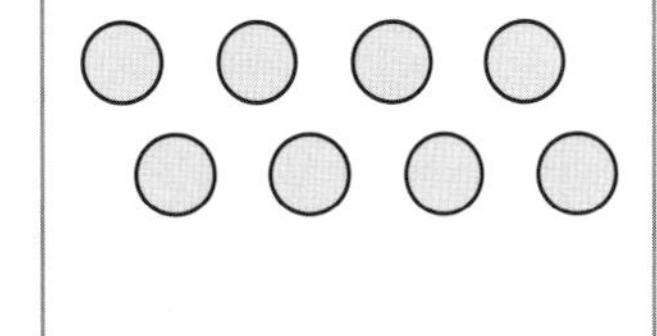

⑫ △ : ○ = 3:5

⑬ ☆ to ♡ = 2:1

⑭ ♤ : ♧ = 3:4

⑮ ◇ to ♡ = 4:5

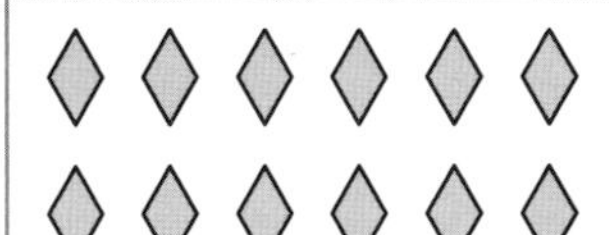

Write the totals and each ratio in simplest form to complete the table. Then answer the questions.

⑯

Hockey Team	Wins	Losses	Ties	Total No. of Games	Wins to Games	Losses to Games	Ties to Games
Terminators	11	4	2				
Bears	12	3	3				
Raiders	10	2	3				
Wizards	15	1	1				

⑰ Which two teams have the same ratio of wins to games?

____

⑱ If the Wizards played 1 more game and lost, what would be their new ratio of

a. wins to games? __________

b. losses to games? __________

Find the unit rates and check the better buy. Round your answers to the nearest cent.

⑲ (A) $62.19 for 3 kg $______ /kg

(B) $119.70 for 5 kg __________

⑳ (A) $39.99 for 12 cups __________

(B) $69.99 for 18 cups __________

㉑ (A) $77.31 for 9 bags __________

(B) $20.16 for 3 bags __________

㉒ (A) $2.76 for 2.4 m

$______ /m

(B) $2.31 for 150 cm

$______ /m

㉓ (A) $15 for 0.5 L

$______ /mL

(B) $12 for 240 mL

$______ /mL

Find the hourly pay for each person. Check who has the best pay in each group.

㉔ John earns $600 in 40 h, Mary earns $200 in 8 h, and Adam earns $18 each hour.

(A) John: $_____ /h

(B) Mary: _________

(C) Adam: _________

Tips: Make sure that the ratios are in the same unit before comparing.

㉕ Tiffany earns $16 in 90 min, Joshua earns $35 in 2 h, and Kenneth earns $24 in 1.5 h.

(A) Tiffany: _________

(B) Joshua: _________

(C) Kenneth: _________

㉖ Michael earns $10 in 30 min, Nicole earns $18 in 1.5 h, and Doris earns $432 in three 8-h work days.

(A) Michael: _________

(B) Nicole: _________

(C) Doris: _________

Find the ratios or rates.

㉗ 20% of the balls are green and the rest are red. What is the ratio of green balls to red balls?

____

㉘ $\frac{3}{7}$ of the cupcakes are vegan. What is the ratio of vegan cupcakes to all cupcakes?

____

㉙ There are green and red balls in a box. If the ratio of "red to all" is 2:5, what is the ratio of red balls to green balls?

____

㉚ The ratio of forks to knives is 1:1. If 4 forks are added, the ratio is 2:1. What is the ratio if 8 forks and 4 knives are added?

____

㉛ Lea ran 120 m in 3 min. What was Jay's speed in m/min if it took him 20 s less than Lea to run the same distance?

____

㉜ 24 cartons of milk are sold for $30.99. If there is a 15% discount, what is the unit price?

____

7 Volume and Surface Area

- finding the volume and surface area of 3-D shapes

Read This

Volume is measured using cubic units (cm^3, m^3, etc.) while surface area is measured using square units (cm^2, m^2, etc.).

Find the volume and surface area.

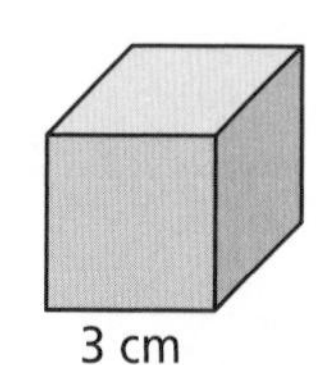

Volume: $3 \times 3 \times 3 = 27$ (cm^3)

Surface Area: $3 \times 3 \times 6 = 54$ (cm^2)

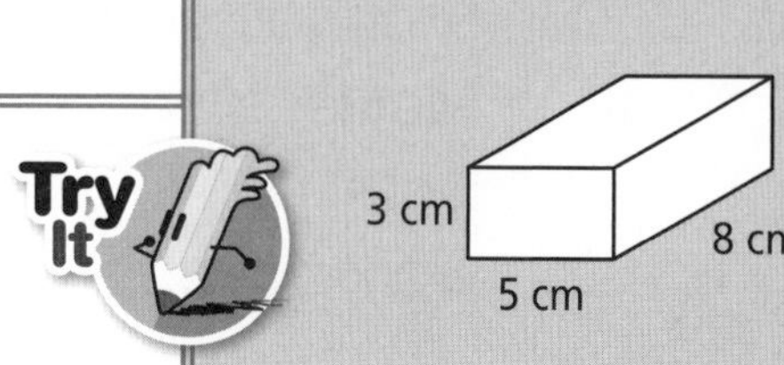

Volume: ______ cm^3

Surface Area:  cm^2

Find the volume and surface area of each solid.

①

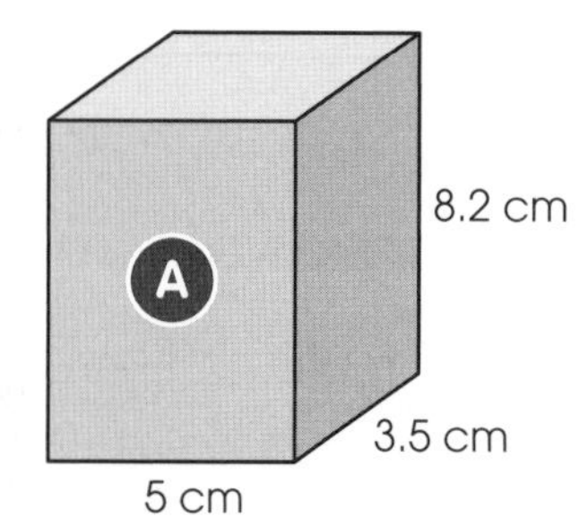

6 cm
B
4 cm
3.5 cm
3 cm

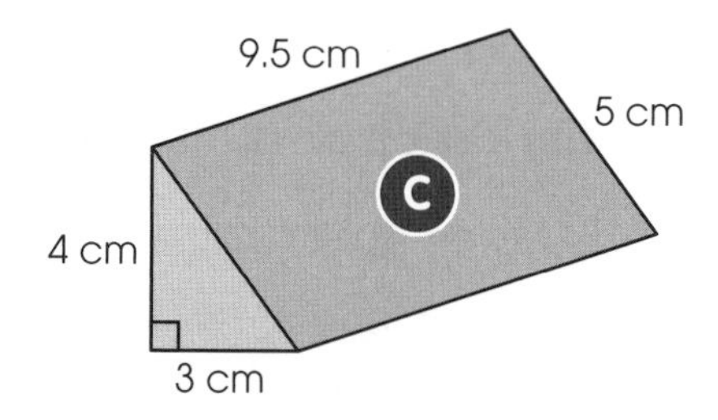

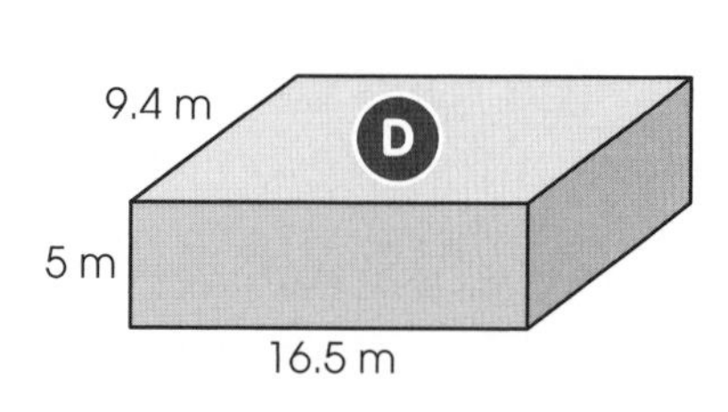

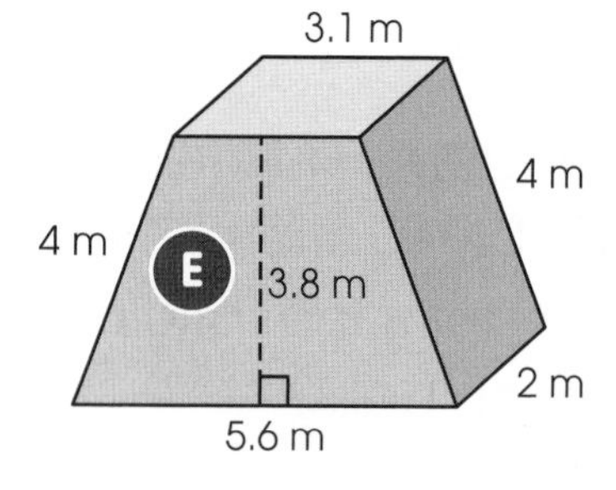

	A	B	C	D	E
Volume	______	______	______	______	______
Surface Area	______	______	______	______	______

Serena builds different solids using unit cubes. Find the volume (V) and the surface area (S.A.). Then read what she says.

② **Unit Cube**

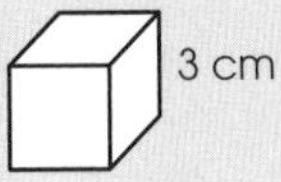

3 cm

V: ________

S.A.: ________

a.

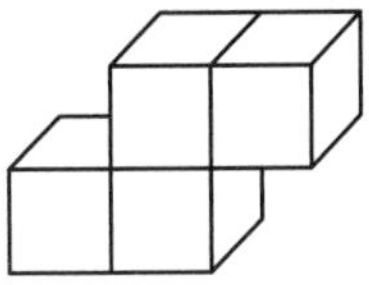

b.

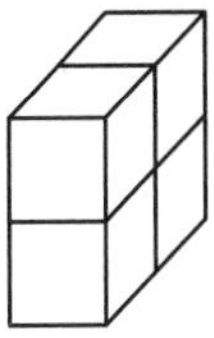

c.

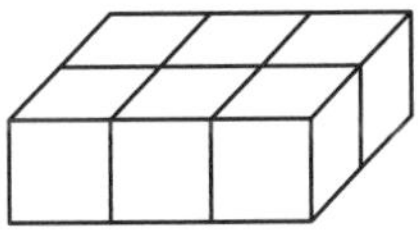

d.

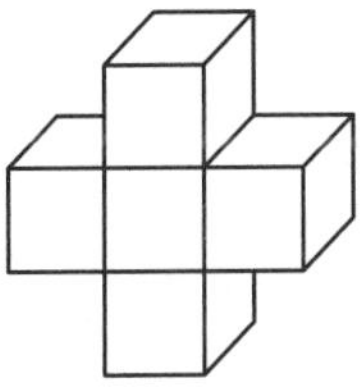

Use these answers to help you complete the activity below.

③ Circle "T" for the true statements and "F" for the false ones.

a. Solids that have the same volume always have the same surface area. **T / F**

b. Solids that have the same surface area always have the same volume. **T / F**

c. If the volume of a solid is greater than that of another solid, its surface area is also greater. **T / F**

d. When two solids are joined, the total surface area is smaller than the sum of their individual surface areas. **T / F**

e. When two solids are joined, the total volume is the sum of their individual volumes. **T / F**

Find the volume and surface area of each combined solid. Show your work.

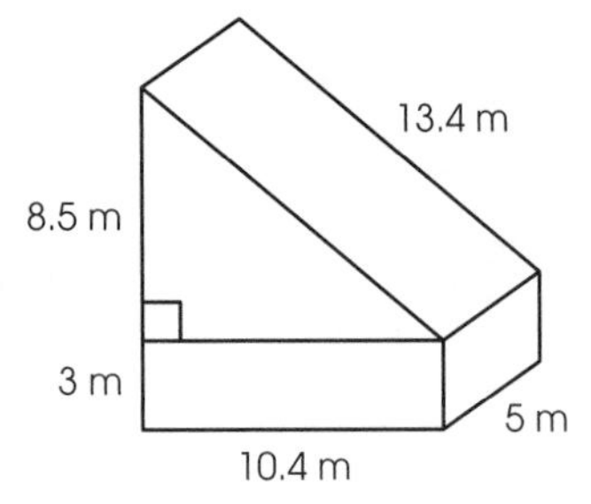

⑤

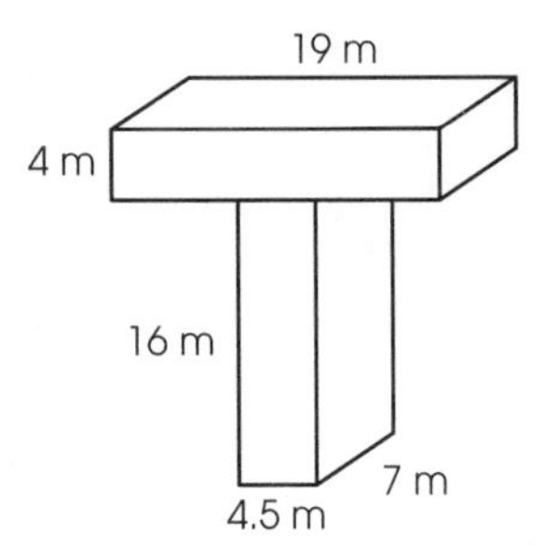

⑥

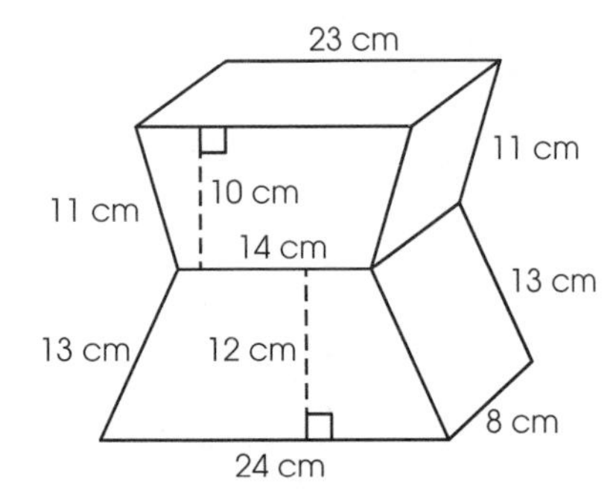

⑦

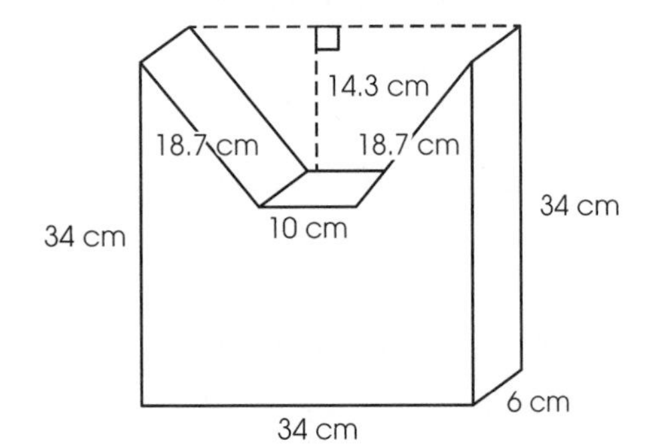

Use the given information to find the answers.

⑧

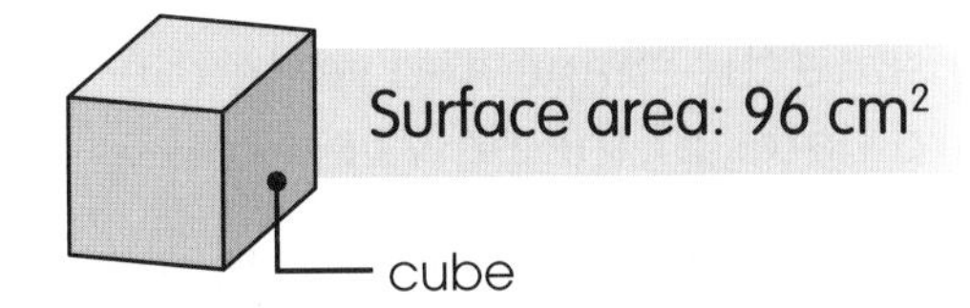

a. Side length: ________

b. Volume: ________

⑨

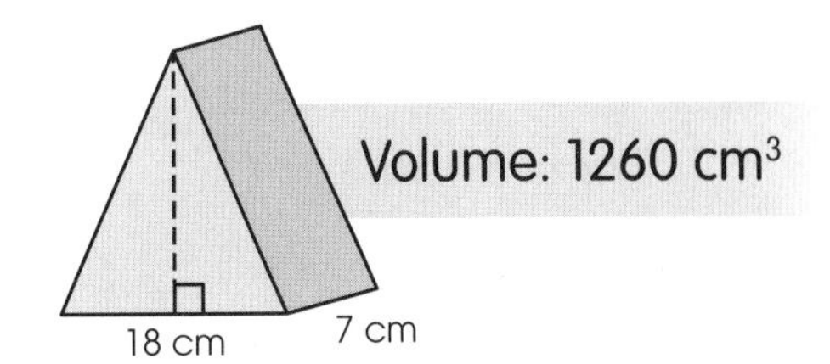

a. Area of base: ________

b. Height of triangle: ________

⑩

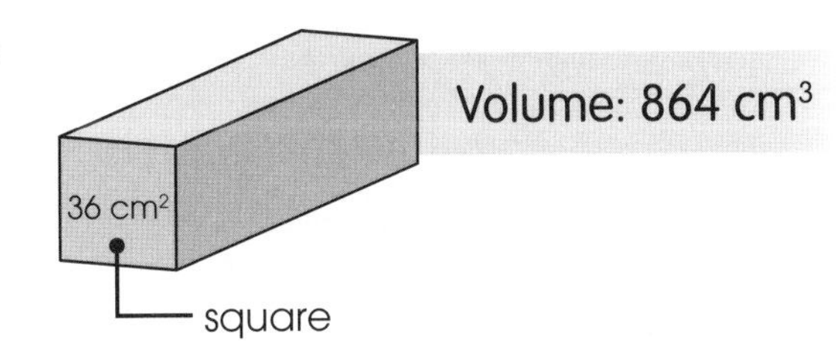

a. Height: ________

b. Side length of base: ________

c. Surface area: ________

Do the conversions.

⑪ $4\ 000\ 000\ cm^3$ = ________ m^3

⑫ $2\ 356\ 000\ cm^3$ = ________ m^3

⑬ $1\ 730\ 000\ cm^3$ = ________ m^3

⑭ $1.5\ m^3$ = ________ cm^3

⑮ $0.007\ m^3$ = ________ cm^3

⑯ $0.6\ m^3$ = ________ cm^3

$1\ m^3 = 1\ 000\ 000\ cm^3$

$m^3 \xrightarrow{\text{multiply}} cm^3$

e.g. $12\ m^3 = 12 \times 1\ 000\ 000$
$= \underline{12\ 000\ 000\ cm^3}$

$cm^3 \xrightarrow{\text{divide}} m^3$

e.g. $2\ 400\ 000\ cm^3$
$= 2\ 400\ 000 \div 1\ 000\ 000$
$= \underline{2.4\ m^3}$

Write ">", "<", or "=" in the circles.

⑰ $3\ m^3$ ◯ $300\ 000\ cm^3$

⑱ $20\ m^3$ ◯ $2\ 000\ 000\ cm^3$

⑲ $1\ 200\ 000\ cm^3$ ◯ $12\ m^3$

⑳ $5\ 000\ 000\ cm^3$ ◯ $50\ m^3$

㉑ $0.4\ m^3$ ◯ $4\ 000\ 000\ cm^3$

㉒ $2.05\ m^3$ ◯ $2\ 050\ 000\ cm^3$

㉓ $700\ 000\ cm^3$ ◯ $0.7\ m^3$

㉔ $10\ 000\ cm^3$ ◯ $1\ m^3$

8 Coordinates and Transformations

• relating coordinates and transformations

Read This

Translation

• move up/down
➡ change the y-coordinate

• move to the left/right
➡ change the x-coordinate

e.g.

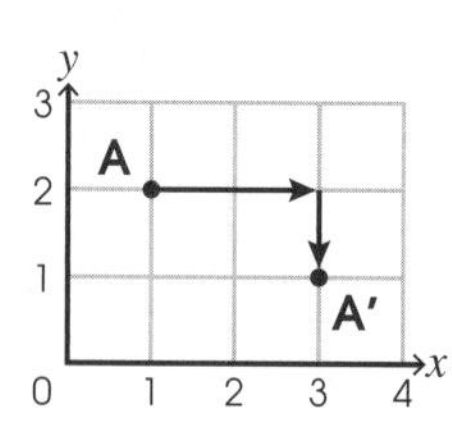

Translate A

• 2 units to the right

• 1 unit down

A(1,2)
+2 ↓ ↓ -1
3 1

A'(3,1)

Example Translate the shape and write the coordinates of the new vertices.

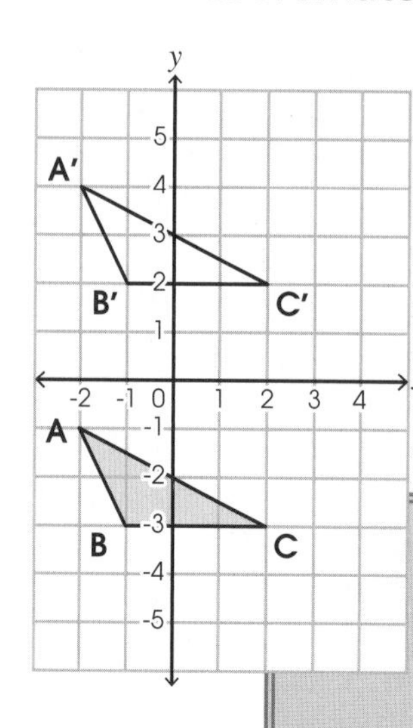

Translate the shaded triangle 5 units up.

A(-2,-1) A'(-2,4)
B(-1,-3) ➡ B'(-1,2)
C(2,-3) C'(2,2)

Try It

Translate the shaded triangle 2 units to the right.

A(-2,-1) A"( )
B(-1,-3) ➡ B"( )
C(2,-3) C"( )

Write the coordinates of the vertices. Then do the transformations and write the coordinates of the new vertices.

①

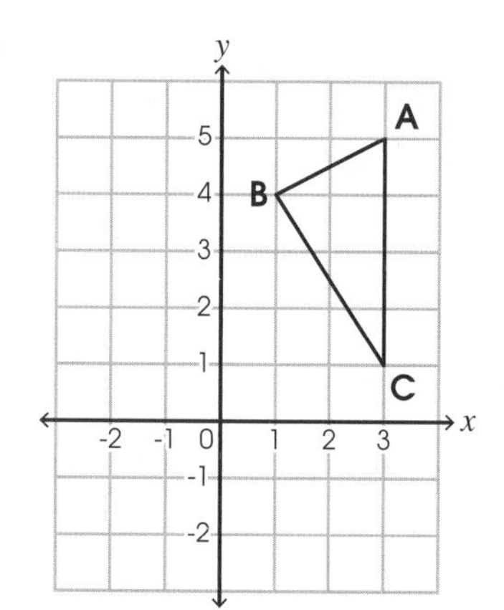

A( ) A'( )
B( ) ➡ B'( )
C( ) C'( )

Translate △ABC 3 units down and 2 units to the left.

②

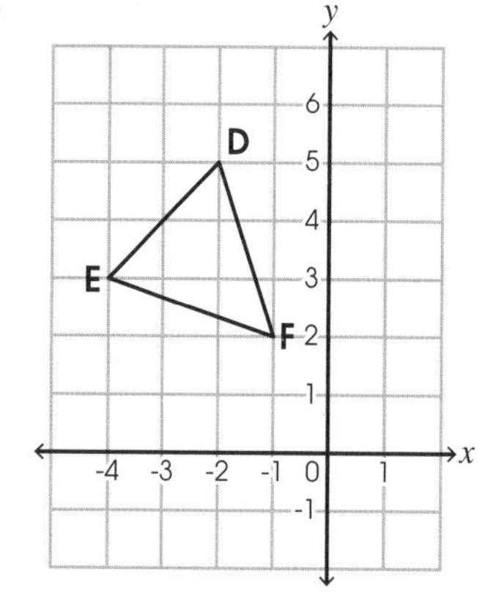

D( ) D'( )
E( ) ➡ E'( )
F( ) F'( )

Rotate △DEF by a $\frac{1}{2}$ turn about (-1,2).

③

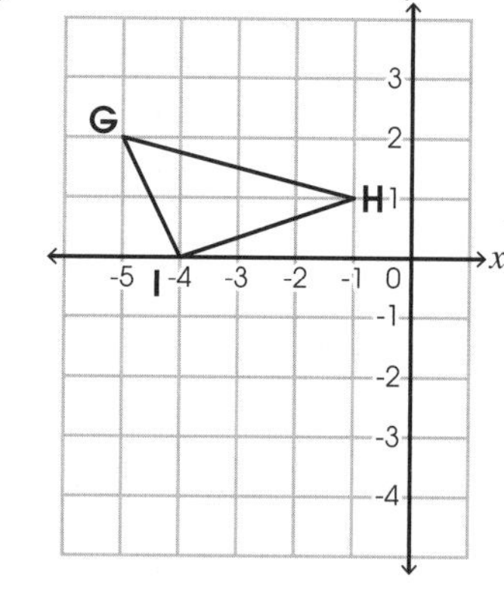

G( ) G'( )
H( ) ➡ H'( )
I( ) I'( )

Reflect △GHI in the x-axis.

For each shape, write the coordinates of the vertices. Then do the transformations and write the coordinates of the new vertices.

④

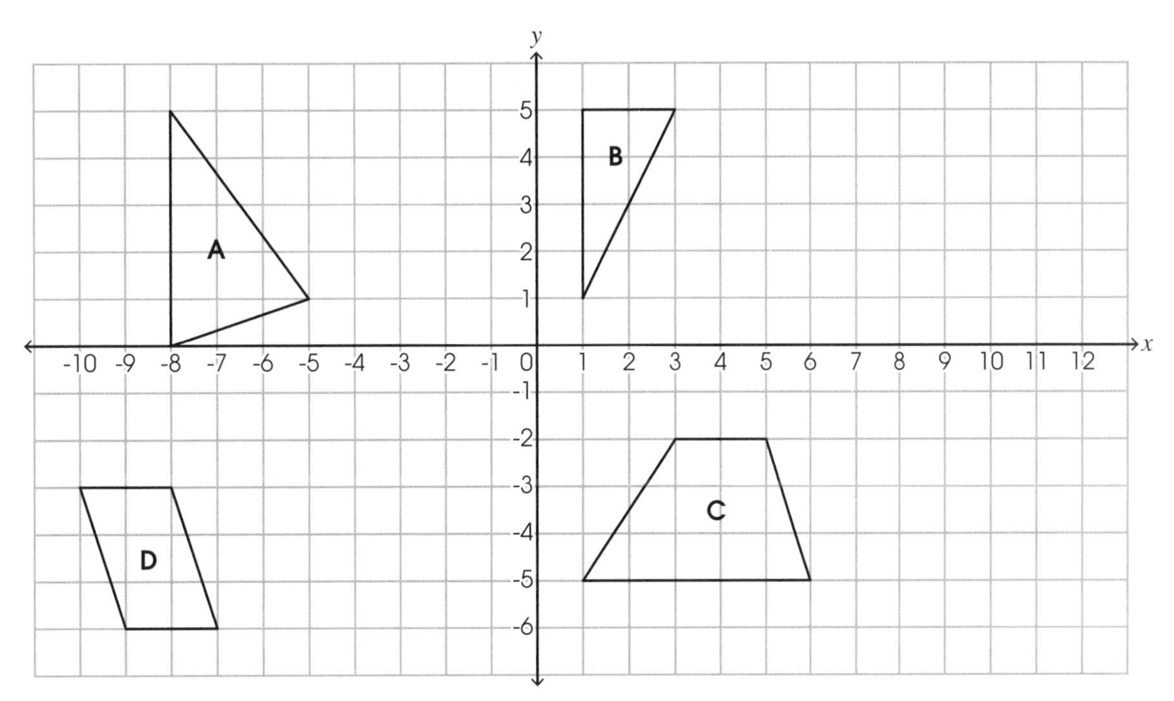

a. **Shape A** → **Shape A′**

Rotate Shape A $\frac{1}{4}$ clockwise about (-8,0).

b. **Shape B** → **Shape B′**

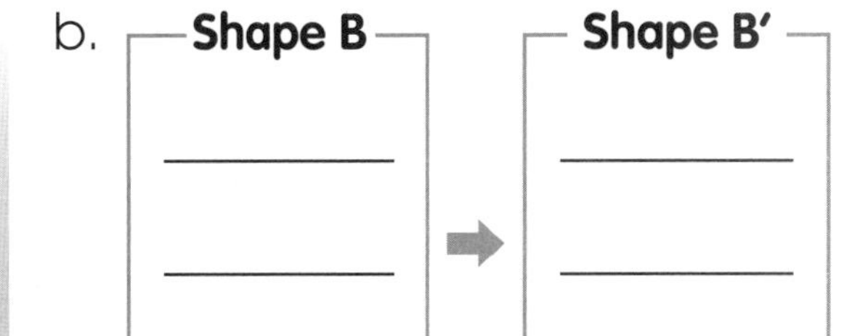

Reflect Shape B in the *y*-axis and translate it 2 units down.

c. **Shape C** → **Shape C′**

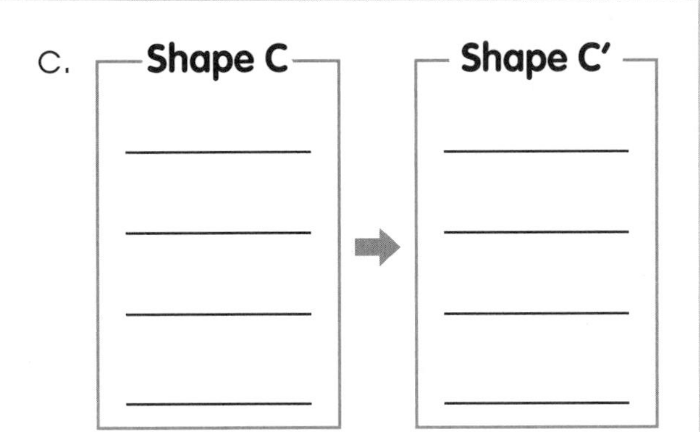

Translate Shape C 4 units up and 3 units to the right.

d. **Shape D** → **Shape D′**

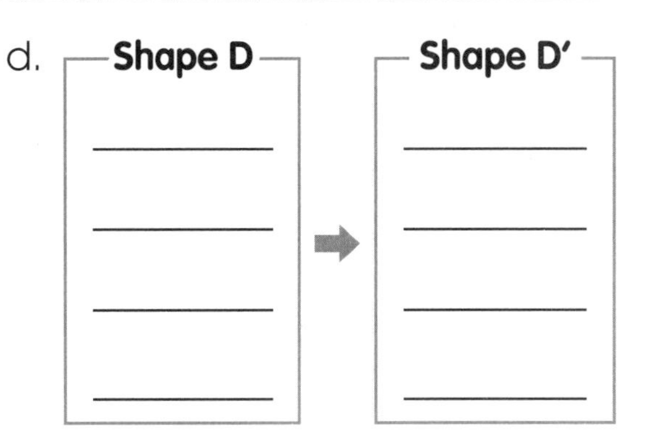

Reflect Shape D in the *y*-axis; rotate it $\frac{1}{4}$ counterclockwise about (10,-3).

Plot the points to draw the images of each shaded shape on the grid. Then write the letters to match the images with how they were transformed from the original shape.

⑤

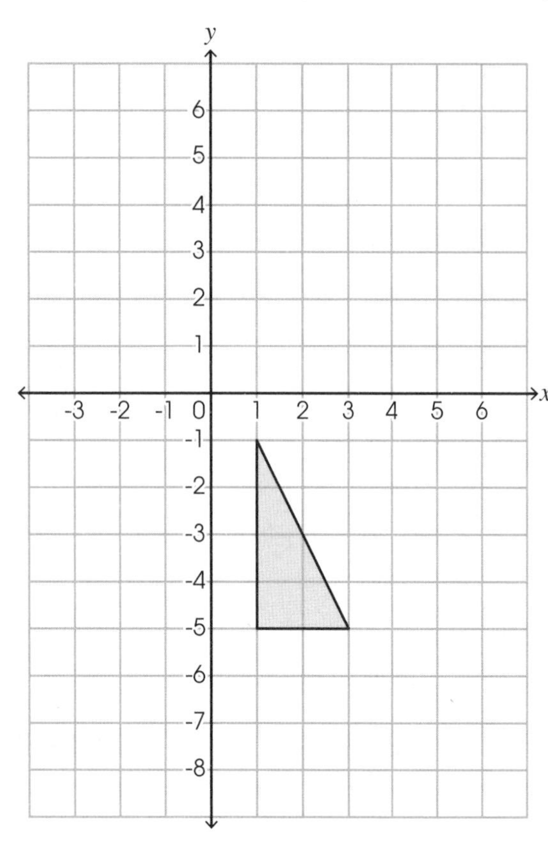

Coordinates of the Images' Vertices

A (4,-1) (4,-5) (6,-5)

B (1,-1) (5,-1) (5,1)

C (-1,-1) (-3,-5) (-1,-5)

D (-3,3) (-3,-1) (-1,-1)

Transformations

◯ a reflection in the y-axis

◯ a translation of 3 units to the right

◯ a $\frac{3}{4}$ clockwise rotation about (1,-1)

◯ a translation of 4 units up and 4 units to the left

⑥

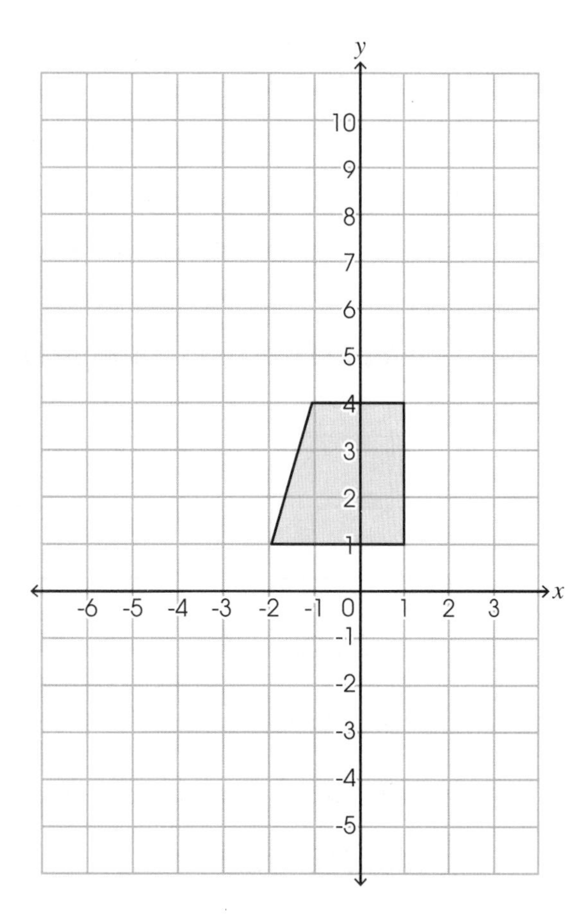

Coordinates of the Images' Vertices

A (-2,1) (-2,4) (-5,4) (-5,2)

B (1,-1) (1,-4) (-1,-4) (-2,-1)

C (1,7) (4,7) (1,4) (3,4)

D (-3,5) (-3,8) (-5,8) (-6,5)

Transformations

◯ a $\frac{1}{4}$ counterclockwise rotation about (-2,1)

◯ a reflection in the x-axis

◯ a translation of 4 units up and 4 units to the left

◯ a $\frac{1}{2}$ rotation about (1,4)

Plot the points for each shape and its image. Then describe the transformations.

⑦

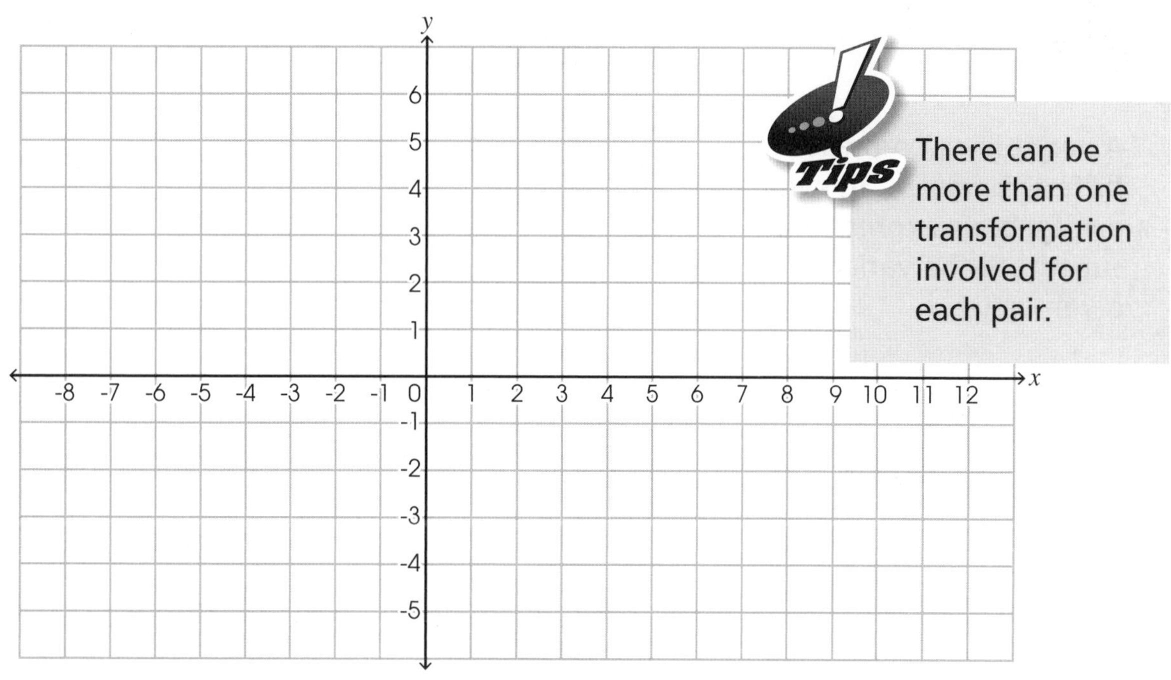

y
x
Tips
There can be more than one transformation involved for each pair.

	Shapes		Images	
a.	(4,3) (1,4) (6,5)	→	(-6,-3) (-3,-4) (-8,-5)	____________
b.	(-1,3) (-7,3) (-4,5)	→	(-1,1) (-1,-5) (-3,-2)	____________
c.	(-6,1) (-4,1) (-8,-3) (-6,-3)	→	(9,-1) (11,-1) (13,-5) (11,-5)	____________

Match the equivalent transformations.

⑧

rotate $\frac{1}{4}$ counterclockwise •	• rotate $\frac{1}{2}$ counterclockwise
translate 3 units up and 3 units down •	• rotate 360° clockwise
reflect in the x-axis and then reflect in the y-axis •	• reflect in the y-axis and then reflect in the x-axis
rotate $\frac{1}{2}$ clockwise •	• rotate $\frac{3}{4}$ clockwise

LEVEL 2 – FURTHER YOUR UNDERSTANDING

9 Algebraic Expressions

- evaluating algebraic expressions

A variable can represent different values. To evaluate an algebraic expression, substitute the variable for the given value.

Example Evaluate $3b + 5$ with the given values of b.

$b = 1$	$3b + 5$ $= 3 \times 1 + 5$ $= 8$

$b = 2$	$3b + 5$ $= 3 \times 2 + 5$ $= 11$

Evaluate $a + 2$.

$a = 3$	$a + 2$ $= \square + 2$ $= \square$

$a = 5$	$a + 2$ $= \square + 2$ $= \square$

Evaluate. Show your work.

① $2y + 5$

$y = 3$ | $y = 2$

② $(d - 1) \div 2$

$d = 3$ | $d = 8$

③ $9(a - 2)$

$a = 6$ | $a = 12$

④ $10 - 2(b + 1)$

$b = 1$ | $b = 2$

⑤ $4 + (6 - b) \div 2$

$b = 2$ | $b = 4$

$b = 6$ | $b = 0$

⑥ $3d + 10 \div d$

$d = 2$ | $d = 5$

$d = 1$ | $d = 10$

Evaluate the algebraic expressions.

⑦

$p = 2$
$q = 5$
$r = 3$
$s = 4$
$t = 1$

a. $6p - 2q$

b. $5(r - t)$

c. $9 - 3r + 2s$

⑧

$v = 6$
$w = 5$
$x = 2$
$y = 3$
$z = 1$

a. $x(y + 5)$

b. $v + 2w$

c. $8z + 4(3 - x)$

⑨

$g = 3$
$h = 2$
$i = 9$
$j = 1$
$k = 10$

a. $9 \div (k - j)$

b. $ij + 10$

c. $3(g - 1) + 2(h - 1)$

Write an algebraic expression for each case. Then evaluate using each set of values.

⑩ Find the sum of double m and triple n.

algebraic expression

a. $m = 8, n = 5$

b. $m = 6, n = 2$

⑪ Find the product of 6 and the difference of x and y.

algebraic expression

a. $x = 8, y = 5$

b. $x = 6, y = 1$

For each scenario, check the correct algebraic expression.

⑫ Rodney read $3x$ books last month and $2y$ books this month.

a. The total number of books he read:

Ⓐ $3y + 2x$

Ⓑ $3x + 2y$

b. How many more books he read last month than this month:

Ⓐ $3x - 2y$

Ⓑ $3(x - 2y)$

⑬ Each week, Jim drinks x cups of juice and Tyler drinks y cups.

a. The total number of cups of juice they will drink in 3 weeks:

Ⓐ $3(x + y)$

Ⓑ $3x + y$

b. How many more cups Jim will drink than Tyler after 2 weeks:

Ⓐ $2(x - y)$

Ⓑ $2x - y$

⑭ Jocelyn gets 3 candies each day. She eats x candies every day and saves the rest.

a. The total number of candies saved after y days:

Ⓐ $(x - 3) + y$

Ⓑ $y(3 - x)$

b. The total number of candies ate in y days:

Ⓐ xy

Ⓑ $x + y$

⑮ Kenneth pays $\$x$ for bus fare on weekdays and \$2 less on weekends.

a. The total bus fare paid for 3 Saturdays:

Ⓐ $3(x - 2)$

Ⓑ $2(x - 3)$

b. The total bus fare paid in a week:

Ⓐ $7(x - 2)$

Ⓑ $5x + 2(x - 2)$

⑯ Allison saved \$20 in her piggy bank. She wants to buy lollipops that are $\$x$ each with her money.

a. The amount of money left after getting y lollipops:

Ⓐ $(20 - x)y$

Ⓑ $20 - xy$

b. The number of lollipops she can get if her mom gives her $\$y$ more:

Ⓐ $(20 + y) \div x$

Ⓑ $(20 \div x) + y$

For each scenario, write an algebraic expression and answer the questions.

⑰

I made a cookies yesterday and b more cookies today than yesterday.

a. Write an algebraic expression to show the number of cookies made today.

____

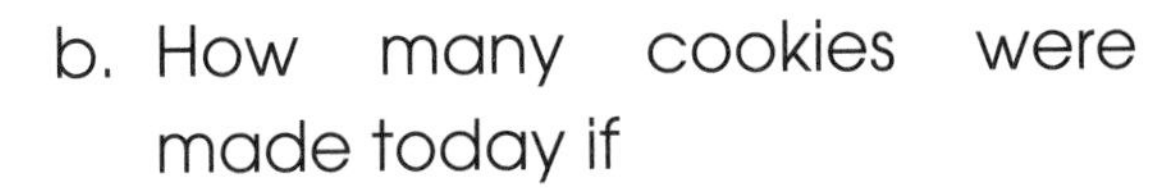

b. How many cookies were made today if

- $a = 20$ and $b = 5$?

  ________ cookies

- $a = 15$ and $b = 10$?

  ____

⑱

I planted $(n - 1)$ trees per week for m weeks.

a. Write an algebraic expression to show the total number of trees planted.

____

b. How many trees were planted if

- $m = 4$ and $n = 1$?

  ____

- $m = 5$ and $n = 2$?

  ____

⑲

I ran $(p \div 2)$ laps per minute for q minutes.

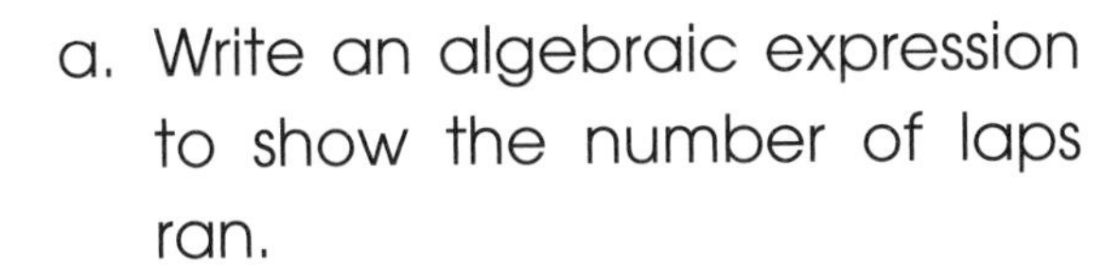

a. Write an algebraic expression to show the number of laps ran.

____

b. How many laps did he run if

- $p = 2$ and $q = 5$?

  ____

- $p = 4$ and $q = 3$?

  ____

⑳

I sew $(i \div 3)$ quilts per month for $(j - 1)$ months.

a. Write an algebraic expression to show the number of quilts sewn.

____

b. How many quilts were sewn if

- $i = 3$ and $j = 4$?

  ____

- $i = 6$ and $j = 5$?

  ____

10 Equations

- setting up equations and solving them

Follow the steps below to solve an equation.

1. Think of how to get rid of any values around the unknown.
2. Apply opposite operations to both sides of the equation to solve for the unknown.

Example Solve the equation.

$x + 2 = 5$ ← Get rid of 2 to isolate x.

$x + 2 - 2 = 5 - 2$ ← Subtract 2 on both sides (opposite operation).

$x = 3$

$y - 1 = 3$

$y - 1 + \square = 3 + \square$

$y = \square$

Solve the equations.

① $a + 2 = 7$

② $d - 4 = 11$

Tips When solving an equation, the goal is to isolate the variable.

③ $3y = 18$

④ $x \div 6 = 2$

⑤ $n + 4 = 24$

⑥ $s - 16 = 2$

⑦ $4i = 12$

⑧ $\frac{k}{9} = 10$

⑨ $14n = 18$

⑩ $\frac{3}{4}y = 24$

⑪ $2x = 16$

⑫ $20 - d = 11$

⑬ $11 - e = 4$

Check the correct equation for each description. Then solve it.

⑭ The sum of a and 8 is 15.

(A) $a + 8 = 15$ (B) $a = 8 + 15$

$a =$ ______

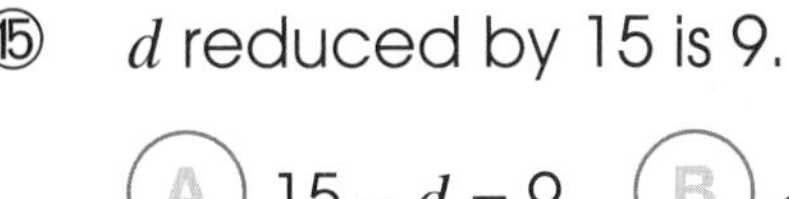

⑮ d reduced by 15 is 9.

(A) $15 - d = 9$ (B) $d - 15 = 9$

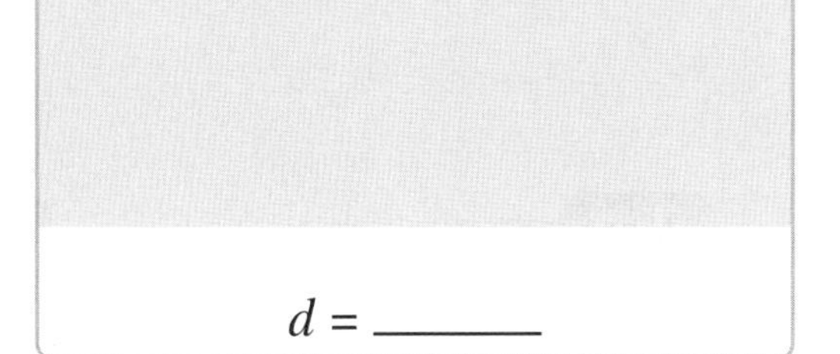

$d =$ ______

⑯ The product of i and 3 is 15.

(A) $3i = 15$ (B) $3 + i = 15$

$i =$ ______

⑰ x divided by 2 is 8.

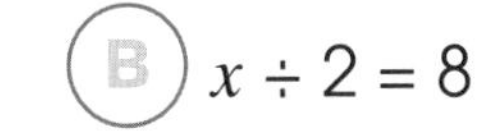

(A) $2 \div x = 8$ (B) $x \div 2 = 8$

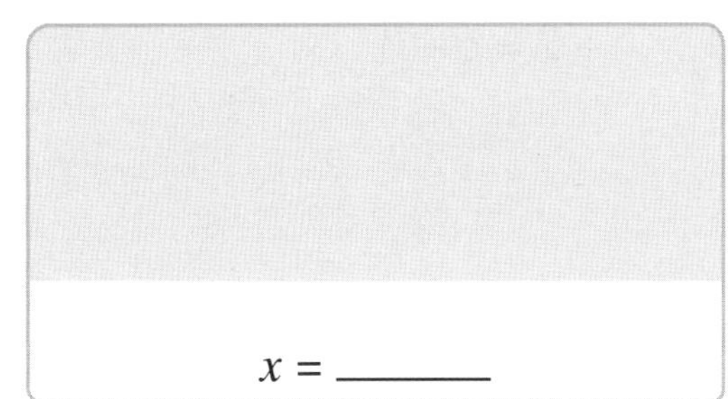

$x =$ ______

Use the guess-and-check method to solve for the variables.

⑱ $2a + 6 = 10$

$a =$ ______

Guess	Check

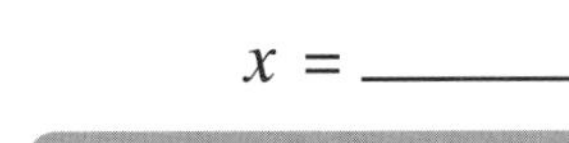

⑲ $x \div 2 - 4 = 8$

$x =$ ______

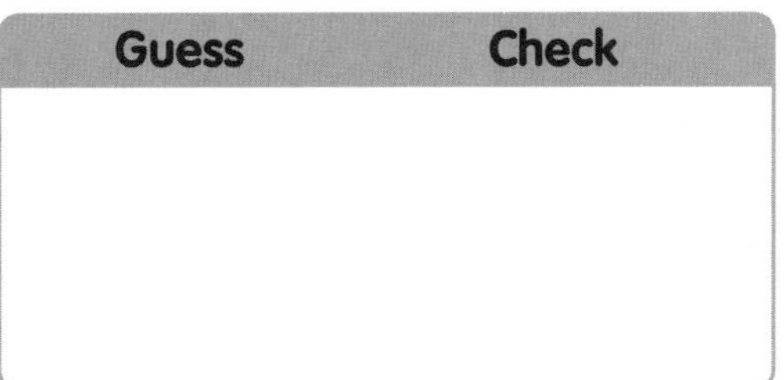

Guess	Check

⑳ $15 - 4d = 3$

$d =$ ______

Guess	Check

㉑ $10 + 8 \div y = 14$

$y =$ ______

Guess	Check

Solve each equation. Use substitution to check your answers.

㉒ $b + 4 = 10 + 3$

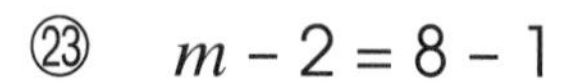

㉓ $m - 2 = 8 - 1$

Check

Check

Substitution is one way to check your answer to an equation.

e.g. $x - 1 = 4$

$x = 5$

Check Substitute 5 into the left side of the equation.

$5 - 1 = 4$ ← equal to the right side of the equation

So, $x = 5$ is correct.

㉔ $k \div 2 = 6 - 3$

㉕ $p \times 4 = 16 \div 2$

㉖ $q - 5 = 12 \div (2 + 1)$

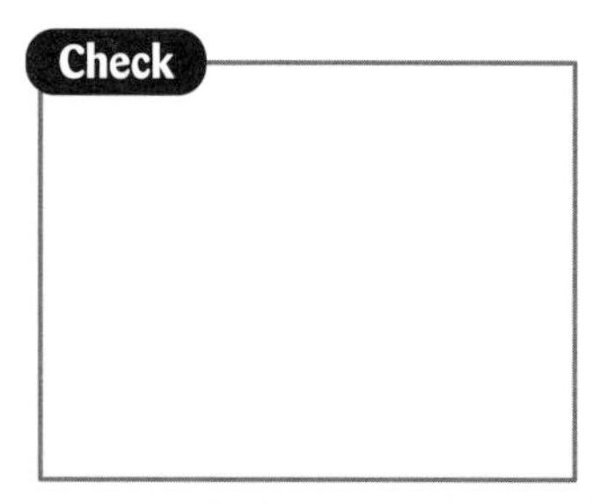

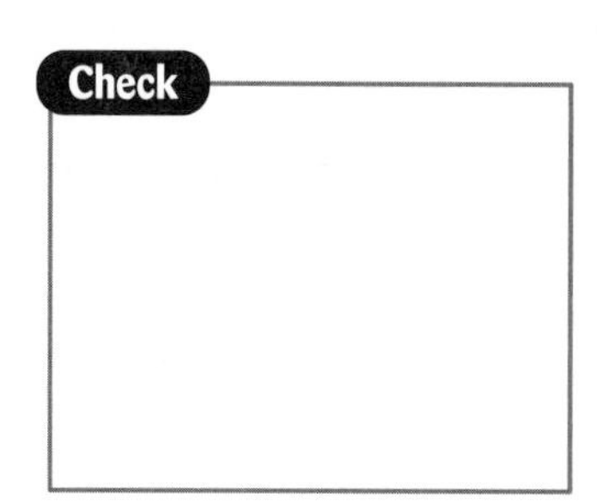

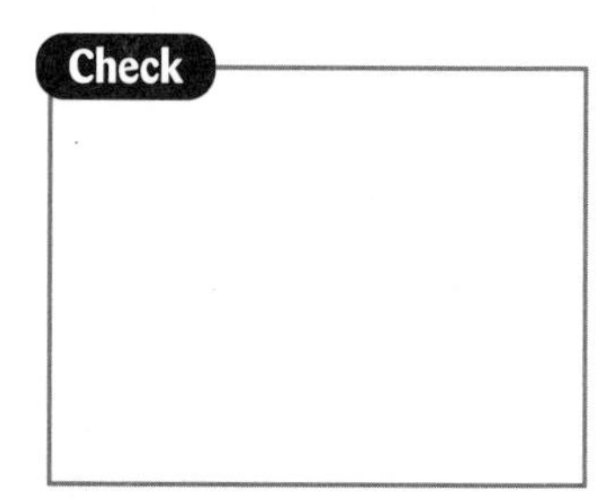

For each scenario, check the correct equation.

㉗ Alan read 3 books in July and b books in August. He read 7 books in total.

(A) $3 + b = 7$ (B) $b + 7 = 3$

㉘ A dog gets 3 treats every week. It gets 12 treats in w weeks.

(A) $3 + w = 12$ (B) $3w = 12$

㉙ 8 flowers bloomed on Sunday and f flowers bloomed on Friday. 2 more flowers bloomed on Sunday than Friday.

(A) $8 - f = 2$ (B) $f - 8 = 2$

㉚ Zed drinks 2 cups of tea every day. He drank a total of 32 cups in the past d days.

(A) $\frac{d}{2} = 32$ (B) $2d = 32$

Write an equation for each scenario, letting x represent the unknown. Then solve it using any method.

31. Sophia has a total of 11 T-shirts. 5 of them are blue and the rest are yellow. How many yellow T-shirts does she have?

equation

She has ______ yellow T-shirts.

32. Gavin drinks 2 L of water every day. After how many days will he have drank 18 L of water?

equation

He will have drank 18 L of water after ______ days.

33. Judy had a bag of candies. She gave the candies out to 5 friends and each of them got 6 candies. How many candies were in the bag at the beginning?

equation

______ candies were in the bag.

34. Alan had a carton of eggs. 4 of them broke and he used half of the remaining to bake a cake. If he used 7 eggs to bake the cake, how many eggs were in the carton at the beginning?

equation

______ eggs were in the carton.

35. Beth had 3 sheets of stickers. After getting 5 more stickers from her brother, she had 35 stickers. How many stickers were on each sheet?

equation

______ stickers were on each sheet.

11 Data Management

- interpreting graphs and answering questions about them

Read This

Double line graphs have the data points connected. They are useful for comparing the trends of two sets of data over the same time period.

Remeber to label each line after graphing it.

Example Which town had 200 kg of bottles collected on Monday?

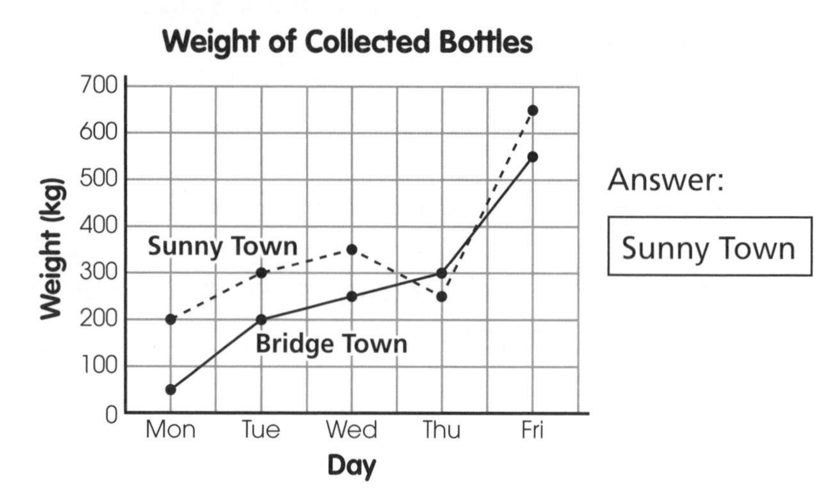

How many kilograms of bottles were collected in each town on Tuesday?

Sunny Town: ______ Bridge Town: ______

Use the graph above to answer the questions.

① How many more kilograms of bottles were collected in Sunny Town than Bridge Town on Friday? ______

② How many fewer kilograms of bottles were collected in Sunny Town than Bridge Town on Thursday? ______

③ Which day had the greatest increase in weight of bottles collected in Sunny Town? ______

④ Which day had the greatest difference in the weight of bottles collected between Bridge Town and Sunny Town? ______

⑤ Which town collected more kilograms of bottles during the week? ______

A circle graph is made to show the weights of bottles collected in nearby towns. Read the information and check the correct circle graph. Then answer the questions.

⑥

- **Sunny Town collected 25% of all bottles.**
- **Silver Town collected the least amount of bottles by weight.**

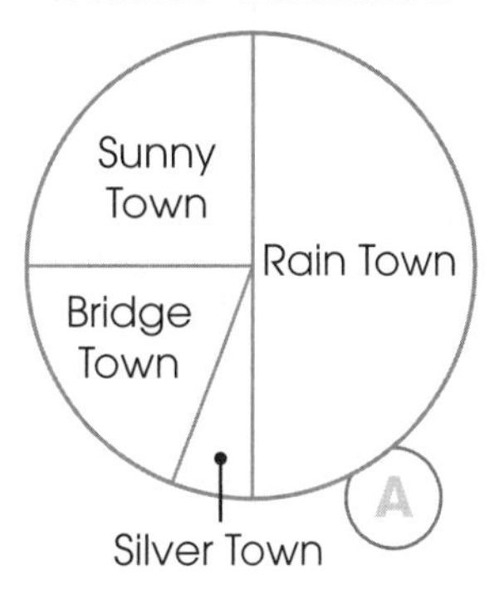

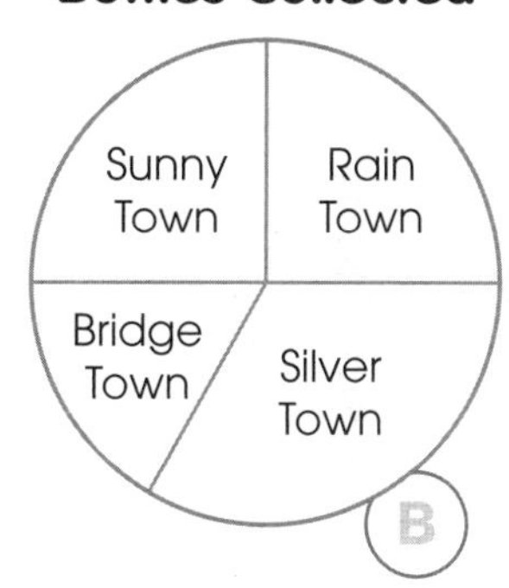

a. Which town collected about half of all bottles? __________

b. Which two towns collected about 25% of all bottles? __________

Read the information about the recycling materials collected in Sunny Town. Complete the table and the circle graph. Then answer the questions.

⑦

Material	Percent	Angle
Paper	30%	108°
Glass	20%	
Plastic	25%	
Metal	15%	
Others	10%	

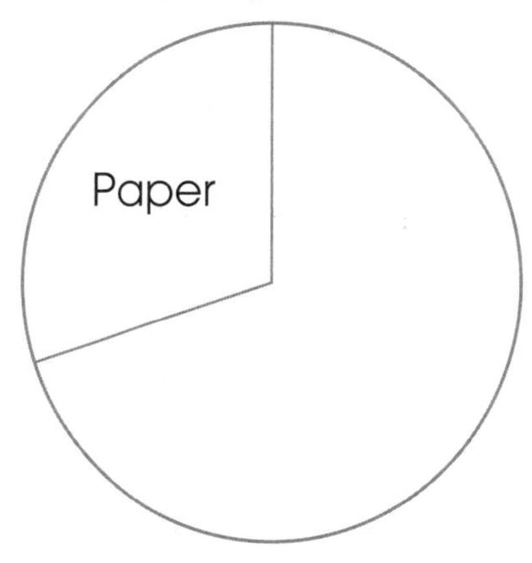

⑧ The total weight of the recycling materials collected is 5000 kg. What is the weight of the

a. paper? __________ b. glass? __________

c. plastic? __________ d. metal? __________

Use the graph to answer the questions.

⑨ **Boxes of Waste Paper Collected**

Weight (kg)	Number of Boxes
0 – 2	5
2 – 4	20
4 – 6	15
6 – 8	30
8 – 10	10

Tips: Histograms are similar to bar graphs but have no space between the bars.

a. Check the type of graph this is.

◯ bar graph ◯ circle graph

◯ line graph ◯ histogram

b. Write two differences between a histogram and a bar graph.

- ______________________
- ______________________

c. How many boxes of waste paper weighed between 6 and 8 kg? ______

d. How many boxes of waste paper weighed under 4 kg? ______

e. How many boxes of waste paper weighed over 8 kg? ______

f. How many boxes of waste paper were collected? ______

g. How many more boxes of waste paper weighed 6 – 8 kg than 2 – 4 kg? ______

h. How many fewer boxes of waste paper weighed 8 – 10 kg than 2 – 4 kg? ______

i. How many boxes of waste paper weighed over 10 kg? ______

j. What is the median weight of a box of waste paper? ______

The amount of plastics collected weekly in a neighbourhood was recorded. Complete the table and the histogram. Then answer the questions.

⑩ **Weights (kg) of Plastics Collected Weekly**

68	77	61	80	71
70	78	56	78	75
62	67	73	77	80
79	69	78	75	66
70	80	59	80	72
65	77	68	79	79

Weight (kg)	Frequency
56 – 60	
61 – 65	
66 – 70	
71 – 75	
76 – 80	

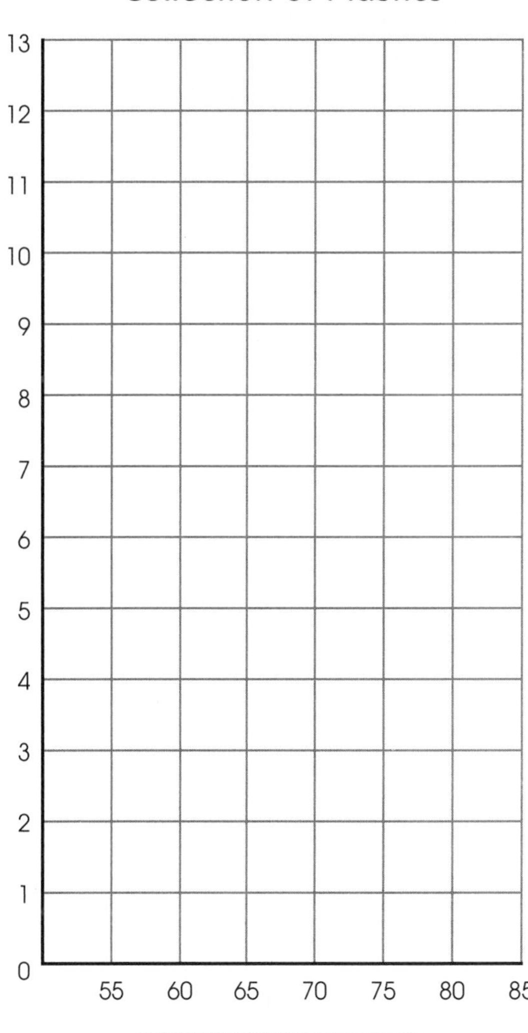

a. What is shown on the vertical axis? __________

b. What is shown on the horizontal axis? __________

c. How long did it take to collect the data set? __________

d. What is the median? __________

e. What range of weight is the most frequent? What does this imply?

____

____

LEVEL 2 – FURTHER YOUR UNDERSTANDING

12 Probability

- finding probabilities

Read
This

Experimental probability is the probability of an event occurring based on the results of actual experiments.

Theoretical probability is the probability of an event occurring in theory.

Example Determine the probabilities.

Janice flipped a coin 10 times and got tails 4 times. Find the probability of getting tails.

Experimental probability: $\frac{4}{10}$ or 40%

Theoretical probability: $\frac{5}{10}$ or 50%

Probability can be expressed as a fraction or a percent.

Try It

Find the probability of getting heads.

Experimental probability: ______

Theoretical probability: ______

Read what each child says. Determine and write whether it involves experimental probability or theoretical probability.

① 
The probability of rolling an even number on a dice is 50%.

____

② 
I rolled a dice 20 times. The probability of getting a "2" was 10%.

____

③ 
A 5-card deck has 1 blue card. The probability of picking a blue card is $\frac{1}{5}$.

____

④ 
I flipped a coin 10 times and the probability of flipping heads was $\frac{3}{10}$.

____

⑤ 
There are 4 buckets in different colours. The probability that a ball will land in a red bucket is 25%.

____

Read each scenario and answer the questions.

⑥ There are 8 cards. A card is drawn at random each time.

a. Find the probabilities.

- P(1) = ______
- P(2) = ______
- P(3) = ______
- P(7) = ______
- P(not 7) = ______
- P(1 or 7) = ______

b. Three cards are taken out. Which three cards are taken out if

- the probability of getting each number is the same?

____

- the probability of getting 2 is 60%?

____

There is more than one answer.

⑦ The spinner is spun.

a. Find the probabilities.

- P(2) = ______
- P(4) = ______
- P(5) = ______
- P(not 2) = ______
- P(an even number) = ______

b. A coin is tossed after the spinner is spun. Check the correct tree diagram and find the probabilities.

(A)

Spinner	Coin	Outcome
2	H	2, H
	T	2, T
4	H	4, H
	T	4, T
5	H	5, H
	T	5, T

(B)

Spinner	Coin	Outcome
2	H	2, H
	T	2, T
4	H	4, H
5	T	5, T

P(2, H) = ______

P(4, H) = ______

P(5, T) = ______

P(2, 5) = ______

Kate is playing with a dice. Answer the questions.

⑧ This is a fair dice with its 6 sides labelled from 1 to 6.

Kate rolls the dice once. Find the probability of each event.

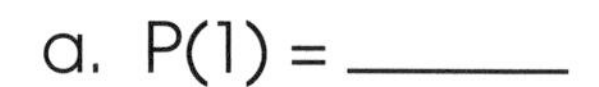

a. P(1) = ______ b. P(6) = ______

c. P(not 5) = ______ d. P(an odd number) = ______

⑨ Kate rolls the dice twice and adds the numbers.

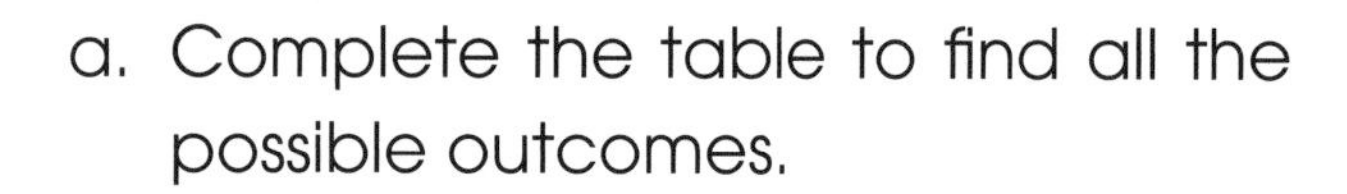

a. Complete the table to find all the possible outcomes.

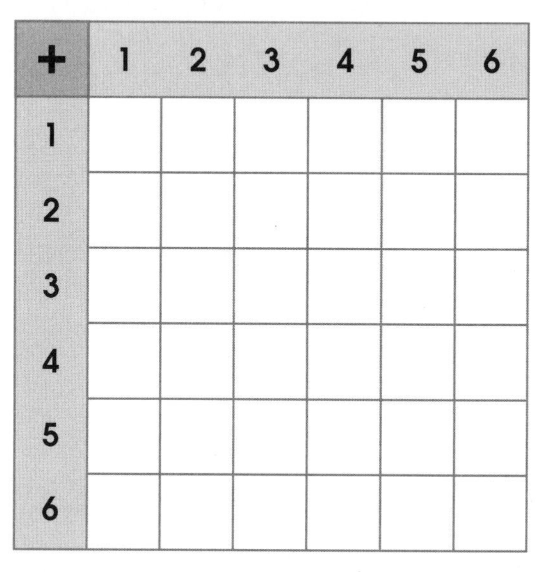

+	1	2	3	4	5	6
1						
2						
3						
4						
5						
6						

b. Find the probabilities of getting each sum.

- P(5) = ______
- P(10) = ______
- P(not 6) = ______

c. Are all the possible sums equally likely to occur? Explain.

____

____

d. Which sum is the most likely? What is the probability?

____

e. What is the probability that Kate rolls "1" twice in a trial?

____

f. What is the probability that Kate gets a "2" for one of her rolls in a trial?

____

Answer the questions.

⑩ In a board game, you can get an extra turn if you roll a double with 2 dice.

a. What is the theoretical probability that you will roll a double?

b. In Rex's 20 turns, he rolled a double 5 times. What is his experimental probability of getting an extra turn?

c. Has Rex been lucky? Explain.

⑪ Anthony's mom made 2 spinners to help decide what to make for Anthony's lunch.

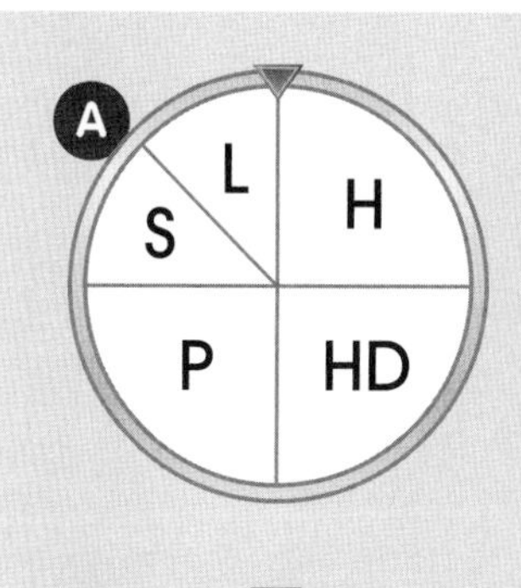

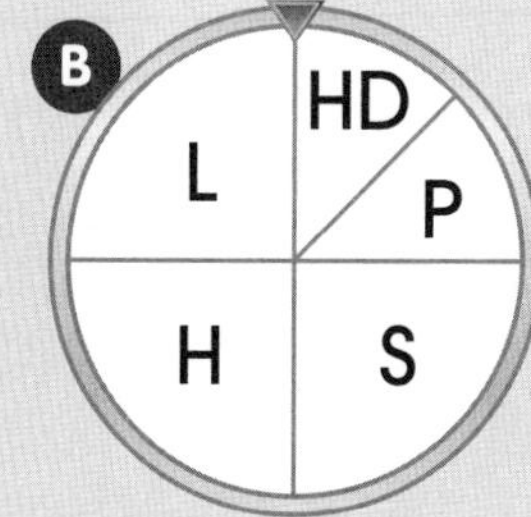

S - spaghetti

P - pizza

H - hamburger

HD - hot dog

L - lasagna

a. What is the probability of getting hamburger on each spinner?

b. If Anthony wants spaghetti, which spinner does he want his mom to use?

c. Spaghetti was spun yesterday. Does the probability of spinning spaghetti today change based on yesterday's result? Explain.

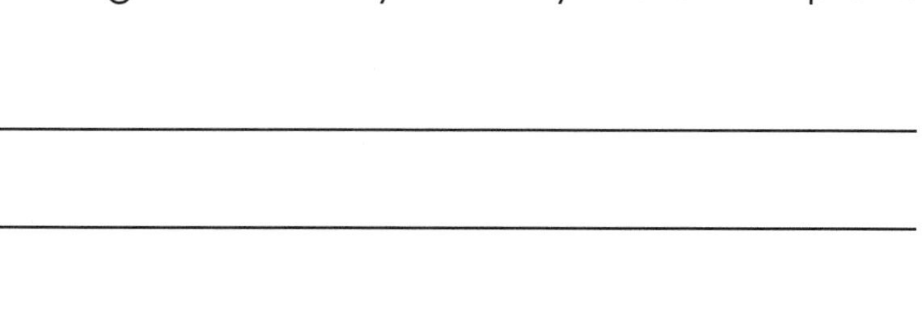

LEVEL 3

APPLICATIONS

1 Exponents

- solving problems involving exponents

A music video has 2×10^7 views online. How many views does the music video have in standard form?

2×10^7

$= 2 \times$ ______

$=$ ______

The music video has ______ views.

Powers of 10 are useful for expressing very large or very small numbers.

e.g.

$800\ 000 = 8 \times 10^5$

$0.00006 = 6 \times 10^{-5}$

Answer the questions.

① What is the difference between 5^4 and 5×4? Explain.

______

______

______

② Is any power a multiple of its base? Illustrate your answer with an example.

______

______

______

③ Which is greater, 3^4 or 4^3?

______

______

______

④ Is it possible for a number to be expressed as a power with different bases? If so, give an example.

______

______

______

⑤

Can you tell which power is the greatest by inspection (without comparing them in standard form)? Explain.

______

______

______

Check the correct expression for each problem. Then solve it.

⑥ There are 100 ducks at Lake Princess. If the number of ducks at the lake doubles every 6 months, how many ducks will there be in 2 years?

Ⓐ 100×2^4

Ⓑ 100×2^6

There will be ________ ducks.

⑦ Patricia follows a pattern when putting money into her piggy bank.

Date	Sep 1	Sep 2	Sep 3	Sep 4
Amount ($)	2^0	2^1	2^2	2^3

How much will she put into her piggy bank on Sep 10?

Ⓐ 2^9

Ⓑ 2^{10}

She will put $________ into her piggy bank.

⑧ The population of Pleasantville doubles every 20 years and is currently 100 000.

a. What will the population be in 60 years?

Ⓐ $100\,000 \times 3^2$

Ⓑ $100\,000 \times 2^3$

The population will be ____________ .

b. What was the population of Pleasantville 20 years ago?

Ⓐ $100\,000 \div 20$

Ⓑ $100\,000 \div 2$

The population was ____________ .

c. What was the population of Pleasantville 60 years ago?

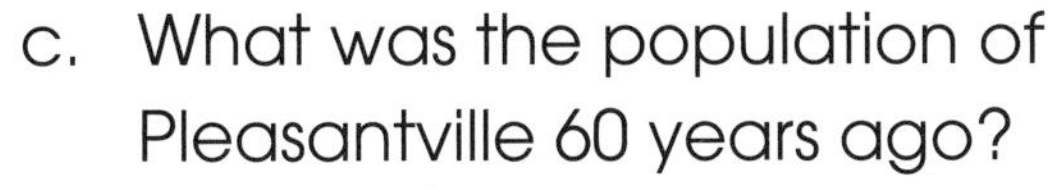

Ⓐ $100\,000 \div 2^3$

Ⓑ $100\,000 \div 2^6$

The population was ____________ .

Solve the problems. Show your work.

⑨ A scientist created a special type of foam that expands when it comes in contact with water. Its size increases to 10 times its previous size every hour. The volume of the foam is 3 cm^3 when it is placed in the water.

a. What will its volume be after 3 hours?

Its volume will be ________ cm^3.

b. What will its volume be after 5 hours?

Its volume will be ________ cm^3.

c. After how many hours will its volume be 3 000 000 cm^3?

It will be after ________ hours.

⑩ There were 200 visitors at an amusement park when it opened at 9:00 a.m.

a. The number of visitors tripled every hour until noon. How many visitors were there at noon?

Tips: Noon is 12 p.m. There are 3 hours from 9:00 a.m. to 12:00 p.m.

There were ________ visitors at noon.

b. The number of visitors reduced by half every hour from noon until close. How many visitors were there at 3:00 p.m.?

There were ________ visitors at 3:00 p.m.

c. At what time in the afternoon were there 1350 visitors at the park?

1350 visitors were at the park at ________ .

⑪ Drew made a cube with a side length of 6 cm. Andy made a cone with a volume of 2×10^2 cm^3. Whose solid has a greater volume?

⑫ Grace's cube has a volume of 27×10^3 cm^3. Which cube is hers?

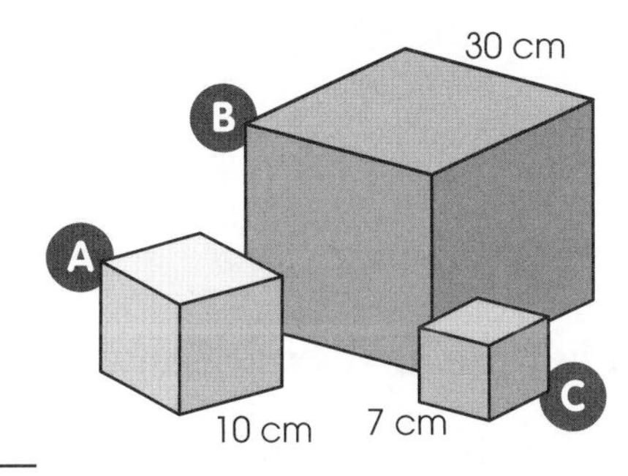

⑬ Tammy created 3 programs to send out e-mails. Beginning at 1, the number of e-mails sent grows each minute according to the rates in the table.

a. How long will it take for each program to send at least 500 e-mails?

Program	Growth
A	doubles
B	triples
C	quadruples

b. How many more e-mails can Program C send than Program B in 8 minutes?

⑭ The value of a painting doubles every 10 years. If it was worth $1000 in 1920, what would its value be in 2020?

2 Order of Operations

- solving word problems involving the order of operations

$12 + (7 - 2) \times 3^2$ ← brackets

$= 12 + \square \times 3^2$ ← exponents

$= 12 + \square \times \square$ ← multiplication

$= 12 + \square$ ← addition

$= \square$

Follow the order of operations with **BEDMAS**:

Brackets
Exponents
Division
Multiplication
Addition
Subtraction

Find the answers. Show your work.

① $(18 + 7) \times 4$

② $60 \div (7 + 5)$

③ $(28 - 7) \div (2^2 - 1)$

④ $5^2 \times 4 - 9^2 + 6$

⑤ $16 \div (2 \times 3^2 - 10)$

⑥ $81 \div 3^2 \times (5 - 2)$

⑦ $(25 - 18) + 7 \times 2^2$

⑧ $30 + (18 \div 3^2)$

⑨ $24 \div 6 \times (2 + 4^2)$

⑩ $4 - 2 + 5 \times 4^2 \div (5 + 3)$

⑪ $19 - 2 + (6^2 \div 3) + 3 \times 4$

⑫ $65 \div (3^2 - 2^2) + (5^2 \times 3)$

⑬ $(7 - 2^2) \times 4^2 \div (6 \times 10 \div 5)$

Answer the questions.

⑭ Pat has an old calculator. She punched in $10 + 12 \div 2$. The calculator did its calculation following the sequence of the input.

a. What answer did the calculator give?

b. Was the calculator's answer correct? If not, what is the correct answer?

⑮ Meghan wrote the expression: $(5 \times 3) + 1$. Her brother walked by and insisted that the brackets in the expression are not necessary. Is he correct? Explain.

⑯ Jim says, "The sum of the squares of two numbers is the same as the square of the sum of the numbers." Is he correct? Give an example.

⑰ Write two different expressions that involve multiplication for finding the perimeter of the rectangle, one with brackets and the other without. Show that the answers are the same.

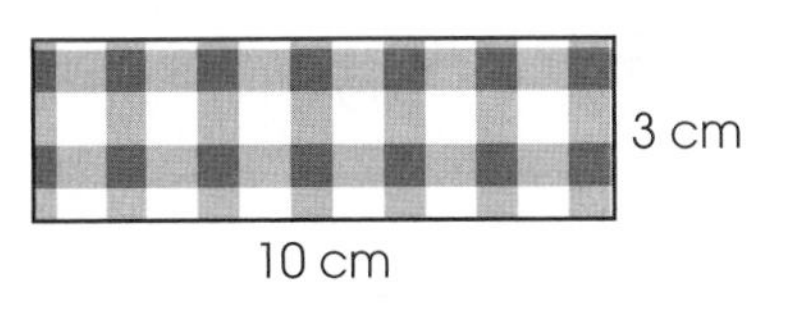

⑱ Nick wrote the mathematical statement: $35 + 3 \div 19 + 7 \times 22 - 13 = 65$.

a. Is the statement correct? Explain.

b. Nick realized that he forgot to add brackets. Help him add brackets to make the statement correct.

LEVEL 3 – APPLICATIONS

Check the correct expression for each problem. Then solve it. Show your work.

⑲ Debbie bought 3 pairs of earrings on sale. The original price was \$6 per pair but they were on sale for \$1 off. What was the final cost of her purchase?

(A) $(6 - 1) \times 3$

(B) $6 \times 3 - 1$

The final cost was \$__________ .

⑳ Oliver bought 4 ties for \$23 each and paid with \$100. What was his change?

(A) $100 - 23 \times 4$

(B) $4 \times 23 + 100$

His change was \$__________ .

㉑ Joy bought 3 CDs at \$19 each. She had a \$5-off coupon and her father agreed to pay half the cost. How much did Joy pay?

(A) $(19 - 5) \times 3 \div 2$

(B) $(19 \times 3 - 5) \div 2$

Joy paid \$__________ .

㉒ Judy has a weekend job that pays \$17 per hour and \$20 per hour overtime. Last week, Judy worked 6 hours on Saturday and 5 hours on Sunday, plus 4 hours overtime. How much did she earn?

(A) $17 \times (6 + 5) + 20 \times 4$

(B) $20 \times (6 + 5) + 17 \times 4$

She earned \$__________ .

㉓ Jane made 2 long distance calls. One lasted 10 minutes longer than the other. The 2 calls lasted a total of 46 minutes. How long was each call?

(A) $46 \div 2 - 10$

(B) $(46 - 10) \div 2$

The calls were __________ and __________ minutes long.

Solve the problems. Show your work.

㉔ At Hill School, there are 200 Grade 7 students and half of them have brown eyes. 90 have both brown hair and brown eyes. How many brown-eyed Grade 7 students do not have brown hair?

㉕ Bill bought 2 pairs of jeans. He paid $15 more for one pair than the other. He paid with a $100 bill and got $25 change. What was the price of the less expensive pair?

㉖ Tom bought 3 T-shirts on sale. Each one had been marked down by $5. He also bought a pair of jeans for $55. The total cost before tax was $106. What was the price of each T-shirt before the sale?

㉗ A plane is flying at 1200 m below the clouds. Another plane at a lower altitude is flying at 1800 m above the ground. The clouds are 4000 m above the ground. What is the difference in altitude between the 2 planes?

㉘ Allisa made some bracelets. She sold 5 of them for $6 each and the remaining 8 bracelets for $7 each. She paid $50 for the materials.

a. What was her profit?

Profit can be found by subtracting the amount spent from the amount earned.

b. If she had sold all bracelets for $7 each, how much more would she have earned?

LEVEL 3 – APPLICATIONS

3 Squares and Square Roots

- solving word problems involving squares and square roots

A square-shaped town has an area of 225 km^2. What is the length of the town?

$\sqrt{225} = \square$

The length of the town is ______ km.

The square root of a square's area gives the side length of the square. Likewise, the square of a side length gives the area of the square.

Solve the problems. Show your work.

① A square picture frame has a length of 30 cm. What is the area of the frame?

____

② Brad pours a square foundation for a garden shed. The area of the foundation is 9 m^2. How many metres of lumber does Brad need to frame the whole foundation?

____

③ Sara has 100 cm of string that she plans to wrap around a square. What is the area of the largest square that Sara can wrap the string around?

____

④ A rectangular rug can be divided into 2 equal squares. What are the dimensions of the rug if its area is 3200 cm^2?

____

⑤ What is the difference in side length of two squares if their areas are 121 m^2 and 289 m^2?

⑥ Cameron's living room is square-shaped. A square section of the room is tiled as shown. What is the area of the untiled section?

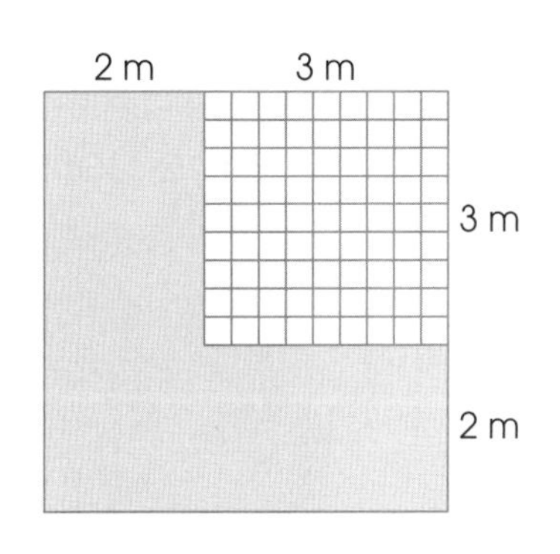

⑦ A square picture frame has an outer width of 20 cm and an inner width of 18 cm. What is the area of the frame?

⑧ A tiling pattern consists of identical big and small squares.

a. What is the area of a small square?

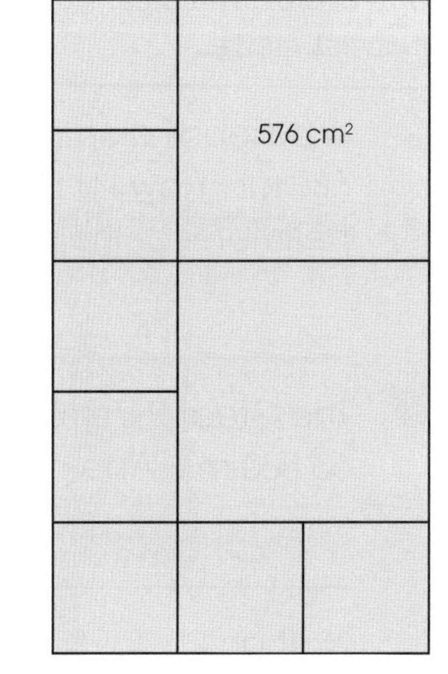

b. What is the area of the entire tiling pattern?

c. What is the perimeter of the tiling pattern?

LEVEL 3 – APPLICATIONS

⑨ A 48-km² rectangular neighbourhood can be divided into 3 identical squares. What are the dimensions of the neighbourhood?

---

⑩ Two square-shaped play pens are next to each other. What is the perimeter of the play pens together?

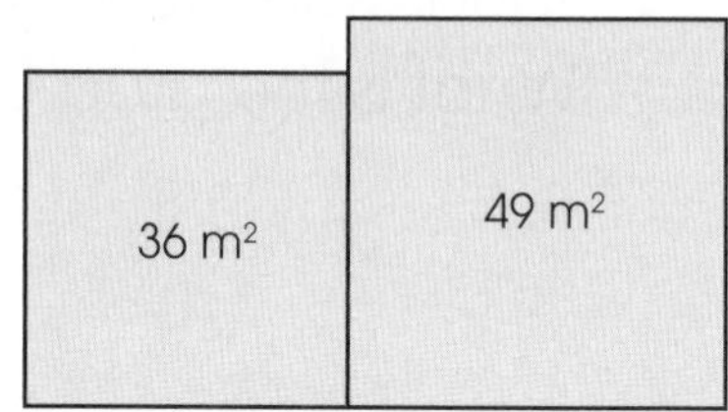

---

⑪ A pool has the given dimensions on all sides. What is the area of the pool?

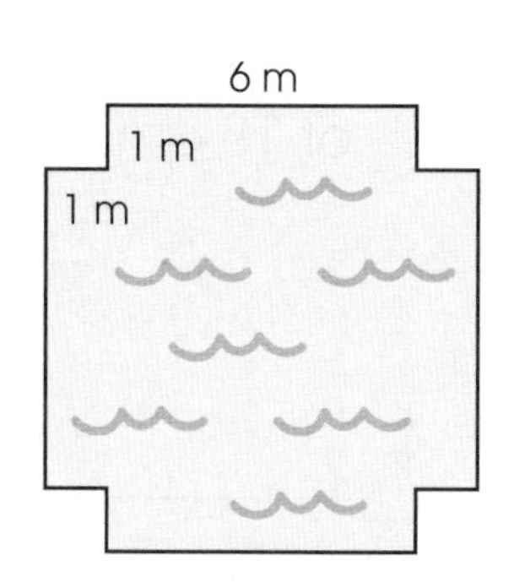

Solve the problems. Use a calculator to find the answers and round them to the nearest metre.

⑫ A baseball diamond has a square infield. The area of the infield is about 680 m². How far does a player need to run to complete a home run (the perimeter of the infield)?

---

⑬ The Great Pyramid of Giza in Egypt has a square base that covers about 53 000 m². About how long is each side of the base?

---

⑭ Nathan Phillips Square is Canada's largest city square. It has an area of 48 500 m². If it is in the shape of a square, what is the approximate perimeter?

Answer the questions about perfect squares.

A perfect square is a number that has a whole number square root.

e.g.

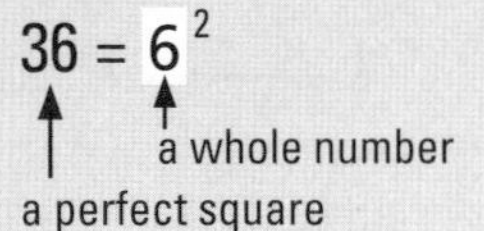

⑮ a. List the first 10 perfect squares.

____

b. List the difference between each consecutive pair of perfect squares.

____

c. Describe the pattern in the differences.

____

⑯ a. If you multiply a perfect square by a perfect square, is the answer always a perfect square? Give an example.

____

b. Tiffany thinks that a perfect square multiplied by a non-perfect square always gives a perfect square as a product. Is she correct? Explain with an example.

____

c. Joshua thinks that the product of 2 non-perfect squares can be a perfect square. Is he correct? Explain with an example.

____

d. Allison thinks that the square root of the sum of 2 perfect squares cannot be a whole number. Is she correct? Explain with an example.

____

4 Integers

- solving word problems involving integers

On one day in February in Toronto, the temperature was -12°C in the afternoon and -22°C at night. What was the change in temperature from day to night?

(______) – (______) = ______

There was a drop of ☐ °C.

The sum of a positive integer and a negative integer may be positive, negative, or zero.

e.g. $(+5) + (-9)$
$= +5 - 9$
$= \underline{-4}$

Complete each statement with "always", "sometimes", or "never". Then match it with an example from the box.

2 > -3   -2 > -3   2 + 3 = 5
-2 + (-3) = -5   2 + (-3) = -1

① The sum of 2 positive integers is a. ______ positive.

e.g. ______

The sum of 2 negative integers is b. ______ positive.

e.g. ______

A positive integer is c. ______ smaller than a negative integer.

e.g. ______

The sum of a positive integer and a negative integer is d. ______ negative.

e.g. ______

If 2 integers are negative, then the one closer to zero is e. ______ greater.

e.g. ______

Check the correct expression for each problem. Then find the answer.

② Lisa's kite was 26 m off the ground. It then descended 18 m. How high is the kite?

(A) 26 + (-18)

(B) -26 + 18

The kite is ______ m high.

③ In a football game, the running back gained 6 m, lost 3 m, lost 2 m, and then lost 4 m. How much did he gain or lose?

(A) 6 + (-3) + (-2) + (-4)

(B) 6 – (-3) – (-2) – (-4)

He ______ ______ m.
gained/lost

Solve the problems. Show your work.

④ A vulture is flying at a height of 8000 m and a goose is flying at 1000 m. Meanwhile, a turtle is swimming at a depth of 1200 m.

a. Locate the missing values on the diagram.

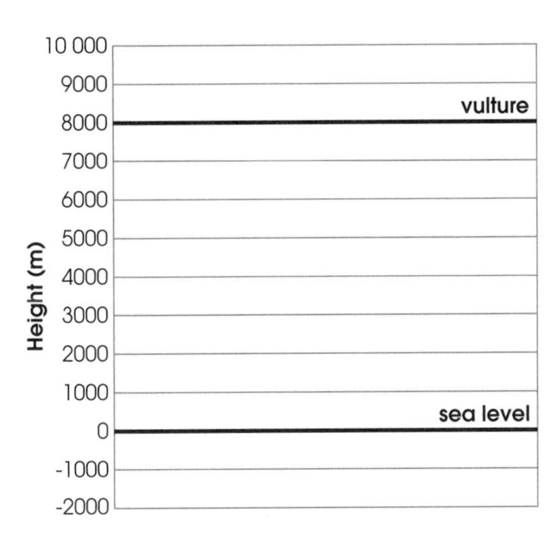

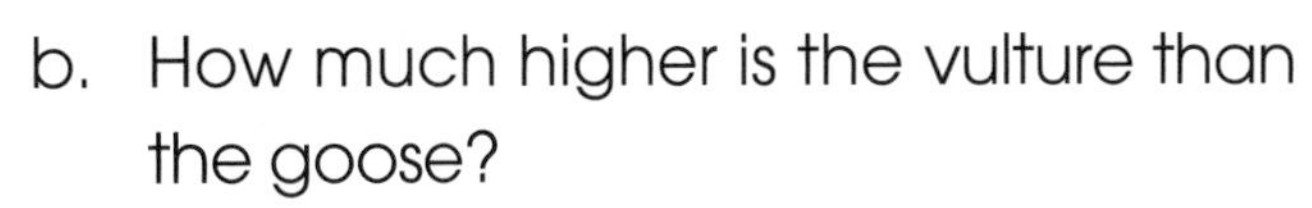

b. How much higher is the vulture than the goose?

____

c. How much higher is the vulture than the turtle?

____

d. If the vulture descends 1500 m to catch its prey, how high was the prey?

____

⑤ Mount Logan in Yukon is 5900 m high. Under the Pacific Ocean, the Mariana Trench reaches 11 000 m below sea level.

a. Write each measurement as an integer.

____

b. Would Mount Logan be seen as an island if it rose from the floor of the Mariana Trench? Explain with an integer.

____

⑥ The initial price of MBank stock was $23 per share. The price changes from Monday to Friday last week are recorded in the table.

a. What was the overall price change?

____

b. Mr. Smith bought 1000 shares of MBank stock last Monday and sold them all on Friday. How much did he gain or lose?

____

MBank Stock

Day	Change ($)
Mon	-2
Tue	+2
Wed	-3
Thu	+2
Fri	-3

⑦ The minimum surface temperatures of some planets are given in the chart.

a. Which planet can get the coldest?

Planet	Minimum Temperature
Mercury	-184°C
Earth	-90°C
Mars	-123°C

b. What is the difference between the minimum temperatures of Earth and Mars?

c. List the temperatures in order using ">".

⑧ Tony has a total of -$35 in his two bank accounts. He has -$19 in Account A.

a. How much does he have in Account B?

b. What is the difference between the balances of the two accounts?

⑨ The graph shows the average daily temperature of a city from Jan 1 to Jan 7.

a. What was the difference in temperature between Jan 1 and Jan 7?

b. If the temperature on Jan 8 was 7°C lower than the previous day, what was the temperature on Jan 8?

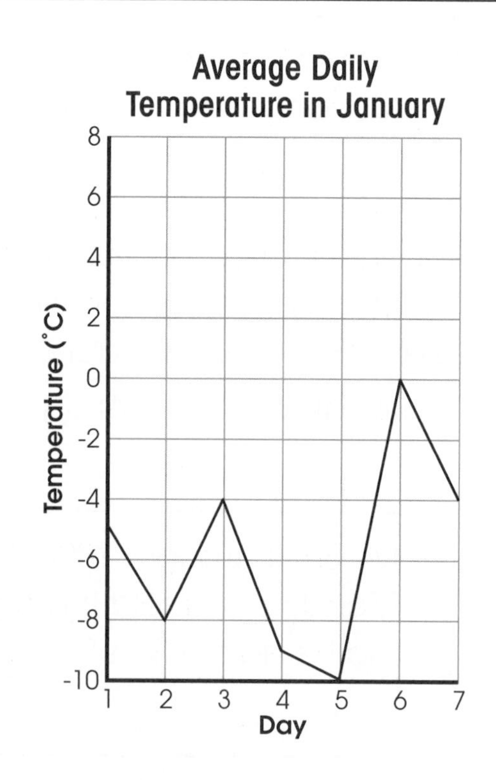

⑩ See the children's scores in the table.

	Round 1	Round 2
Tim	-16	2
Jake	15	
Liz	-9	31

a. How many more points did Tim get in Round 2 than in Round 1?

____

b. How many points did the children get in Round 1 in all?

____

c. If the children got a total of 29 points in Round 2, what was Jake's score in Round 2?

____

⑪ Mr. Smith is a golfer. The table shows his performance in last week's 9-hole tournament.

Hole 1	2	3	4	5	6	7	8	9
par	double bogey	birdie	eagle	par	bogey	bogey	birdie	double bogey

a. Was Mr. Smith's score over or under par?

____

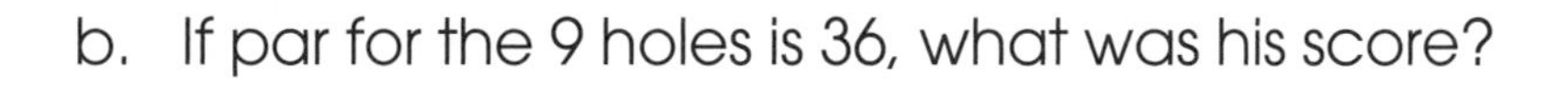

b. If par for the 9 holes is 36, what was his score?

____

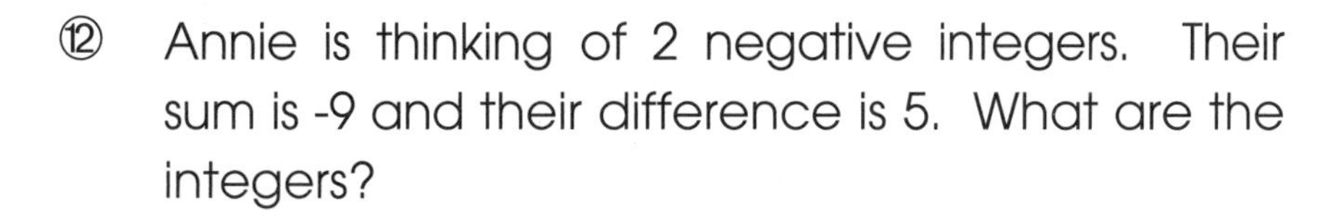

⑫ Annie is thinking of 2 negative integers. Their sum is -9 and their difference is 5. What are the integers?

____

Golfing terms:

- par: the number of strokes one should take to hit the ball into a hole
- bogey: 1 over par
- double bogey: 2 over par
- birdie: 1 under par
- eagle: 2 under par

5 Fractions

- solving word problems involving fractions

A bag has 10 apples. How many apples are there in $\frac{1}{2}$ of the bag?

☐ × ☐ = ☐

There are ☐ apples.

Use multiplication to find the fractional part of a group.

e.g. $\frac{2}{3}$ of 15

$= \frac{2}{3} \times 15$

$= \underline{10}$

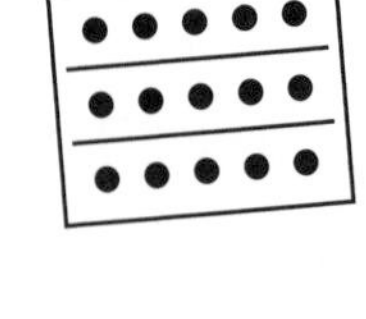

Find the answers. Show your work.

①

How many oranges are there in

a. $\frac{1}{4}$ of the box?

There are ________ oranges.

b. $\frac{3}{8}$ of the box?

There are ________ oranges.

②

How many candies are there in

a. $\frac{2}{5}$ of the box?

There are ________ candies.

b. $\frac{7}{10}$ of the box?

There are ________ candies.

③

How many millilitres of milk did I drink if I drank...

a. $\frac{1}{2}$ of the carton?

He drank ________ mL of milk.

b. $\frac{18}{20}$ of the carton?

He drank ________ mL of milk.

1 L = 1000 mL

Solve the problems. Write the answers in simplest form.

④ Daniel spent $1\frac{1}{2}$ h playing soccer on Saturday and $1\frac{4}{5}$ h on Sunday.

a. How much time did he spend playing soccer altogether that weekend?

b. If he spent $2\frac{3}{4}$ h playing soccer the following weekend, how much less time did he spend on soccer than the previous weekend?

⑤ A pizza costs $8. How much does $\frac{5}{12}$ of the pizza cost?

⑥ A bag of 12 apples weighs 1500 g. Consider $\frac{2}{3}$ of the bag of apples.

a. How many apples are there?

b. How much do they weigh altogether?

⑦ Janet wants to build 12 m of fencing using planks that are $1\frac{1}{4}$ m wide. How many planks does she need?

⑧ Mary has 13 m of ribbon. She needs $3\frac{1}{4}$ m for one dress. How many dresses can she make?

⑨ Mrs. Black wants to put some textbooks on a 90 cm-long shelf. If each book is $4\frac{1}{4}$ cm thick, how many textbooks will fit on the shelf?

⑩ In a parking lot, $\frac{1}{3}$ of the cars are white, $\frac{1}{4}$ black, and $\frac{1}{5}$ red. What fraction of the cars are neither white, black, nor red?

Tips "1" represents all cars in the parking lot.

⑪ The heights of 2 shrubs are $\frac{1}{4}$ m and $\frac{3}{4}$ m. What is the mean height?

⑫ Judy ate $\frac{3}{4}$ of a 240-g chocolate bar and $\frac{2}{3}$ of a 210-g bar. How much chocolate did she eat?

⑬ A classroom is half full with 10 people. How many people are in the classroom if it is $\frac{3}{5}$ full?

⑭ There are 2 parts on this test. Of the 20 questions in Part A, I answered $\frac{3}{5}$ of them correctly. For Part B, I answered $\frac{5}{6}$ of the 30 questions correctly.

How many correct answers did she get on the test?

Read the table about the Grade 7 students at Riverdale Public School. Solve the problems. Show your work.

⑮

Facts about the Grade 7 Students

Fraction of Grade 7 students in the school	$\frac{2}{17}$	
Number of students	Class A: 30 ($\frac{1}{2}$ are boys) Class B: 36 ($\frac{1}{3}$ are girls)	
Average time spent on homework each day	English: $\frac{1}{5}$ h Science: $\frac{1}{4}$ h	Math: $\frac{1}{3}$ h French: $\frac{1}{4}$ h
Number of students who got an "A" on the Math test	Boys: 20 Girls: 15	

There are 66 students in Grade 7.

a. What fraction of the Grade 7 students got an "A" on the Math test?

____

b. How many students are there in the school?

____

c. How many students in Class A are boys?

____

d. How many Grade 7 students are girls in total?

____

e. How many minutes does a Grade 7 student spend on Math on average?

____

f. How many minutes does a Grade 7 student spend on homework each day on average?

____

LEVEL 3 – APPLICATIONS

6 Decimals

- solving word problems involving decimals

Alan paid $17.97 for 3 combo meals.
How much did each combo meal cost?

$17.97 ÷ 3 = ______

Each combo meal cost ______.

When solving problems involving money, we usually round the answer to the nearest hundredth/cent.

e.g. Unit price:
$2 ÷ 3
= $0.66666...
= $0.67

Solve the problems. Show your work.

① Mary and Judy are at a candy store. Each bag of candy costs $7.98 and each box of chocolates costs $8.95.

a. Mary buys 3 bags of candy. How much does she pay?

______

b. Judy buys 2 boxes of chocolates. How much does she pay?

______

c. How much do Mary and Judy pay altogether?

______

d. If the girls pay with a $50 bill, how much change will they get?

______

e. If the girls have $35, how much more money do they need?

______

② You have $12.30. How many comic books can you buy if they cost $2.50 each?

③ Bill commutes 320 km each week for work. His gas consumption averages 7 L per 100 kilometres, and gas costs $1.02/L. Calculate his weekly gas expense.

④ Derek got $100 for his birthday. He bought 5 CDs which cost $15.49 each. How much money did he have left?

⑤ John paid $150 to buy 3 CDs at $16.99 each and 2 T-shirts at $24.25 each, plus 15% tax.

a. How much did he spend?

b. How much change did he get?

⑥ Tony and 7 of his friends shared $4.75 equally. How much money did each of them get?

⑦ Pat has $3.48. What is the maximum number of nickels she can have?

⑧ A map has a scale of 150 km to 1 cm. If two towns are 3.73 cm apart on the map, how far apart are they on land?

⑨ It takes Earth 365.3 days to orbit the sun and complete a year. If it takes Mars 1.88 times as long as Earth to orbit the sun, how long is a Mars year?

⑩ Valentina Tereshkova, the first woman in space, made 48 orbits around Earth in 70.83 hours.

a. How long did each orbit take?

b. How many orbits did she make in 10 hours?

⑪ A number plus $\frac{1}{4}$ of itself is 4.5. What is the original number?

LEVEL 3 – APPLICATIONS

Complete the table to solve the problems. Round the answers to the nearest thousandth.

⑫ **Lightning Baseball Team – Performance of Batters**

(June 1, 2019 – July 10, 2019)

Batter	Times at Bat	Number of Hits	Batting Average
J. Green	14	5	5 ÷ 14 =
S. White	13	4	
C. Brown	12	4	
K. Jones	16	5	

⑬ J. Green was at bat 3 times without a hit in the game on July 11, 2019. What was his new batting average?

____

⑭ The batting average of S. White rose to 0.375 after the match on July 11, 2019. He was at bat 3 times in the game. How many more hits did he get?

____

⑮ The batting average of C. Brown remained the same though he got 1 hit in the game on July 11, 2019. How many times was he at bat that day?

____

⑯ What was the batting average of the four batters as of July 10, 2019?

____

Use the map below to solve the problems. Round the answers to the nearest thousandth.

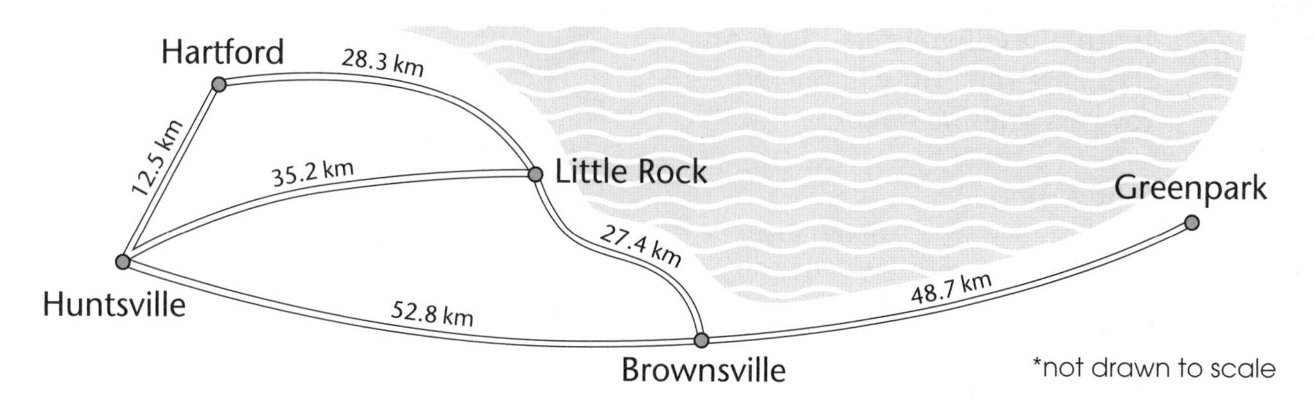

⑰ If Tony drives from Huntsville to Greenpark via Little Rock and Brownsville at a speed of 65.5 km/h, how long will it take him to reach Greenpark?

---

⑱ Everyday, Mr. Wong drives 2.2 h from Huntsville to Greenpark via Hartford, Little Rock, and Brownsville. What is his average speed?

---

⑲ It took Mr. Black 2.9 h to go from Huntsville to Greenpark and back along the same route. He kept an average speed of 70 km/h. Which route did he take?

---

⑳ Ms. Green drove from Brownsville to Huntsville and then Hartford. She then drove to Little Rock and back home to Brownsville. The entire trip took about 1.7 h. What was her average speed?

㉑ Jordan drove from Huntsville to Little Rock. He drove $\frac{3}{5}$ of the trip at 65 km/h and the rest at 40 km/h. How long was his trip?

---

7 Percents

- solving word problems involving percents

Joseph has 50 toy cars. 10% of them are red. How many red toy cars does he have?

50 × 10%

= 50 × ☐ ← convert to a fraction or a decimal

= ☐

He has ☐ red toy cars.

Follow the steps to find the percent of a number mentally.

e.g. Find 20% of 400 kg.

❶ Think: 10% of 400 kg

$400 \times 10\% = 400 \times \frac{1}{10}$

$= 40$

❷ 10% → 40 kg

So, 20% → 80 kg

So, 20% of 400 kg is 80 kg.

Find the answers without using a calculator.

① a. If Uncle Tim sells 20% of the oranges, how many oranges does he sell?

He sells ________ oranges.

b. If 10% of the oranges are rotten, how many oranges are rotten?

________ oranges are rotten.

c. Mrs. Winter buys a box of oranges and pays 80% of the price. How much does she pay?

She pays $________ .

d. Mr. Hill buys a box of oranges and pays 15% of the price for delivery. How much does he pay for delivery?

He pays $________ for delivery.

② A baby at birth is about 30% of its adult height. If Mr. Smith is 170 cm tall, about how tall was he at birth?

He was about ________ cm tall at birth.

③ All items in a shop are 25% off.

a. What is the sale price of a $36 helmet?

Discounts are subtracted from a price; taxes are added to a price.

The sale price is $________ .

b. If there is a 10% tax, what is the final cost of the helmet?

The final cost is $________ .

④ An $80 board game is on sale for 20% off.

a. How much is the board game?

The board game is $________ .

b. If Ken has a coupon for an additional 25% off, what is the final price?

The final price is $________ .

⑤ Tony bought a shirt for $40 and a second one at 50% off.

a. What was the total cost before taxes?

The total cost was $________ .

b. There was a 5% tax. How much did Tony pay in total?

Tony paid $________ in total.

Find the price of each item. Show your work in the corresponding box. Then fill in the price tags and answer the boy's question. Round the answers to the nearest cent.

⑥

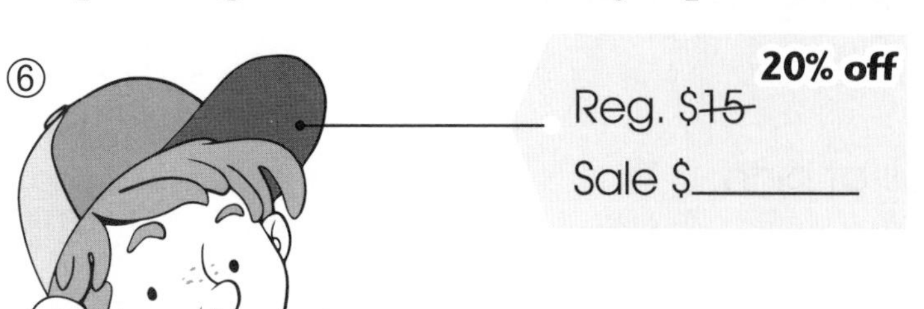

20% off
Reg. $~~15~~
Sale $________

cap

100% – 20% = ________

$15 × ________ = $________

15% off
Reg. $~~28.50~~
Sale $________

T-shirt

30% off
Reg. $~~34.50~~
Sale $________

shorts

35% off
Reg. $~~32.90~~
Sale $________

soccer ball

If the tax rate is 15%, what is the final cost of each item below?

⑦

$42.59 (skateboard)

$22.99 (baseball glove)

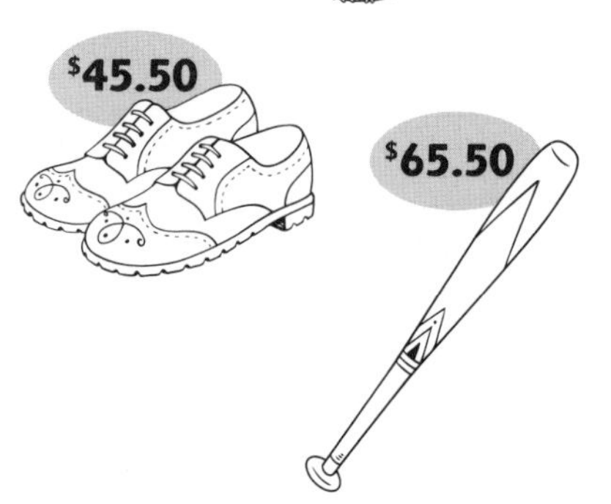

$45.50 (shoes)

$65.50 (baseball bat)

a. skateboard

b. baseball glove

c. shoes

d. baseball bat

Solve the problems. Round the answers to the nearest hundredth.

⑧ Helen bought a $45 skirt. With a tax rate of 9%, how much did she pay?

⑨ Susan bought some music online. She paid $9 in tax at a tax rate of 15%. How much did she pay for the music before tax?

⑩ In 2019, a value-added tax (VAT) of 20% was added to most purchases in the UK. If a tea set cost $8.50 in tax, what was the price of the tea set before tax?

⑪ Ben's parents give him a 10% raise in his weekly allowance. This increases his allowance by $2. What was his old allowance?

⑫ 40% of North America's national parks are located in Canada. If there are 34 national parks in Canada, how many are there in North America?

⑬ Low fat cheese has 30% less calories than regular cheese. If a piece of low fat cheese contains 210 calories, how many calories are there in a piece of regular cheese of the same size?

⑭

John has 20 hockey cards and I have 64. If I give John 25% of my cards, how many cards will each of us have?
Tom

8 Perimeter and Area

- solving word problems involving perimeter and area

The dimensions of a backyard deck are given in the diagram below. What is the area of the deck?

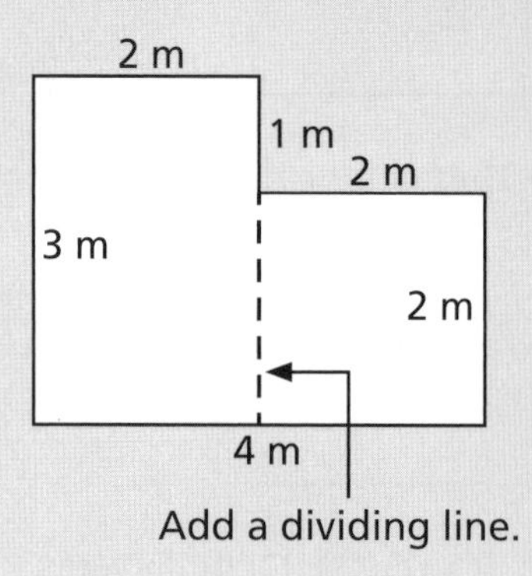

Area:

☐ + ☐

= ☐

The area of the deck is ☐ m^2.

To find the area of a composite shape, add lines to divide it into squares, triangles, parallelograms, or trapezoids. Then find their areas and add.

Solve the problems. Show your work.

① Read the question above again. How many metres of fencing is needed to enclose the deck?

② The design of a flag is a star on a white square between two red rectangles of the same size.

a. What is the perimeter of the flag?

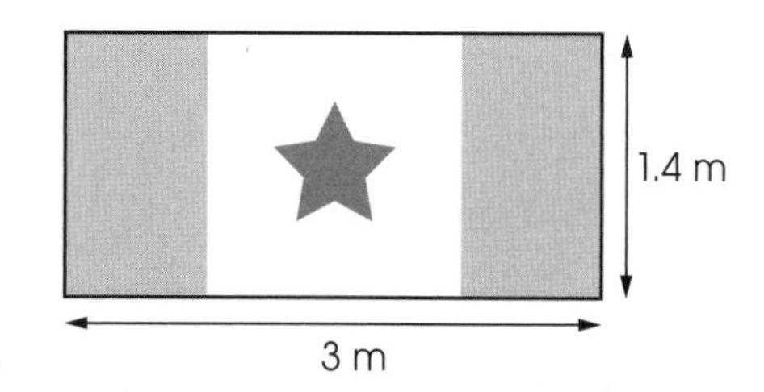

b. What are the areas of the quadrilaterals on the flag?

③ Find the perimeter and area of the shape.

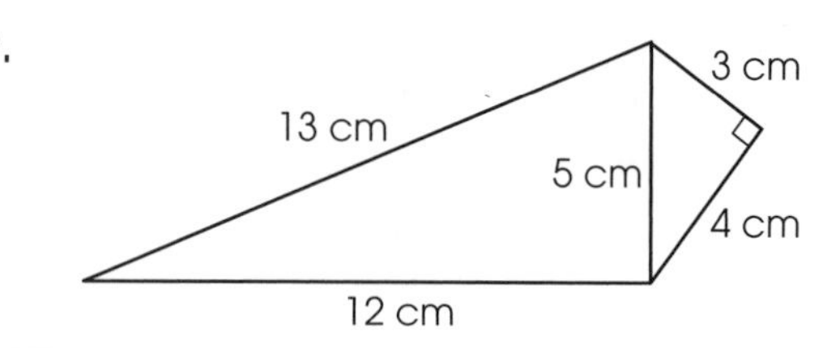

④ 6 circles each with a diameter of 50 cm are drawn in a rectangle as shown.

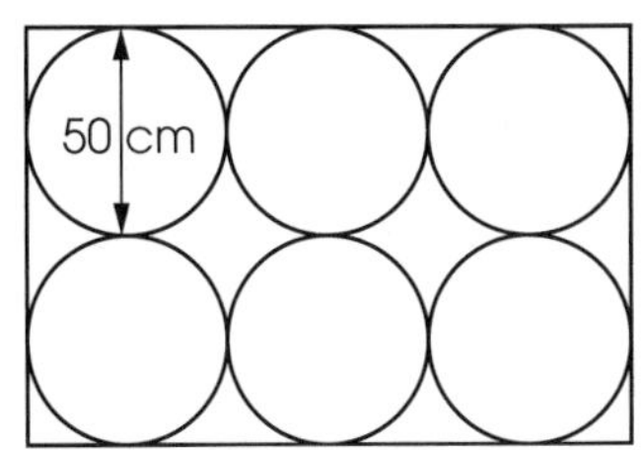

a. What are the length and width of the rectangle?

b. What are the perimeter and area of the rectangle?

d

A diameter is a straight line that joins two points on a circle and passes through the centre.

⑤ A road sign is in the shape of a regular hexagon.

a. What is its perimeter?

50 cm

24 cm

88 cm

b. What is its area?

⑥ The given pattern is part of a quilt design made up of congruent triangles. Determine the total area of the shaded triangles.

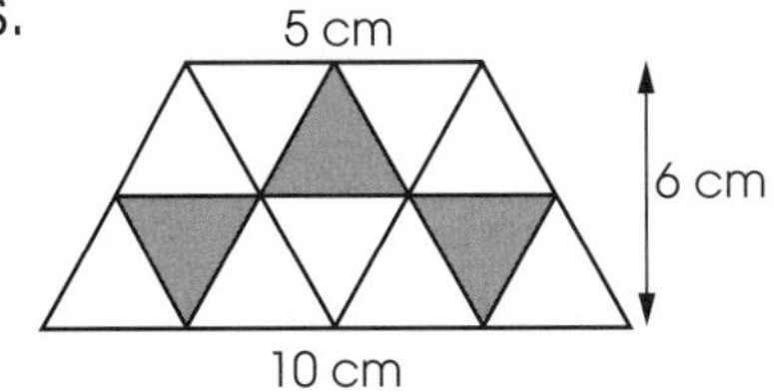

⑦ A carpet contains the hexagonal design shown. Determine the area of the shaded hexagon.

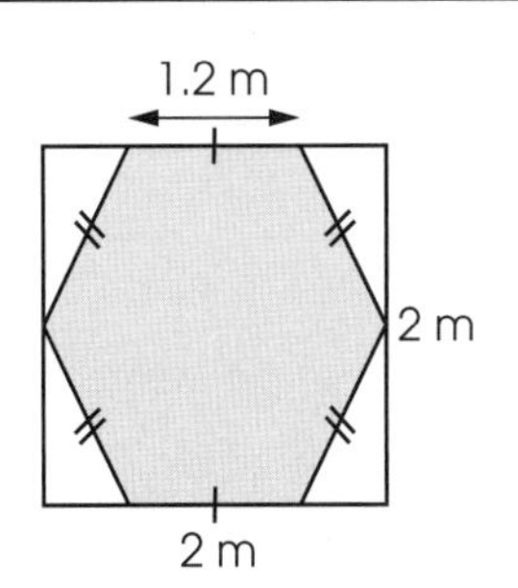

LEVEL 3 – APPLICATIONS

⑧ Tony measured his recycling bin. The front, back, and sides are all in the shape of a trapezoid.

a. Find the total area of its front and back.

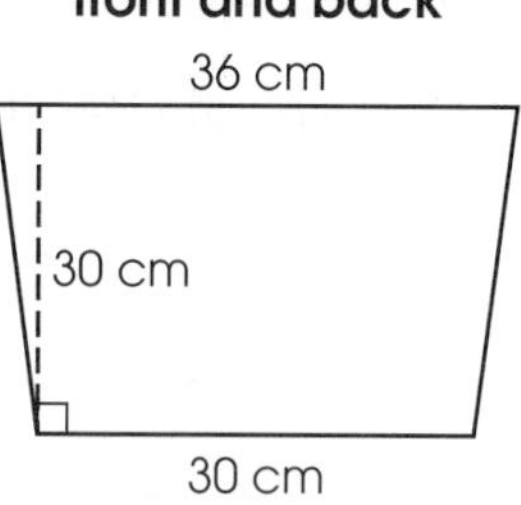

b. Find the total area of its sides.

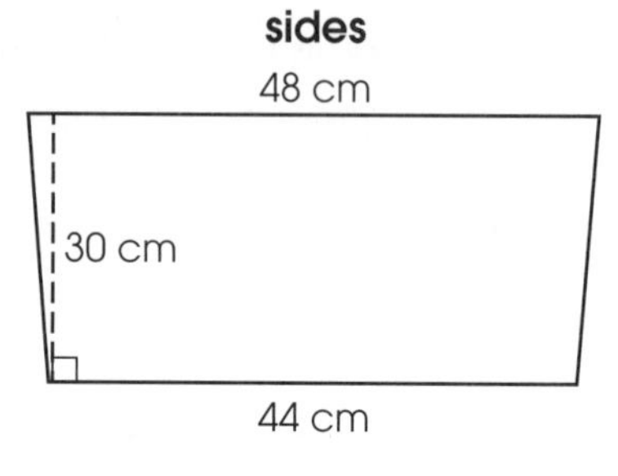

⑨ The design of a quilt is made up of identical parallelograms.

a. How much fabric is needed to make the design?

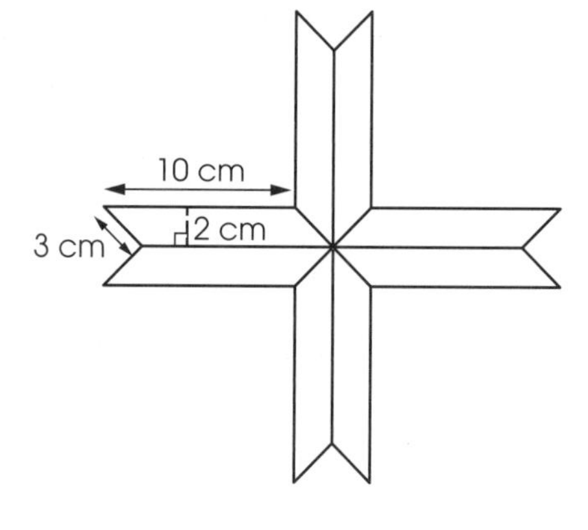

b. How much lace trimming is needed to go around the design?

⑩ A triangle with a base of 10 cm has the same area as a square with sides of 5 cm. What is the height of the triangle?

Tips
Sketch the shapes to help you.

Adam and Nadine have bought a house. The diagrams below show the dimensions of the backyard and the living room. Study the diagrams carefully and solve the problems. Show your work.

⑪

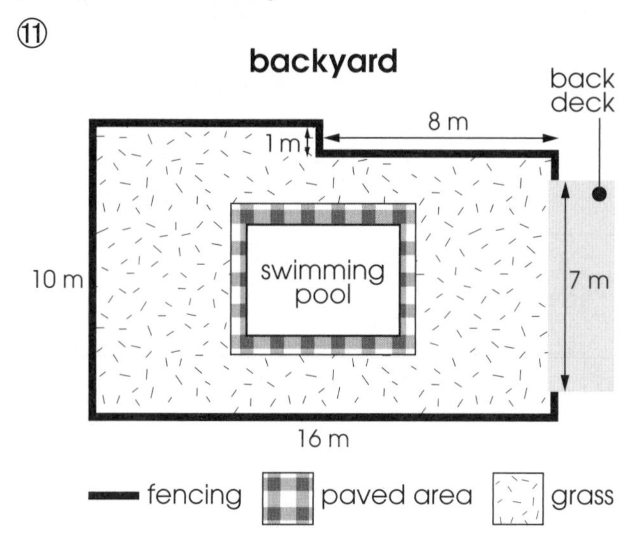

a. The backyard is enclosed by fencing which cost \$15/m. How much did the fencing cost?

____

b. The swimming pool is 5 m by 7 m. The paved area is 1 m wide. What is the area of the paved area?

____

c. The backyard is covered with grass sod which cost \$12/m^2. How much did the sod cost?

____

⑫

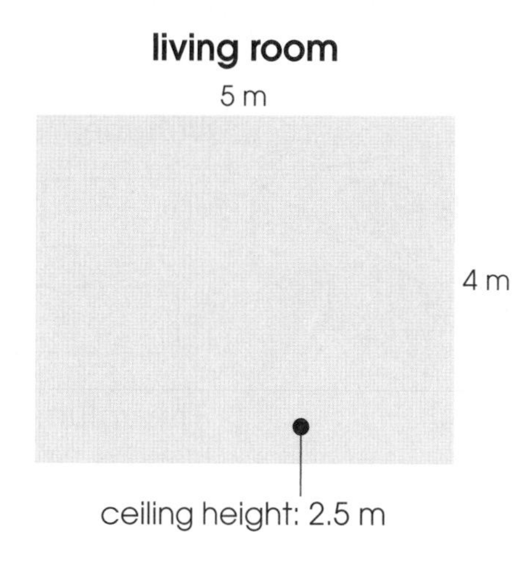

a. Adam and Nadine want to renovate their living room. If the whole floor is to be carpeted and it costs \$70/m^2, how much will it cost in total?

____

b. They also want to paint the walls and ceiling of the living room. The paint costs \$8.99/L and 1 L covers 5 m^2. If the windows occupy 7 m^2, how much will the paint cost? (Round to the nearest cent.)

____

9 Volume and Surface Area

- solving word problems involving volume and surface area

A tissue box has the dimensions of 12 cm by 22 cm by 0.07 m. What is the volume of the box?

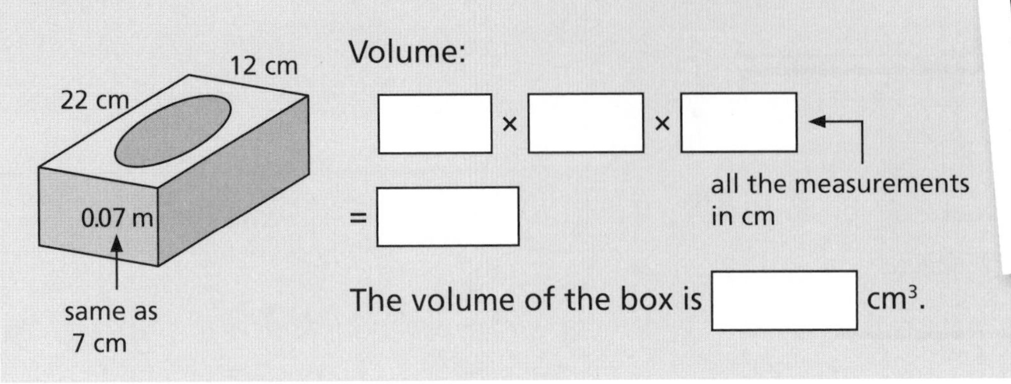

The volume of the box is ____ cm^3.

Make sure all measurements are in the same unit before finding the volume and surface area.

Solve the problems. Show your work.

① A cardboard box is a cube with a side length of 30 cm. How much cardboard is needed to make the box?

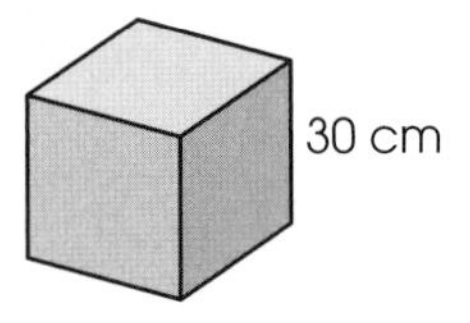

② The total surface area of a cube is 54 cm^2. What is the side length of the cube?

③ A cube-shaped water container has a side length of 12 cm. How many litres of water can it hold?

$1 \text{ cm}^3 = 1 \text{ mL}$

④ A wooden block has the given dimensions. If George is going to paint the entire block, what is the area of the block to be painted?

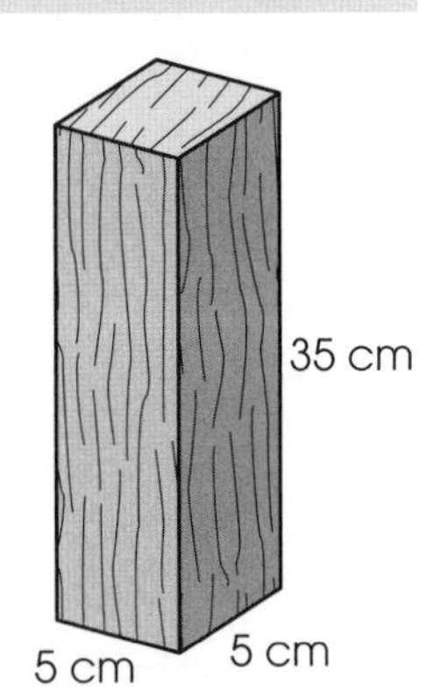

⑤ Jill made a rectangular prism with a piece of cardboard measuring 12 cm by 12 cm. Look at how she cut and folded the cardboard to make the prism.

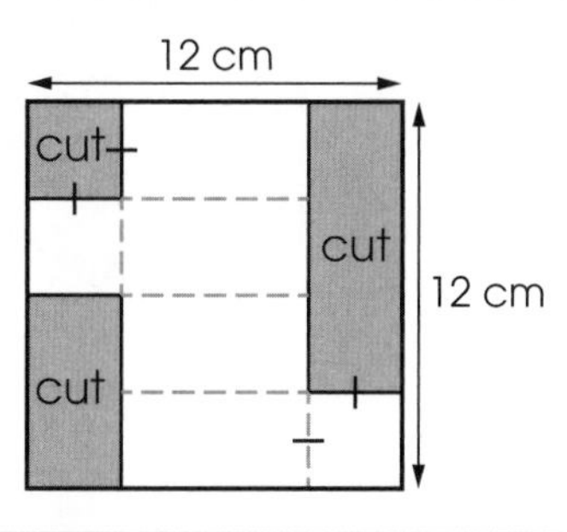

Make a sketch of the folded prism.

a. What are the dimensions of the prism?

b. What is the surface area of the prism?

c. What is the volume of the prism?

⑥ Jennifer wants to wrap a birthday gift for Anna. The dimensions of the box are 20 cm by 30 cm by 8 cm. Jennifer has a piece of 1-m^2 paper. Does she have enough paper?

⑦ If you double the side length of a cube,

a. how does it affect the surface area?

b. how does it affect the volume?

⑧ The surface area of a cube is 216 m^2. What is the volume of the cube?

⑨ Bill constructed a compost bin with the dimensions shown.

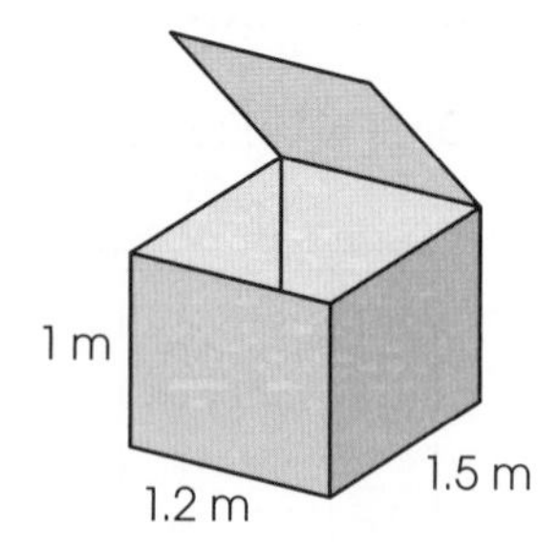

a. Determine the amount of material required to construct the bin.

b. Determine the volume of compost it can hold.

⑩ The dimensions of a room are shown. The total area of the window and door is 3 m². If one can of paint can cover 36 m² and you want to paint the walls with 2 coats each, how many cans of paint are needed?

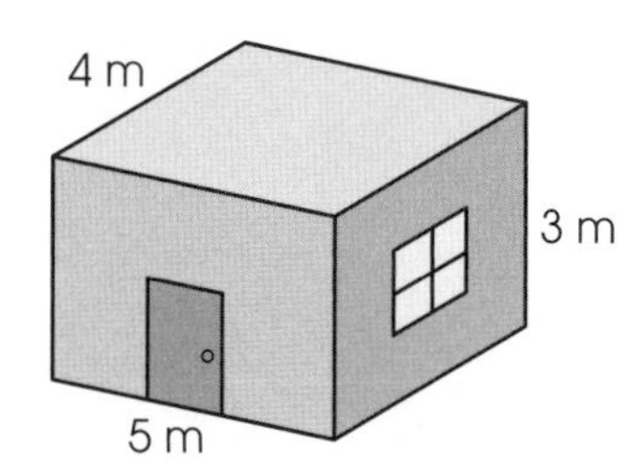

⑪ How many dice, measuring 2 cm by 2 cm by 2 cm each, can be placed in a box measuring 10 cm by 10 cm by 10 cm?

⑫ Ann has a collection of 50 hardcover books. 30 of the books measure 16 cm by 23 cm by 3 cm and the other 20 books measure 16 cm by 23 cm by 1.5 cm. What is the minimum volume of a container that can hold Ann's entire book collection?

⑬ A box of laundry detergent (Box A) measuring 17 cm by 30 cm by 30 cm costs $11.99. Another box (Box B) measuring 15 cm by 25 cm by 25 cm costs $6.99. Which is a better buy? Explain.

⑭ A cereal box measures 31 cm by 20 cm by 7 cm. It is completely filled with cereal. How many servings of cereal does it contain if each serving has a volume of 175 mL?

⑮ It takes one worker one hour to dig a hole that is 3 m by 3 m by 3 m. How long will it take two workers to dig a hole that is 6 m by 6 m by 6 m?

⑯ The volume of the pyramid is $\frac{1}{3}$ of the volume of the water in the rectangular tank. If the pyramid is submerged in the water, how high will the new water level be?

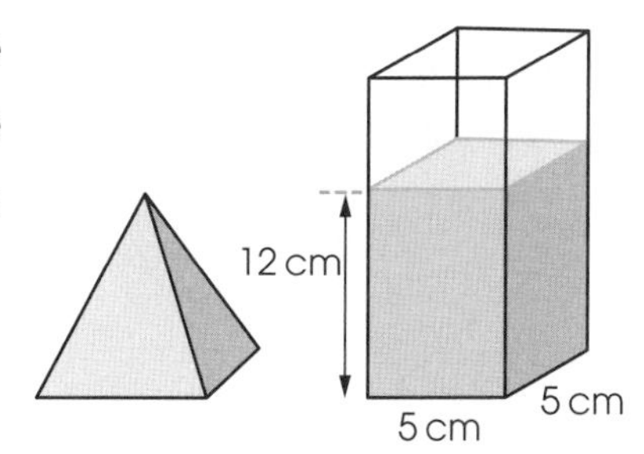

⑰ The concrete steps leading to John's front door have the dimensions shown. Determine the volume of cement needed to build the steps.

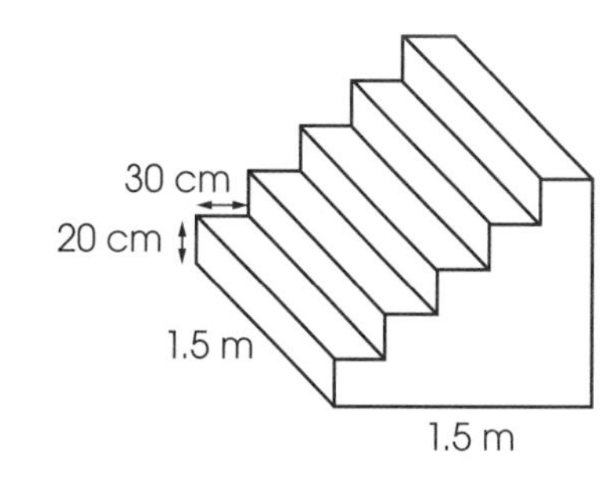

⑱ A square piece of cardboard has an area of 25 cm². A small square of 1 cm² is cut from each corner. The sides are then folded to make an open box. What is the capacity of the box in millilitres?

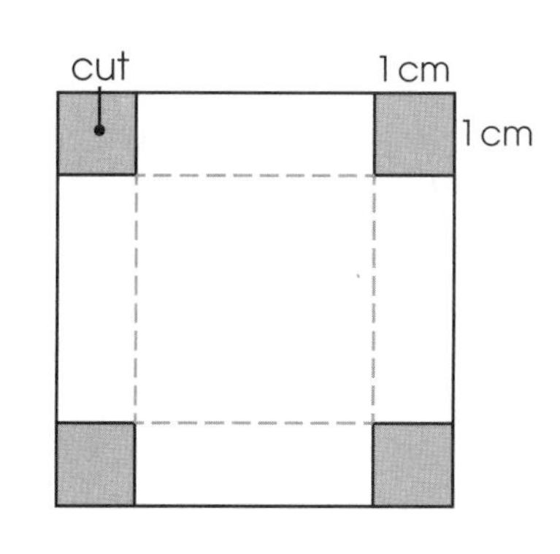

LEVEL 3 – APPLICATIONS

10 Congruence and Similarity

- solving word problems involving congruence and similarity

Determine whether the triangles in each pair are congruent or similar.

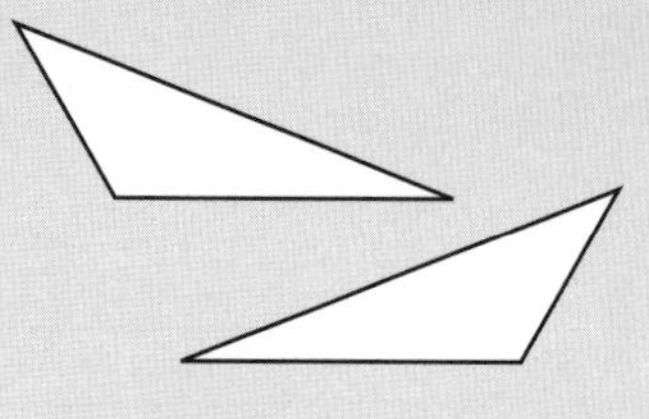

congruent / similar

congruent / similar

Congruent shapes are equal in size and shape. Similar shapes have the same shape but are not the same size.

e.g.

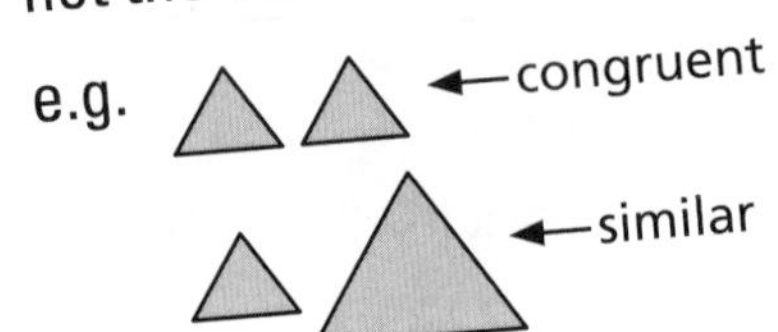

Determine whether the triangles in each pair are congruent, similar, or neither.

①

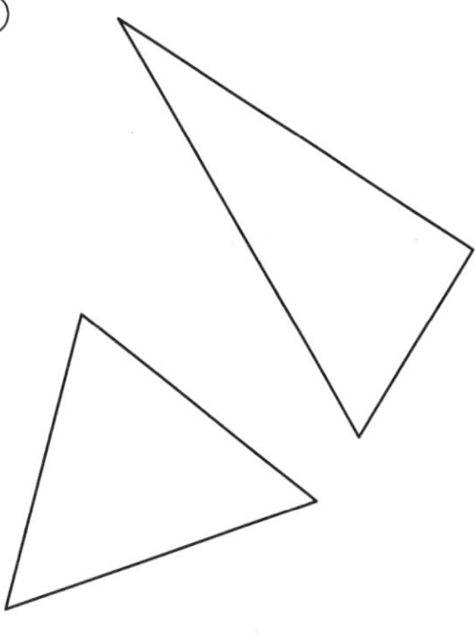

____

②

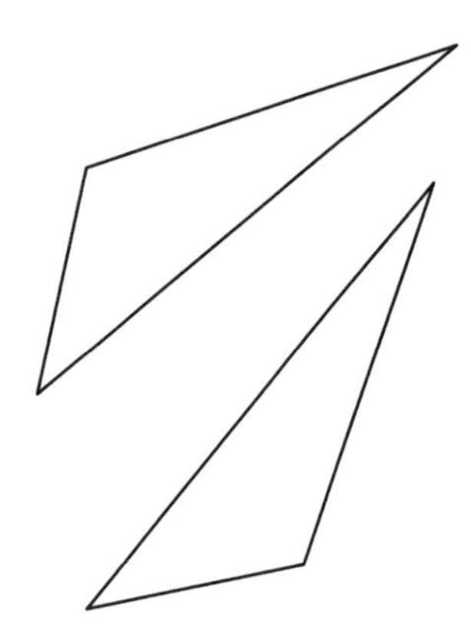

____

③

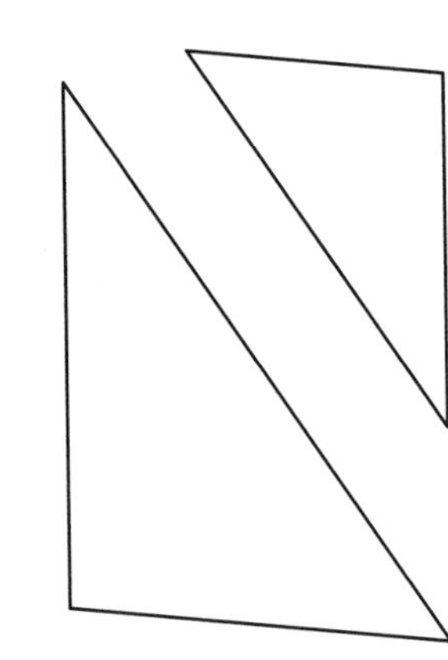

____

④

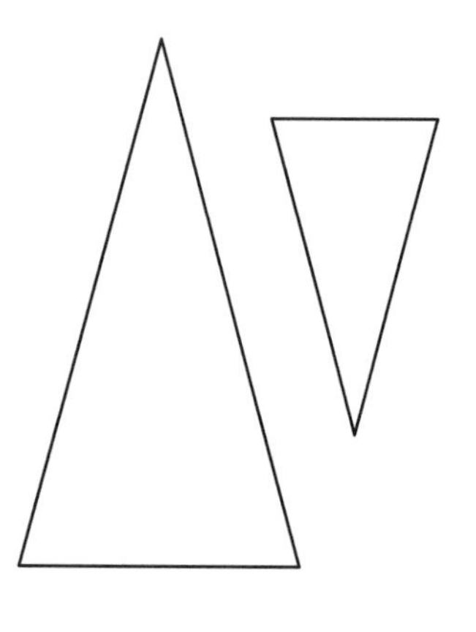

____

⑤

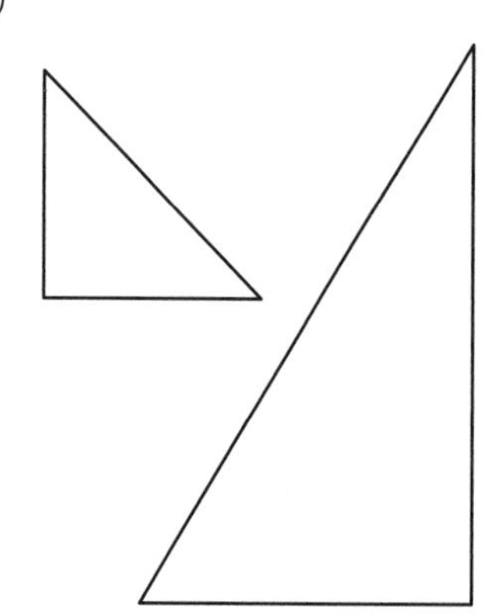

____

⑥

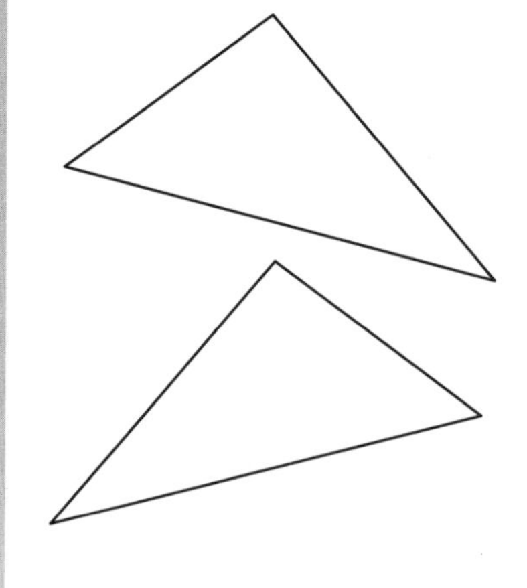

____

Trace each pair of corresponding sides and colour the corresponding angles of the similar triangles with the same colour.

⑦

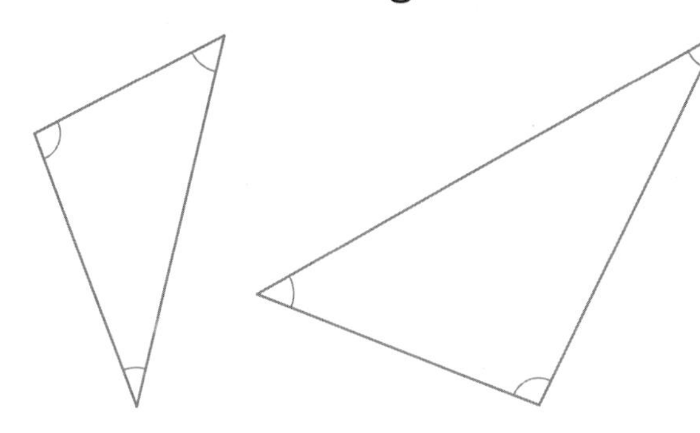

Tips: In similar shapes, the corresponding sides are proportional and their corresponding angles are the same.

⑧

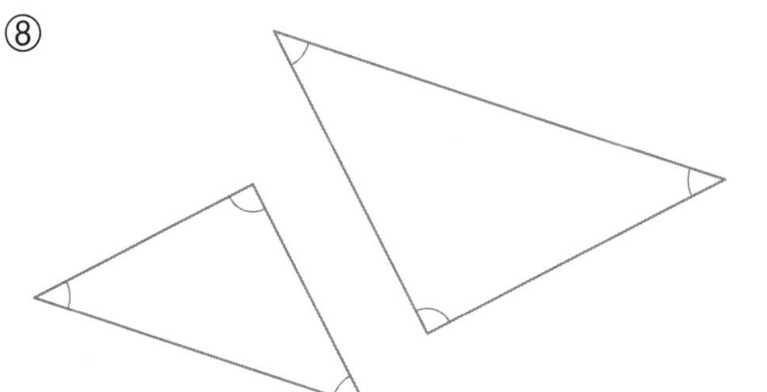

⑨

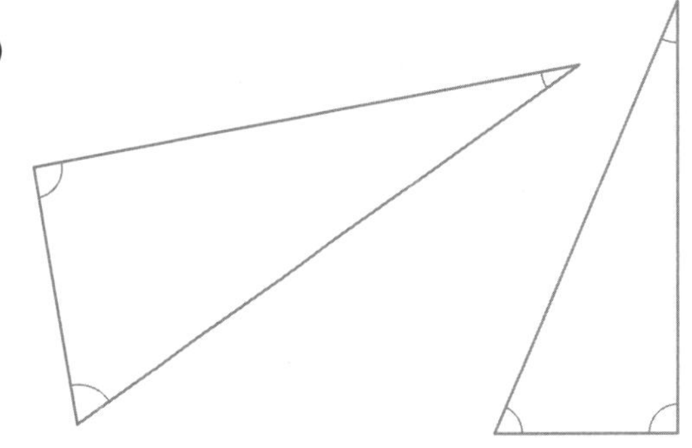

For each pair of triangles, find the ratios of their sides to determine whether they are similar.

Tips: If the ratios of the corresponding sides are the same, then the triangles are similar.

⑩

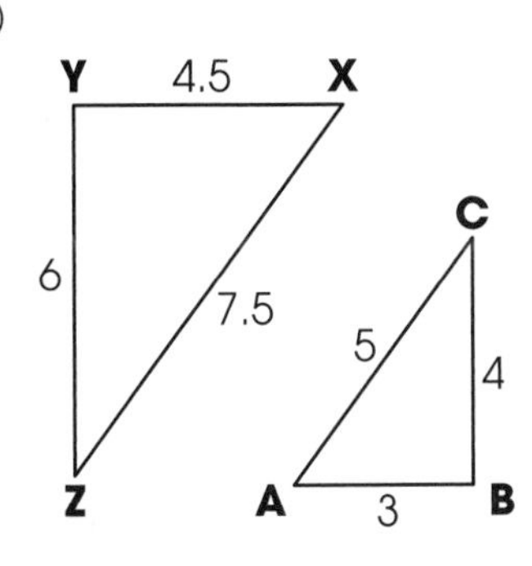

$\frac{AB}{XY} = \frac{3}{4.5} =$ ______

$\frac{AC}{XZ} =$ ______ $=$ ______

$\frac{BC}{YZ} =$ ______ $=$ ______

$\triangle ABC$ ______ (is/is not) similar to $\triangle XYZ$.

⑪

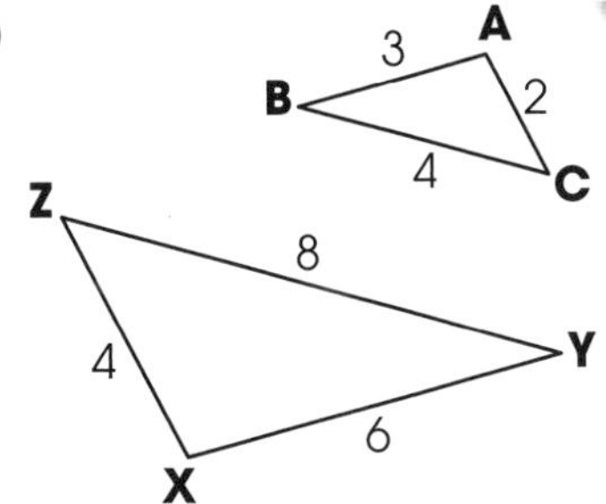

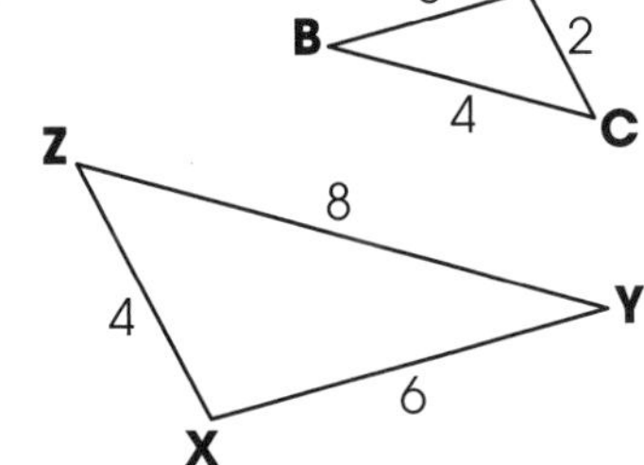

$\frac{AB}{XY} =$ $\frac{AC}{XZ} =$ $\frac{BC}{YZ} =$

$\triangle ABC$ ______ (is/is not) similar to $\triangle XYZ$.

⑫

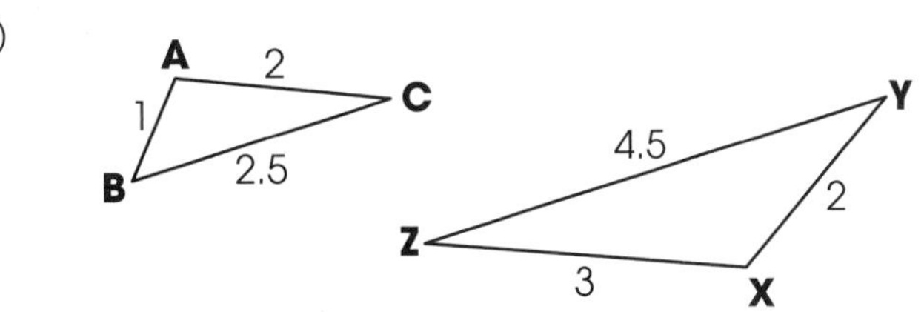

$\frac{AB}{XY} =$ $\frac{AC}{XZ} =$ $\frac{BC}{YZ} =$

$\triangle ABC$ ______ (is/is not) similar to $\triangle XYZ$.

Read the statements. Circle "T" for true and "F" for false.

⑬ If 2 figures are similar, they are also congruent. **T / F**

⑭ 2 equilateral triangles are always similar. **T / F**

⑮ 2 rectangles of the same perimeter are always similar. **T / F**

⑯ If 2 figures are congruent, they are also similar. **T / F**

⑰ 2 squares with the same perimeter are congruent. **T / F**

⑱ A scale model of an object is similar to the object. **T / F**

⑲ 2 circles are always congruent. **T / F**

⑳ 2 squares are always similar. **T / F**

Answer the questions.

㉑ Tom and Jack each construct a triangle with sides of 4 cm, 5 cm, and 6 cm. Are the triangles congruent? Explain.

㉒ Ann and Betty each construct a triangle with angles of 60°, 30°, and 90°. Are the two triangles congruent? Explain.

㉓ An equilateral triangle has sides of 5 cm. Another equilateral triangle has sides of 4 cm. Are the triangles similar? Explain.

Solve the problems. Show your work.

㉔ Bill has a rectangular swimming pool that measures 3 m by 7 m. Bob has a swimming pool that measures 3.6 m by 8.4 m. Are the two swimming pools similar? Give reasons to support your answer.

㉕ There are 2 rectangular pictures on the wall of Bill's living room. One has dimensions of 37 cm by 27 cm and the other 26 cm by 21 cm. Are they similar? Explain.

㉖ 2 rectangular stamps are similar to each other. The smaller stamp has 2 sides of 2 cm and 2 sides of 1.2 cm. The larger stamp has 2 sides of 3 cm. Determine the length of the other 2 sides.

Tips: There are 2 possible values.

㉗ Jonathan's jean pockets are similar. Determine the side length of the larger pocket marked x.

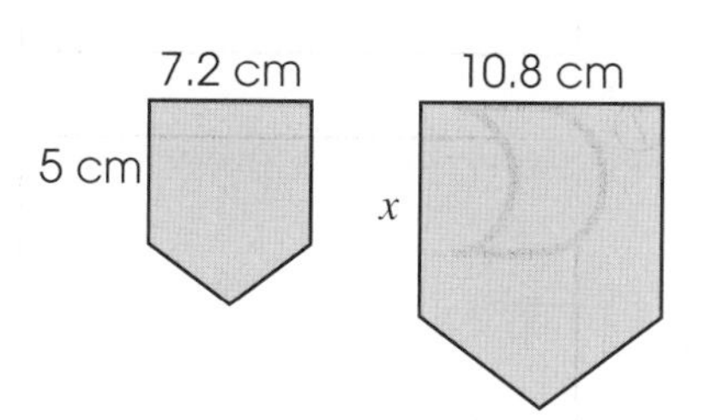

㉘ 2 rectangular pictures are hung side-by-side as shown. Are the pictures similar? Explain.

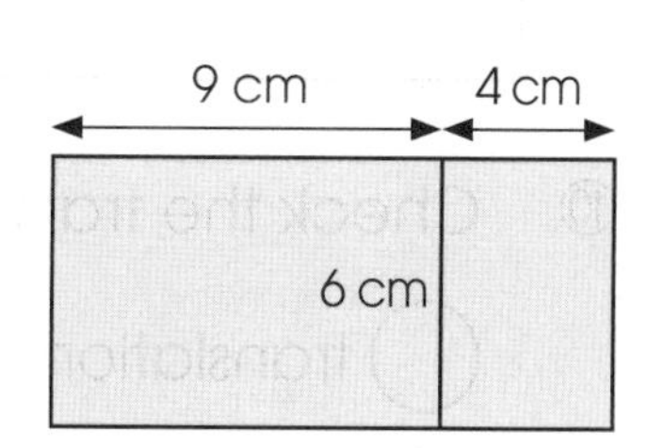

---

11 Transformations and Tiling

- solving word problems involving transformations and tiling

Johannes wants to construct tiling patterns with the following shapes. Which shapes can be used?

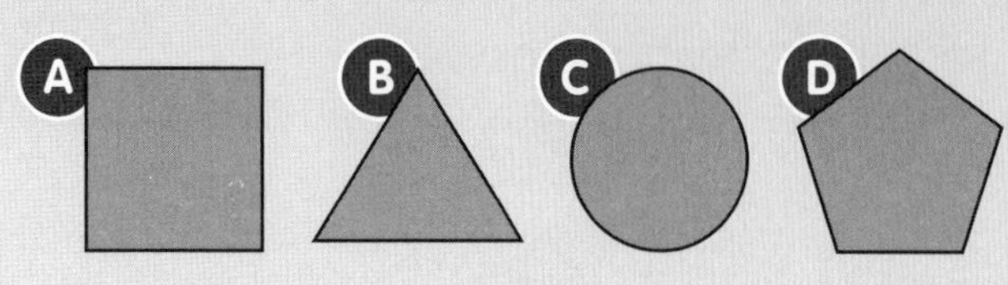

______ and ______ can be used to construct tiling patterns.

A tiling pattern is a pattern made by transforming an identical shape to cover a surface area with no gaps or overlaps.

For each shape, determine whether it can form a tiling pattern. If so, sketch the pattern in the box. Then answer the question.

①
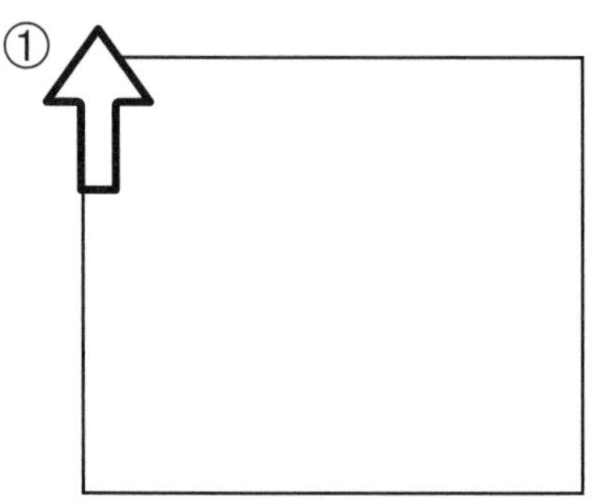

②
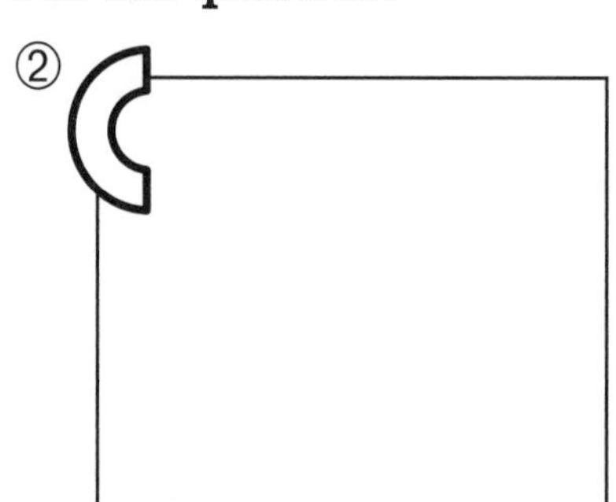

③
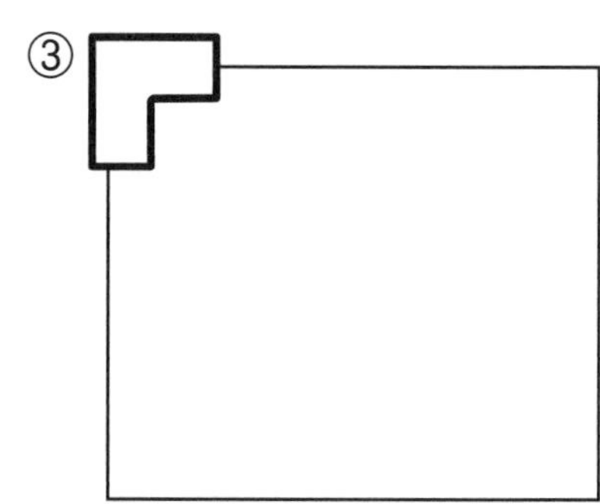

④
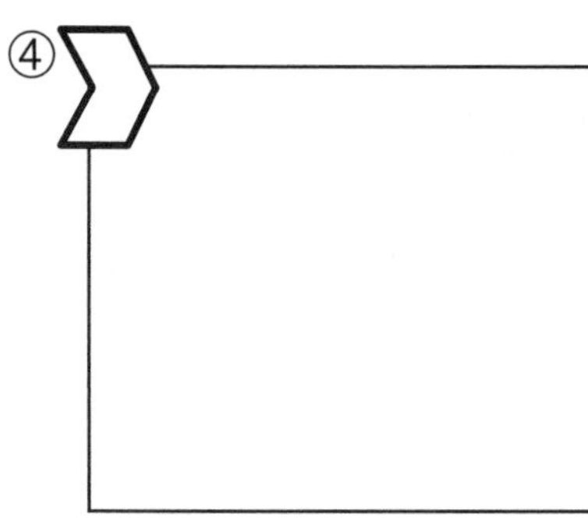

⑤
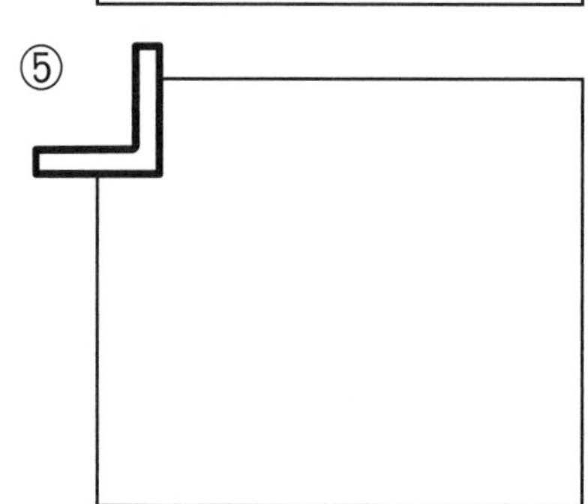

⑥
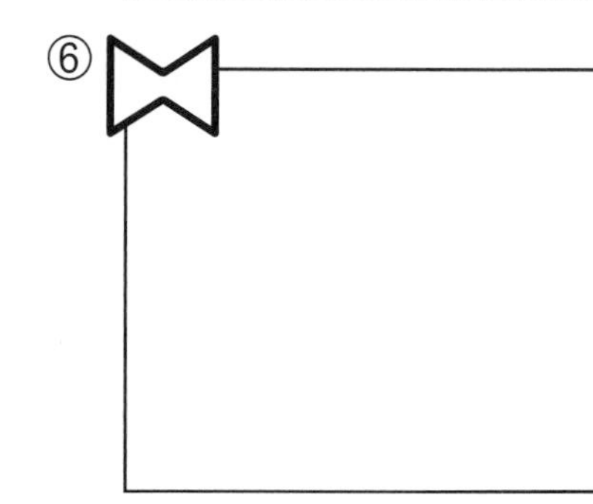

⑦
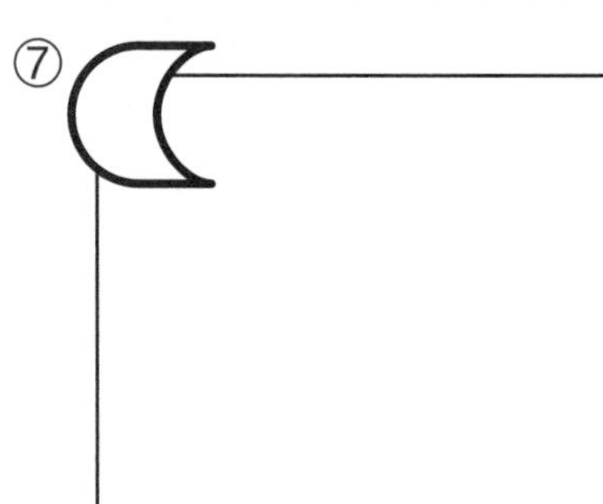

⑧
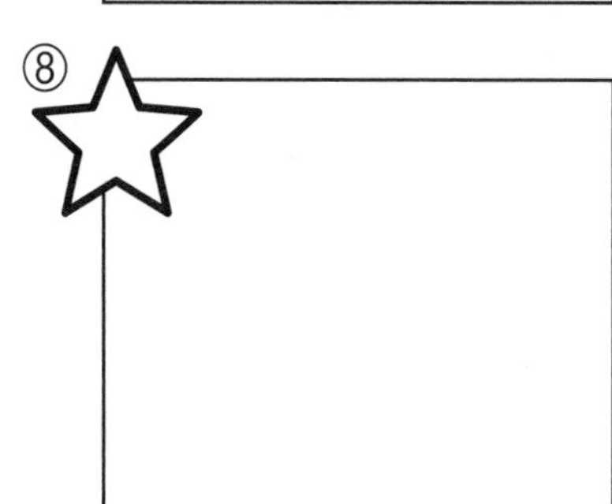

⑨
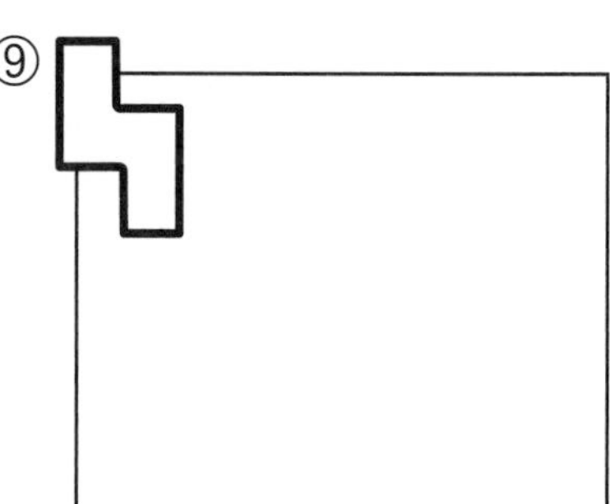

⑩ Check the transformations you can use to form a tiling pattern.

○ translation ○ rotation ○ reflection

○ enlargement ○ reduction

Complete each tiling pattern. Then describe the transformations used.

⑪

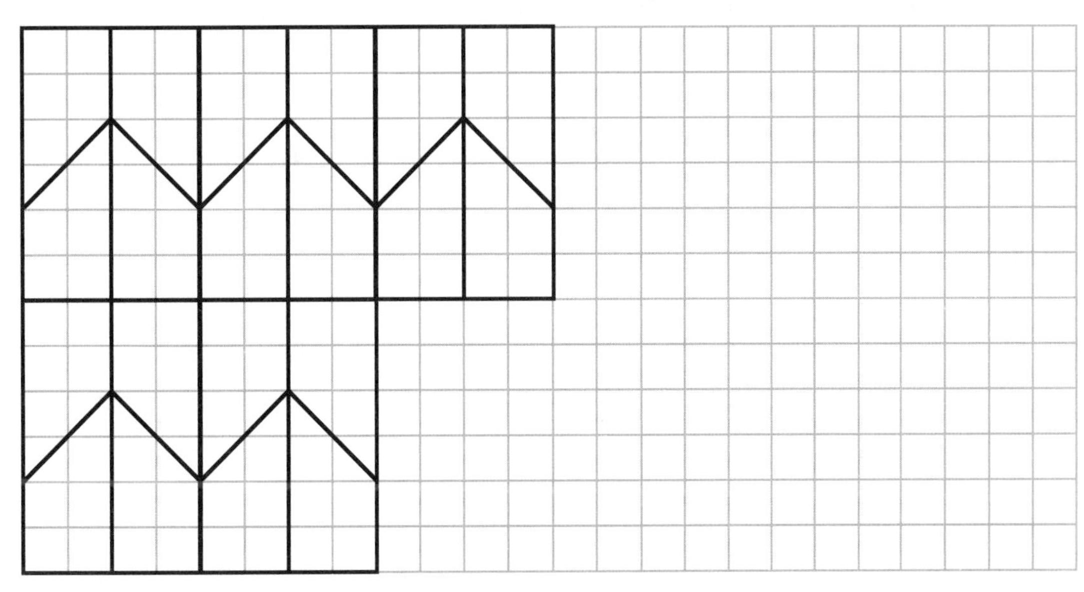

Transformations:

a. 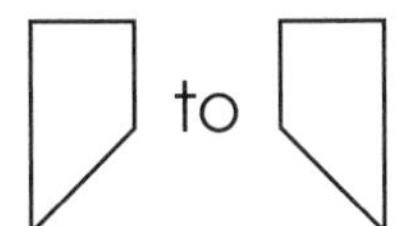

b. 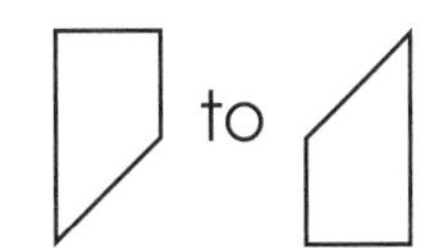

c. to

______ ______ ______

⑫

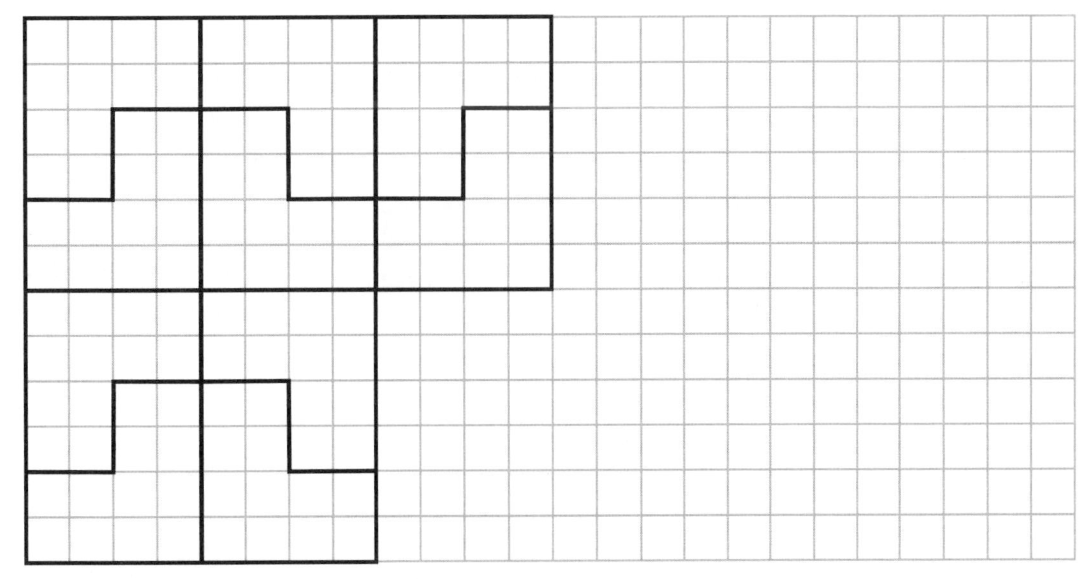

Transformations:

a. 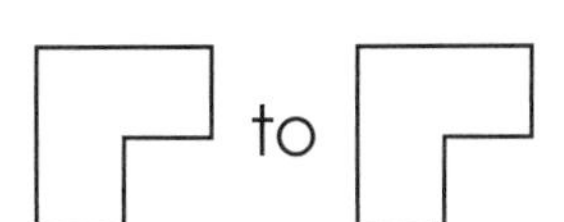

b. 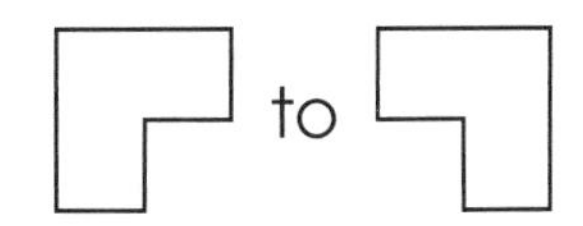

c. to

______ ______ ______

Create a tiling pattern with the given shape using translation, reflection, and rotation. Then draw the transformed shapes in the boxes.

⑬

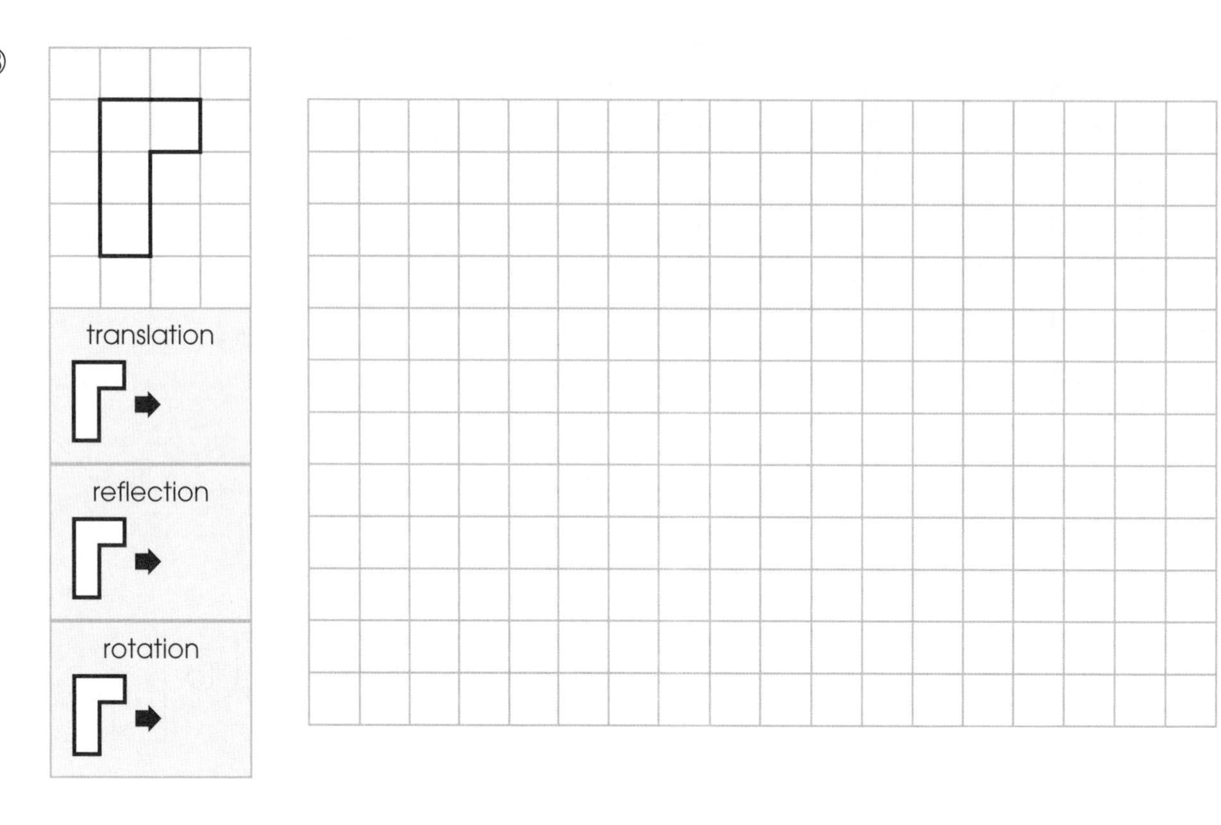

Solve the problems. Show your work.

⑭ See how Bill moves his favourite coaster. Describe the transformations in order.

B → B → B → B

____

⑮ Marc wants to build a tiling pattern on his driveway with triangles. Help him design it by drawing it out. Describe the transformations used.

____

⑯ Each ceramic tile has 2 triangles on it. Casey created a pattern on the wall using the tiles. Describe the transformations used.

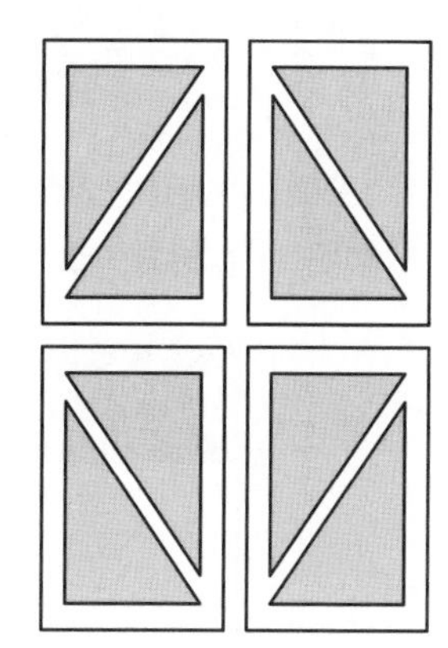

⑰ You are going to tile the floor of your kitchen using tiles in one of the shapes below. Which of the following shapes will you not consider? Explain.

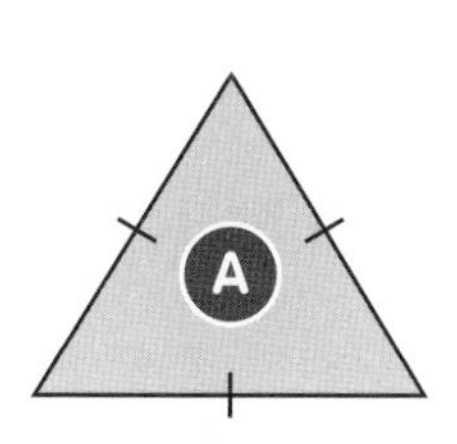

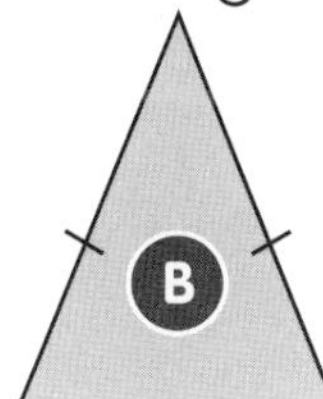

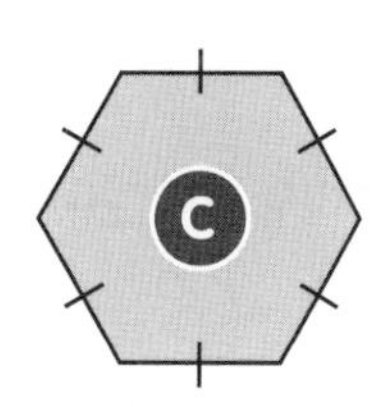

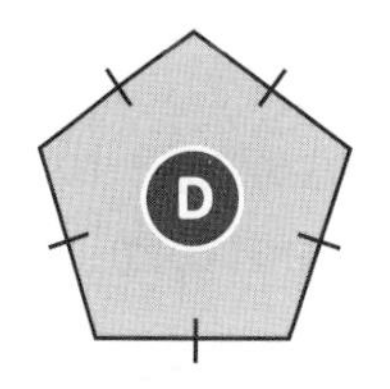

⑱ Ms. Escher made pictures with different figures. Which of the figures can cover a flat surface without gaps or overlaps?

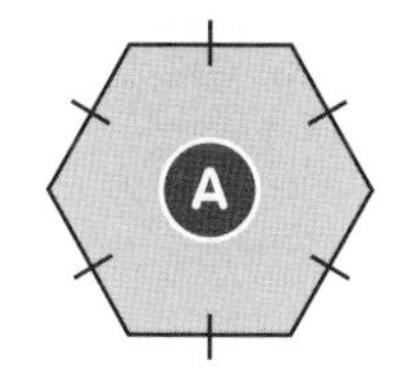

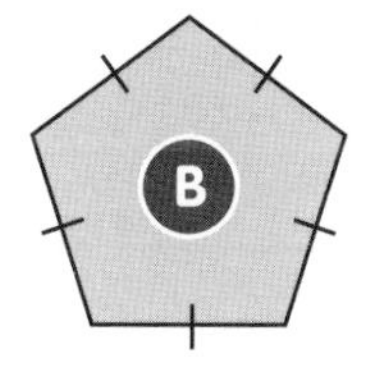

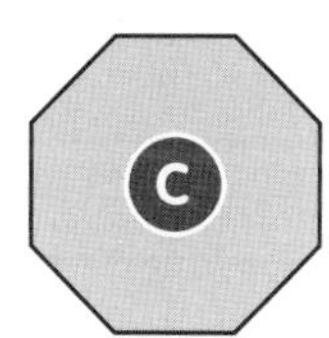

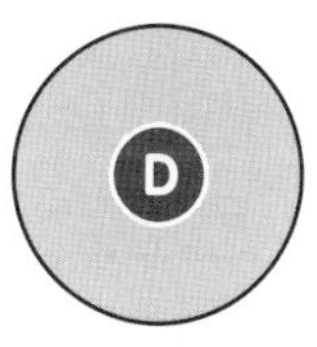

⑲ Mr. Chan has many interlocking bricks which are regular octagons. He cannot make a tiling pattern with the bricks when he uses them to pave his driveway. What should he do? Draw the pattern.

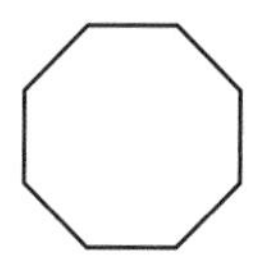

Add a shape to the pattern.

12 Patterning

- solving word problems involving patterning

Write an algebraic expression to describe the pattern and answer the question.

No. of Weeks (Term No.)	No. of Treats (Term)
1	2
2	4
3	6
4	8
5	10

algebraic expression

How many treats are there in 10 weeks?

______ treats

Algebraic expressions can be used to relate term numbers and terms.

They help make it easier to find the term that corresponds to the term number.

Write a pattern rule to describe the pattern. Then write an algebraic expression and complete the pattern.

①

Term No.	Term
1	4
2	5
3	6
4	
5	

- Add ______ to the term number.
- $n +$ ______

②

Term No.	Term
1	3
2	6
3	9
4	
5	

- Multiply the term number by ______.
- $n \times$ ______

③

Term No.	Term
1	1
2	3
3	5
4	
5	

- ______
- ______

④

Term No.	Term
1	0
2	1
3	2
4	
5	

- ______
- ______

⑤

Term No.	Term
1	4
2	8
3	12
4	
5	

- ______
- ______

⑥

Term No.	Term
1	7
2	8
3	9
4	
5	

- ______
- ______

Complete each table and write an algebraic expression to describe the pattern. Plot the points on the graph and connect them. Then answer the questions.

⑦

Term No.	Term
1	3
2	5
3	7
4	
5	

algebraic expression

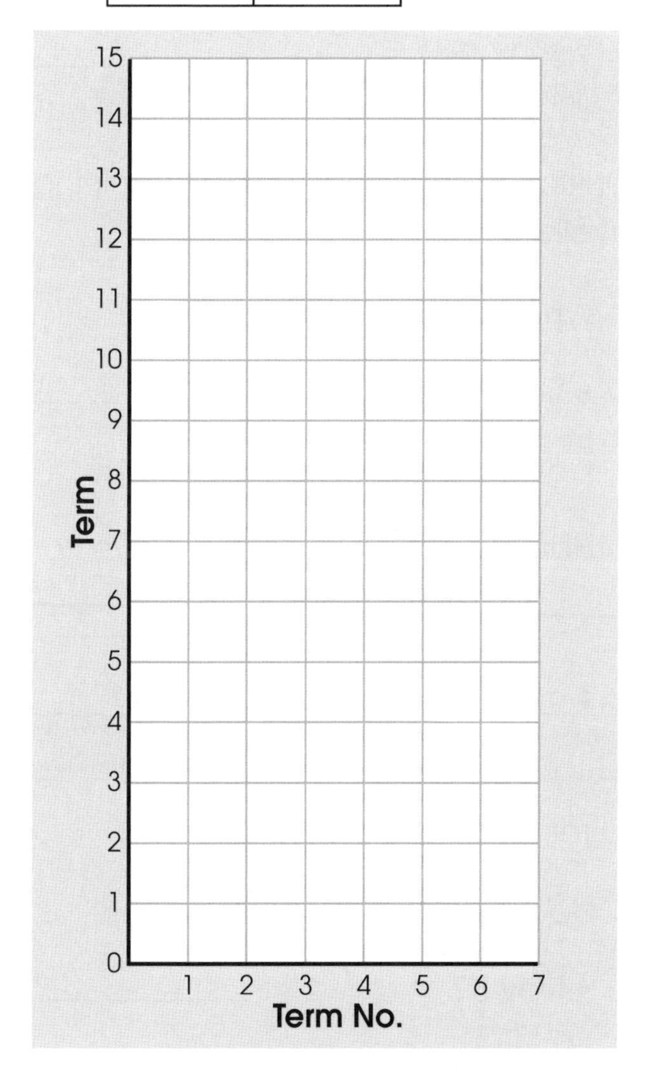

a. Extend the line to continue the pattern.

b. What is the 7th term?

____

⑧

Term No.	Term
1	13
2	11
3	9
4	
5	

algebraic expression

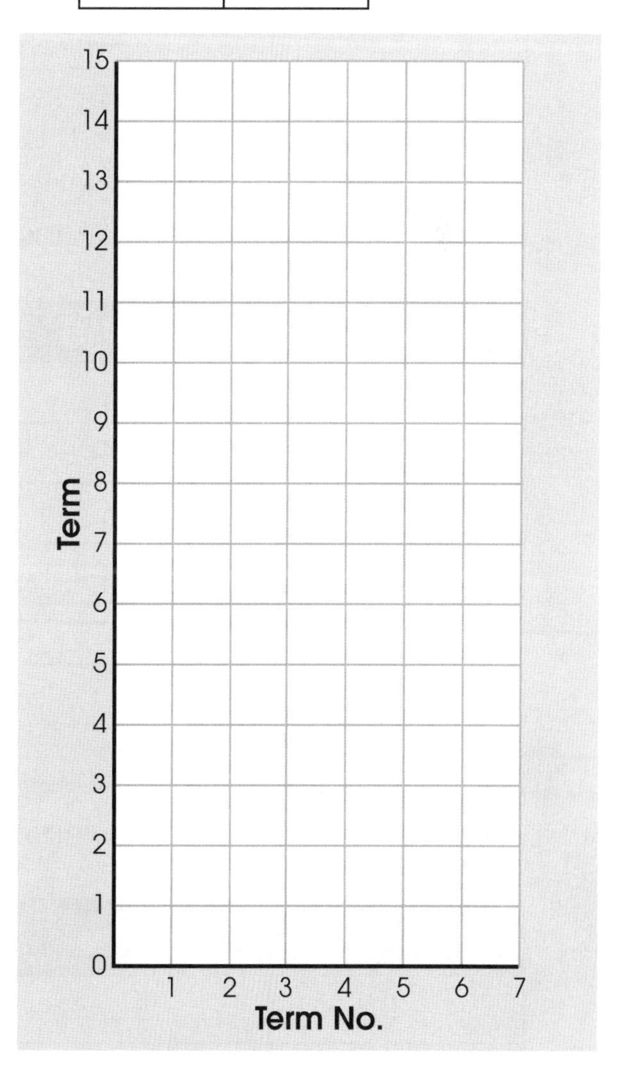

a. Extend the line to continue the pattern.

b. What is the 7th term?

____

LEVEL 3 – APPLICATIONS

Complete each table and solve the problems.

⑨ A baby chick weighs 100 g at birth. Its weight increases by 100 g every week.

Week No.	Weight (g)
1	100
2	200
3	300
4	
5	

a. How much will it weigh by

- Week 4? __________
- Week 5? __________

b. By which week will it weigh

- 600 g? __________
- 800 g? __________

⑩ The cost of a cab ride includes an initial charge of \$2. The remaining charge depends on the distance travelled.

Distance Travelled (km)	Cost (\$)
0	2
1	4
2	6
3	8
4	
5	

a. What is the cost of a

- 5-km ride? __________
- 6-km ride? __________

b. How far can a passenger travel for

- \$16? __________
- \$20? __________

⑪ Adam and David are growing mould for a science project.

Day	Weight (g)
1	5
2	9
3	13
4	
5	

a. How much will it weigh on

- Day 8? __________
- Day 10? __________

b. On which day will it weigh

- 29 g? __________
- 45 g? __________

Solve the problems. Show your work.

⑫ A baseball card is bought for \$2. After the first month, its value will increase to \$5 and then to \$8 after the second month. If the pattern continues, how much will the baseball card be worth after

a. the third month?

b. the sixth month?

⑬ Tania collects stamps. She collected 19 stamps in the first year, 39 in second year, and 59 in the third year. If the pattern continues,

a. how many stamps will she collect in the fifth year?

b. in which year will she collect 139 stamps?

⑭ A teapot increased in value by \$3 every year. Jamie sold the teapot for \$45 after 6 years.

a. How much was the teapot when Jamie bought it?

b. In how many years will the teapot be worth double the value Jamie sold it at?

LEVEL 3 – APPLICATIONS

13 Equations

- solving word problems involving equations

Keith picked 11 apples on a farm. He picked 1 more than double the number of apples Karen picked. How many apples did Karen pick?

Let x represent the number of apples Karen picked.

Karen picked ______ apples.

To solve a problem using an equation, first choose a variable to represent the unknown in the problem.

For each problem, use x as the unknown and state what it represents. Then solve the equation to find the answer. Show your work.

① Two sides of a triangle measure 6 cm and 8 cm. The perimeter of the triangle is 24 cm. What is the length of the third side?

Let ______ represent the length of the third side.

② Ali is thinking of a number. If she multiplies it by 3 and then subtracts 15, she gets 21. What is Ali's number?

Let ______ represent Ali's number.

③ A plumber charges a fixed cost of $30 plus $25 for each hour of service. If he charged $80, how long was the service?

Let ____________________.

④ Snowville has about 10 cm less than 5 times the snowfall of Sunville. If Snowville's snowfall was 360 cm, what was Sunville's?

Let ____________________.

⑤ The Kim family rented a car. They paid \$20 plus \$1 per kilometre driven. If the total cost was \$50, how far did the Kims travel?

⑥ David subtracted 15 from a number. He then doubled the difference and got 22. What was David's original number?

⑦ A diver dives to a depth of 80 m. She then rises at a rate of 2 m/s. How long will it take her to rise to a depth of 14 m?

⑧ John has 2 boxes of marbles, each containing the same number of marbles. If he gets 5 more, he will have 39 marbles in all. How many marbles are there in each box?

⑨ A family of 4 spent \$78 at the exhibition. They spent \$22 on rides and the rest on entrance fees. How much was the entrance fee per person?

⑩ Joey paid for 3 posters and 2 T-shirts with \$100 and got \$13 back. If a T-shirt cost \$21, how much was a poster?

Solve the problems using equations. Show your work.

⑪ Zoe walks 8 km daily, which is about 2 km more than double the distance she jogs. How far does Zoe jog?

⑫ A health club charges a monthly membership fee of $24 plus $2 for each yoga class. Anna spent $40 last month. How many yoga classes did she go to?

⑬ Cindy earns $16 each hour plus a $50 bonus each month. How many hours did she work if she earned $370 last month?

⑭ James has saved all his nickels in a box. The box weighs 500 g and each nickel weighs 4 g. How many nickels has James saved if the total weight of the box of nickels is 700 g?

⑮ A chandelier has a maximum wattage of 600 W. If there are 8 identical bulbs surrounding a 120 W central bulb, what is the maximum allowable wattage of each of the surrounding bulbs?

⑯ Two numbers have a sum of 40 and a difference of 2. What are the numbers?

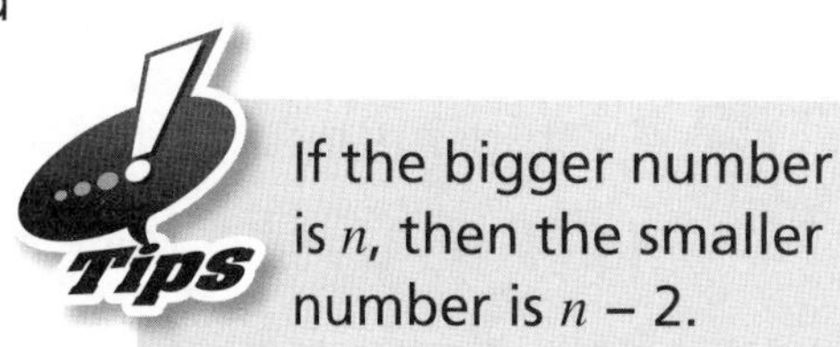

Tips
If the bigger number is n, then the smaller number is $n - 2$.

____

⑰ A local utility company is offering a rebate of $20 to families who install efficient light bulbs. The Jones family spent $50 on efficient light bulbs last month and expect to save $2 on their monthly utility bill. How long will it take for the Jones family to recover the setup cost?

____

⑱ There are 83 boys and girls in Grade 7 at Queenston Junior High School. The number of boys is 31 fewer than 2 times the number of girls. How many boys and girls are there?

____

⑲ A box of 24 tea bags costs $4 to produce and package. The included packaging cost is $1.

a. How much does each tea bag cost to produce?

____

b. If the packaging cost remains as $1, how much does a box of 75 tea bags cost to produce and package?

____

14 Data Analysis

- analyzing data

15 students recorded their heights in centimetres.

150 147 155 154 160 158 148 156

155 162 170 165 163 144 153

Make a stem-and-leaf plot and find the answers.

Stem	Leaf
14	
15	
16	
17	

Mean: ____ cm

Median: ____ cm

Mode: ____ cm

The measures of central tendency help examine a set of data.

- mean: the average of a set of data
- median: the middle value
- mode: the most common value

The stem-and-leaf plot shows the weekly number of hours spent watching TV of a group of 13-year-old children.

①

Stem	Leaf
0	6, 9
1	1, 2, 4, 4, 8, 8, 9
2	1, 3, 4, 6, 6, 8, 8
3	5, 5, 5, 8, 8
4	8, 9

A sample is a set of data from a group that represents a population.

A census is a set of data from an entire population.

a. Find the measures of central tendency.

- mean: ____
- median: ____
- mode: ____

b. The data is used in a study on how much time 13-year-olds in Calgary spend watching TV. Is this set of data a sample or a census? Explain.

____

Answer the questions.

②

I recorded the number of shots on goal a hockey team made each game last month.

a. Determine the measures of central tendency.

- mean:

____

- median:

____

- mode:

____

b. Is this set of data a sample or a census? Explain.

____

Hints

Primary data is a set of data that is collected first-hand.

Secondary data is a set of data that is from other sources.

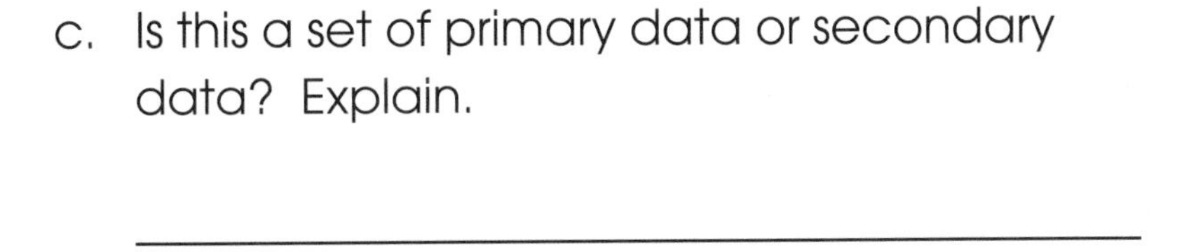

c. Is this a set of primary data or secondary data? Explain.

____

____

d. If one additional game was played with 29 shots on goal, would this affect the mean, median, and/or mode? Explain.

- mean:

____

- median:

____

- mode:

____

For each of the statements, create a set of data with at least three values that makes the statement true.

③ The mean is smaller than the mode. ______________________

④ The mode is smaller than the mean. ______________________

⑤ The median is smaller than the mean. ______________________

⑥ The mean is smaller than the median. ______________________

⑦ The mode and the median are the same. ______________________

⑧ The mean of the values is 7. ______________________

Answer the questions.

⑨ The table shows the annual salary of the employees at ABC Construction Company.

Annual Salary	No. of Employees
\$30 000	12
\$40 000	2
\$45 000	3
\$60 000	2
\$100 000	1 (President)

a. Find the mean, median, and mode salaries.

____

____

b. In an advertisement to attract new employees for the company, would you use the mean, median, or mode? Why?

____

c. If the president got a pay raise, would this affect the mean, median, and/or mode?

____

d. Which measure of central tendency best represents the salaries? Why?

____

Solve the problems. Show your work.

The range is the difference between the highest and lowest data values.

⑩ Determine the range of the data set below.

68, 70, 72, 78, 78

⑪ A class of 25 students averaged 70% on a test. Another class of 30 students averaged 60%. Calculate the average score of all 55 students.

⑫ The average of 3 numbers is 20. One of the numbers is 15. Calculate the sum of the other 2 numbers.

⑬ The line graph shows the production of hard drives at a factory in the past 8 months.

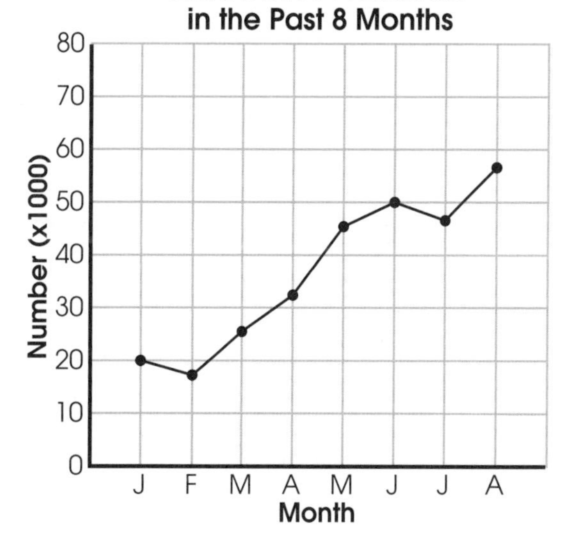

a. What trend do you notice in the number of hard drives produced?

b. Find the approximate mean.

c. What is the range?

d. Predict how many hard drives will be produced in September.

LEVEL 3 – APPLICATIONS

15 Data Management

- interpreting graphs

The circle graphs show the fraction of Canada's population under 19 years old in the years 2004 and 2014. Check the conclusion that can be drawn from the graphs.

Read all the information contained in a graph and draw conclusions based on the data.

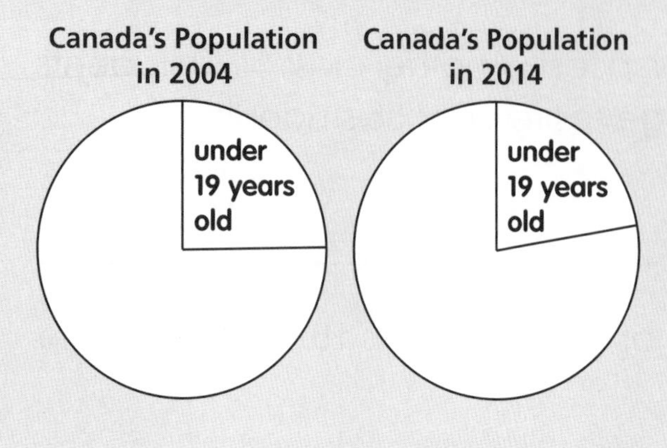

(A) Canada's population remained unchanged from 2004 to 2014.

(B) The number of people under 19 years old decreased from 2004 to 2014.

(C) The percentage of the population under 19 years old decreased from 2004 to 2014.

Check the type of graph that best represents each set of data.

① The scores of all soccer teams in the World Cup

(A) line graph (B) bar graph (C) circle graph

② The percent of Tony's weekly spending on different activities

(A) line graph (B) bar graph (C) circle graph

③ The number of students of different heights in a school

(A) line graph (B) bar graph (C) circle graph

④ The number of cars in different colours in a parking lot

(A) line graph (B) bar graph (C) circle graph

⑤ Toronto's highest and lowest temperatures each day for a week

(A) double line graph (B) double bar graph (C) circle graph

⑥ The number of boys and girls in different parks

(A) double line graph (B) double bar graph (C) circle graph

Answer the questions about the graphs.

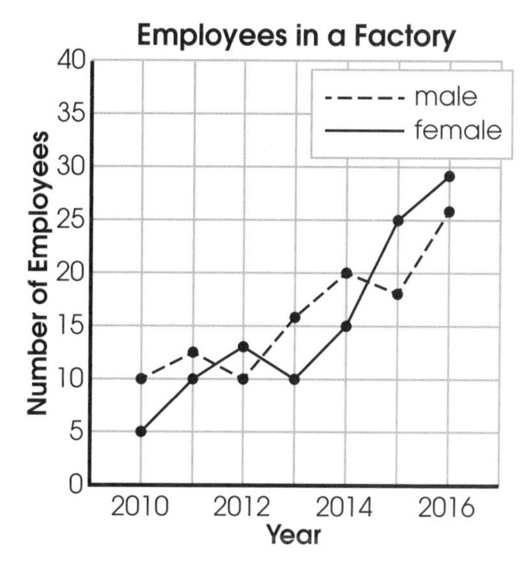

⑦ How many employees were there in total in

a. 2010?

____

b. 2015?

____

⑧ In which years were there more female than male employees? By how many?

____

⑨ The percent of employees of each gender is shown in the graph.

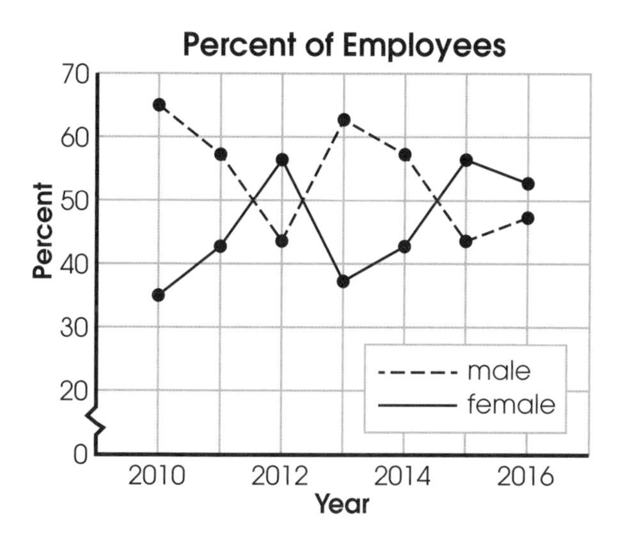

a. What is the sum of the percents each year?

____

b. Find the percent of male employees and female employees for the years below.

- 2010: ______________________

  ____

- 2016: ______________________

  ____

⑩ Which graph above would you use to find the answer to the questions below? Answer with the graph's title.

a. How many female employees will there be in 2019? ______________________

b. In which years were more than half of the employees female? ______________________

⑪ Can a circle graph be used to answer the questions in Question 10? Explain.

____

____

Write the children's names to match the circle graphs that represent their days. Then answer the questions.

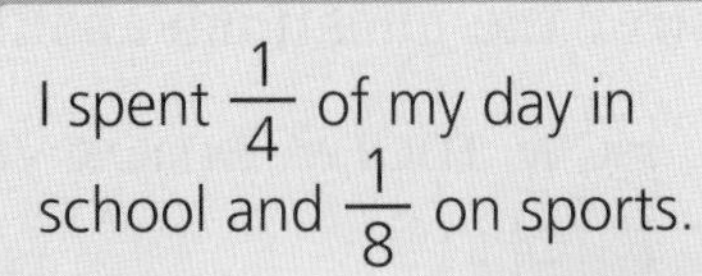

I spent $\frac{1}{4}$ of my day in school and $\frac{1}{8}$ on sports.

I spent $\frac{1}{3}$ of my day sleeping and $\frac{1}{10}$ on homework.

⑫

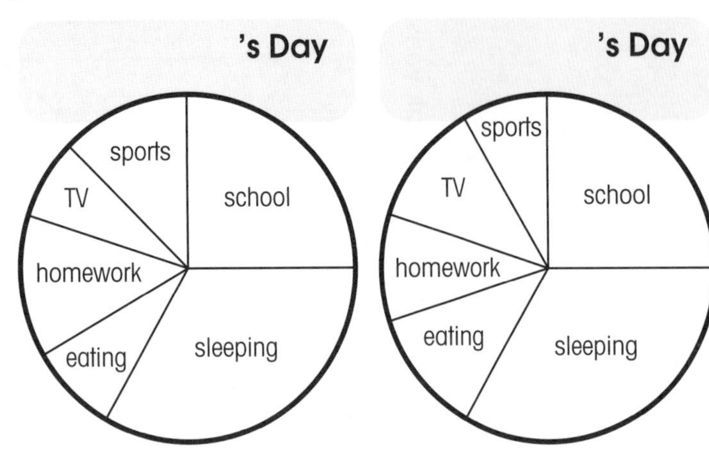

a. Who spent more time watching TV than playing sports?

____

b. How many hours did each child spend

- doing homework? ____________________
- eating? ____________________

c. Which categories did the children spend the same amount of time on? How many hours did they spend on each?

____

⑬ Terrance recorded the number of hours he spent on each activity yesterday. Help him represent his data with a circle graph.

Activity

school: 9 h sleeping: 7 h homework: 3 h
eating: 2 h reading: 1 h sports: 2 h

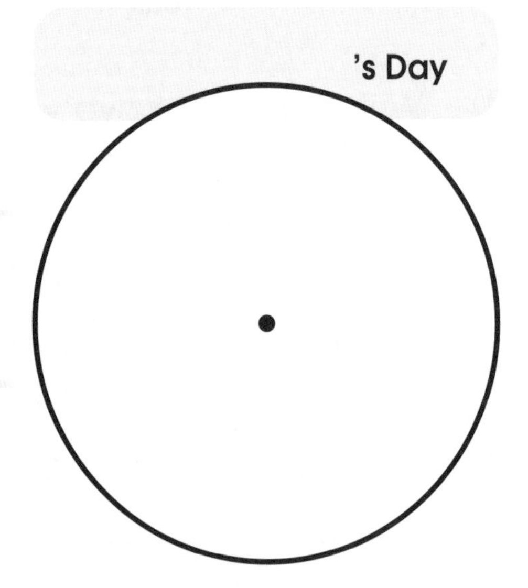

Which two activities took up about half of Terrance's day?

____

The graphs show the profits earned by a transportation company. Answer the questions.

⑭

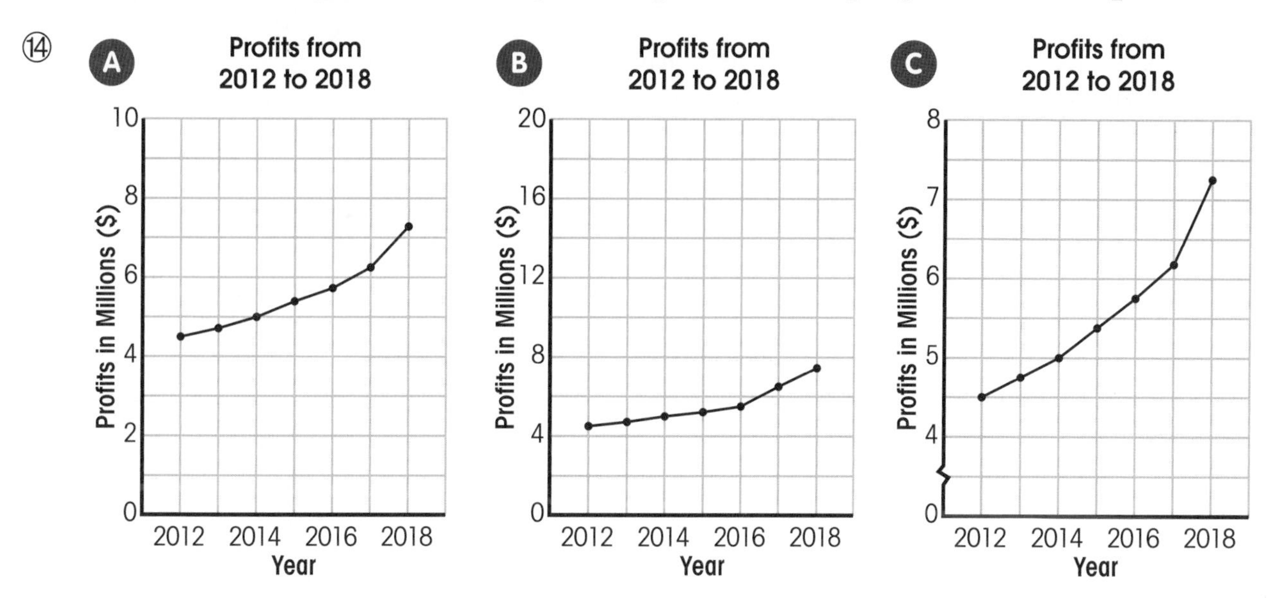

a. Which of these graphs would you use if you were

- trying to encourage investors to invest in the company? Explain.

- a politician talking about excessive profits of the transportation company? Explain.

- the president of the company explaining to the press that the transportation company's profits are not significant?

b. Predict what the company's profit would be in the year 2020.

c. Do you think a bar graph is more appropriate than a line graph for this set of data? Explain.

16 Probability

- solving word problems involving probability

Try It

Janice is going to flip three coins. Draw a tree diagram to find the possible outcomes.

A tree diagram is a helpful tool to find all possible outcomes and theoretical probabilities.

There are ______ possible outcomes.

Look at the tree diagram above. Answer the questions.

① What is the probability that Janice flips

a. 1 head and 2 tails? ______

b. 2 heads and 1 tail? ______

c. more than 1 head? ______

d. fewer than 2 tails? ______

e. all heads? ______

f. no heads? ______

g. all 3 coins with the same result? ______

h. all 3 coins with different results? ______

Solve the problems. Show your work.

② Jimmy wants to buy a new bike. He can choose a mountain bike or a road bike. Both bikes are available in black, white, or gold.

a. How many options does Jimmy have?

b. If he chooses randomly, what is the probability that the new bike will be

- black?

- a road bike?

- a black mountain bike?

③ Ann flipped a coin 50 times and heads turned up 24 times.

a. What fraction of the flips were heads? Is this what you would expect? Explain.

b. If Ann flipped the coin 500 times, how many times should she expect to flip heads?

c. If the first 5 flips were all heads, what is the probability that the next flip will be heads? Explain.

d. Is it possible to flip a coin 20 times and get heads every time? If so, what is the likelihood?

Look at George's spinners and answer the questions.

④

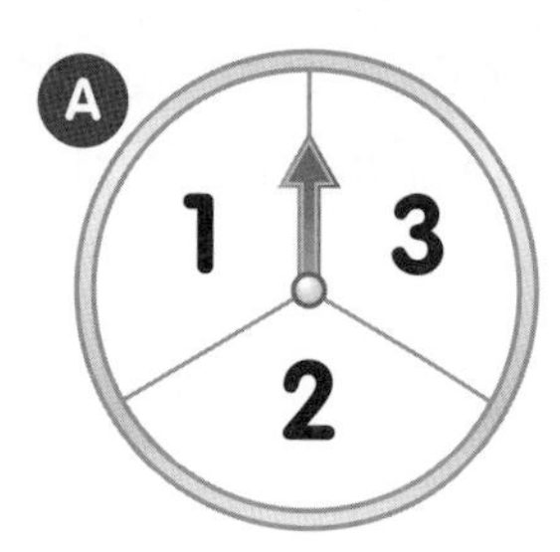

a. What is the probability that the result will be a 3 in each case?

____

b. How many times should George expect each spinner to land on a 3 if he spins each one 30 times?

____

c. Does the probability of landing on a 3 depend on the number of times he spins each spinner?

____

d. Which spinner is fair?

____

⑤ In a TV guessing game, a bucket is filled with coloured balls. One after the other, contestants are asked to guess the colour of the ball that will be drawn. A ball is then drawn and put back into the bucket. The contestant wins a prize if the guess was correct.

a. Does the first or last contestant have a better chance to win? Explain.

____

b. If the bucket contains 10 red, 20 white, and 30 black balls,

- what is the probability that a white ball will be drawn?
- which ball is most likely to be drawn?

______________ ______________

Handy Reference

Order of Operations

BEDMAS

- **B** rackets
- **E** xponents
- **D** ivision
- **M** ultiplication
- **A** ddition
- **S** ubtraction

Squares and Square Roots

$$\sqrt{a \times b} = \sqrt{a} \times \sqrt{b}$$

$$\sqrt{a \div b} = \sqrt{a} \div \sqrt{b}$$

$$\sqrt{a^2} = a$$

$$\sqrt{a}^{\,2} = a$$

Operations with Integers

Addition

+ + ➡ +

+ − ➡ −

e.g. 3 **+** (-2)
= 3 **−** 2
= 1

Subtraction

− + ➡ −

− − ➡ +

e.g. 3 **−** (-2)
= 3 **+** 2
= 5

Multiplication

(+) × (+) ➡ +

(+) × (−) ➡ −

(−) × (+) ➡ −

(−) × (−) ➡ +

e.g. (-2) × (**+**3)
= -6

Division

(+) ÷ (+) ➡ +

(+) ÷ (−) ➡ −

(−) ÷ (+) ➡ −

(−) ÷ (−) ➡ +

e.g. (-8) ÷ (-4)
= **+**2

Perimeter and Area of Polygons

Perimeter

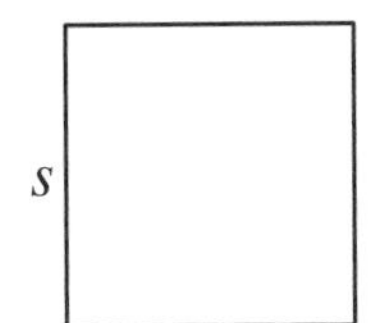

P = $4s$

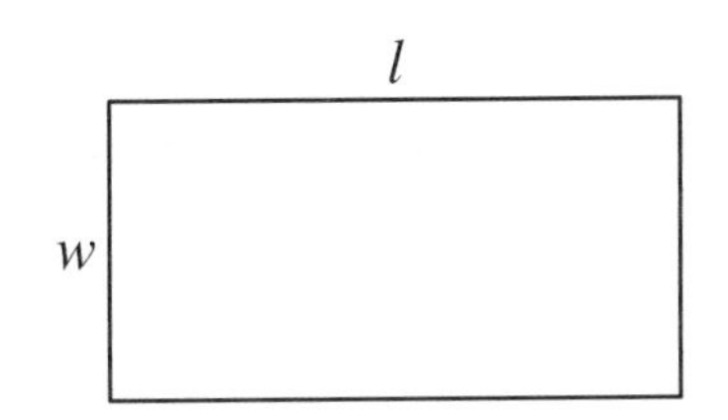

P = $2(l + w)$

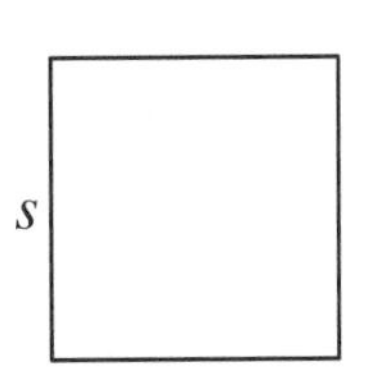

A = s^2

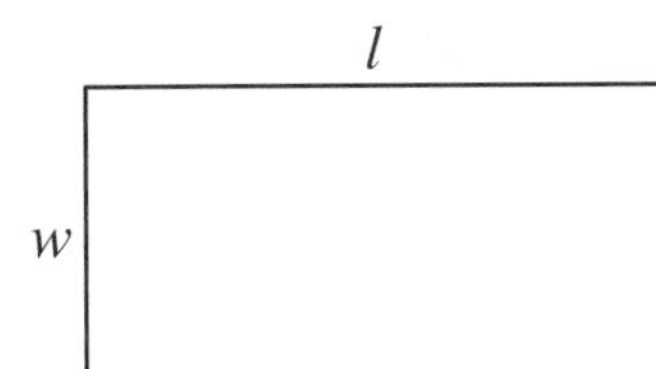

A = lw

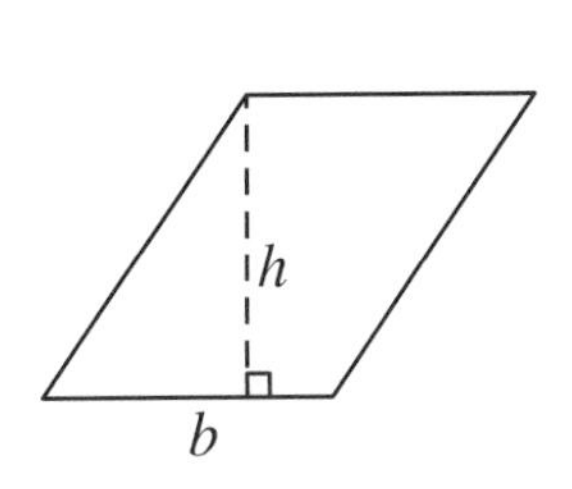

A = bh

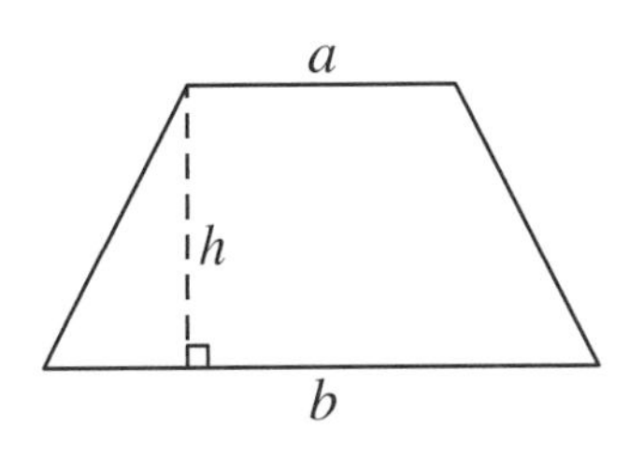

A = $(a + b)h \div 2$

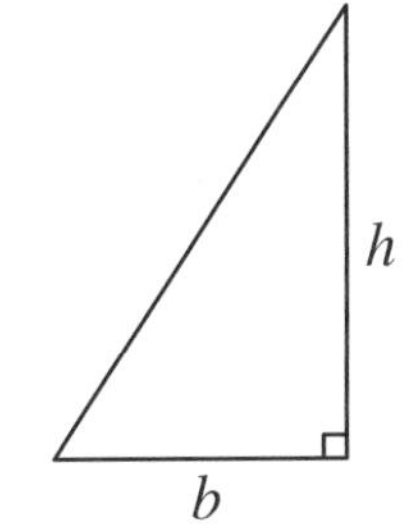

A = $bh \div 2$

Circumference and Area

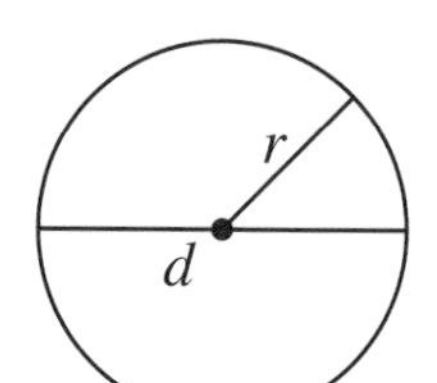

$C = \pi d = 2\pi r$

$A = \pi r^2$

Pythagorean Relationship

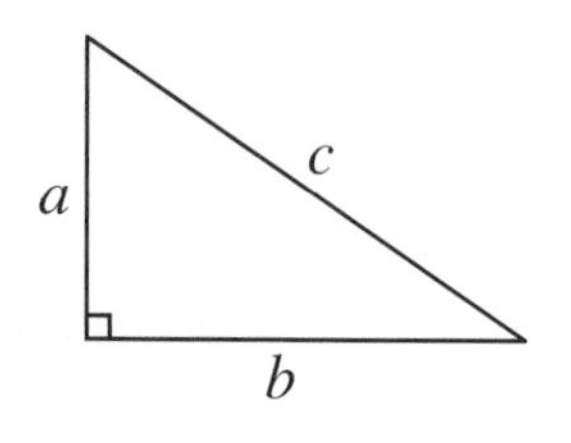

$a^2 + b^2 = c^2$

Volume and Surface Area

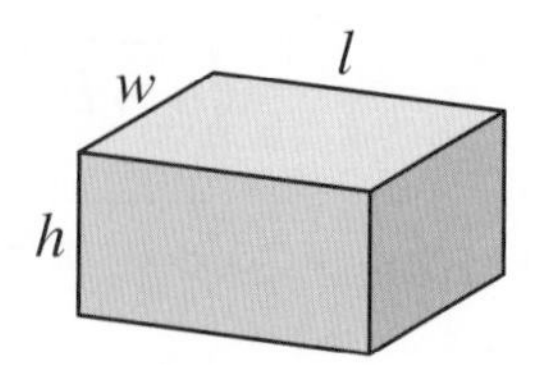

$V = l \times w \times h$

$S.A. = 2lw + 2lh + 2wh$

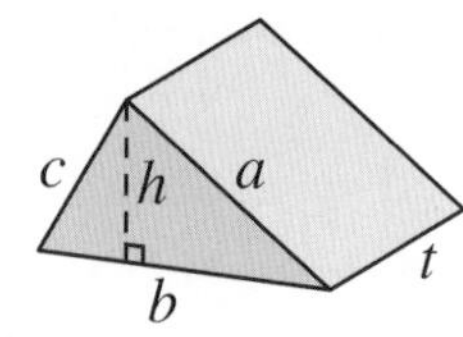

$V = (bh \div 2) \times t$

$S.A. = 2(bh \div 2) + at + bt + ct$

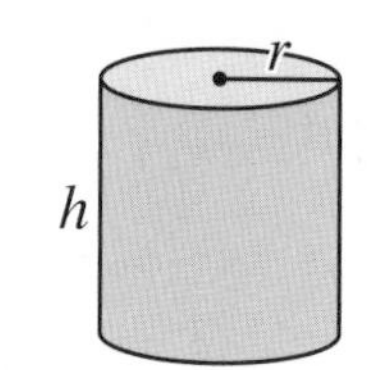

$V = \pi r^2 h$

$S.A. = 2\pi r^2 + 2\pi rh$

Unit Conversions

Length

1 km = 1000 m
1 m = 100 cm
1 cm = 10 mm

Mass

1 kg = 1000 g
1 g = 1000 mg

Capacity and Volume

1 L = 1000 mL
1 L = 1000 cm^3
1 mL = 1 cm^3

Angle Properties

Complementary Angles

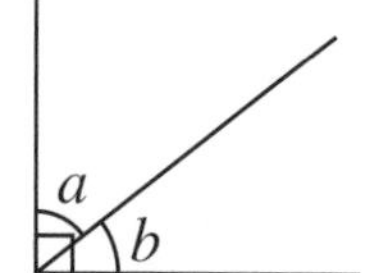

$a + b = 90°$

Supplementary Angles

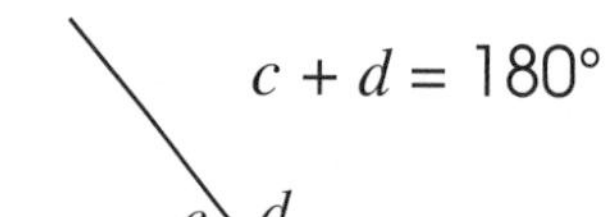

$c + d = 180°$

Opposite Angles

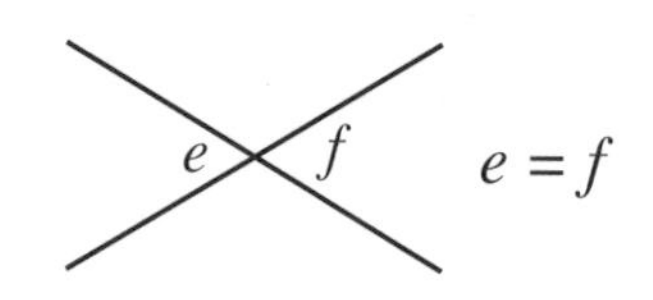

$e = f$

Corresponding Angles

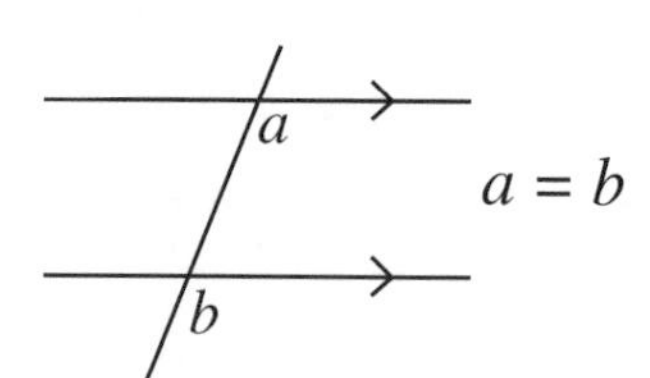

$a = b$

Alternate Angles

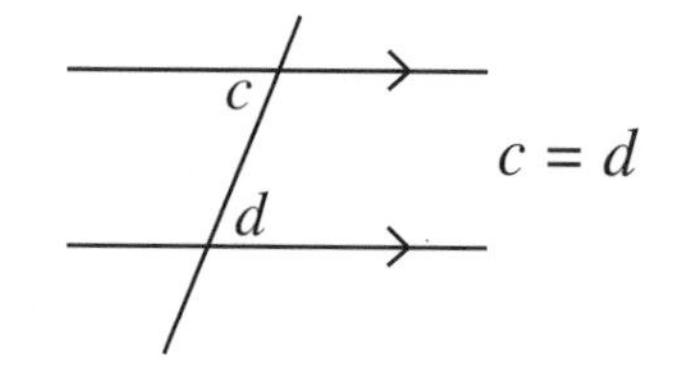

$c = d$

Consecutive Interior Angles

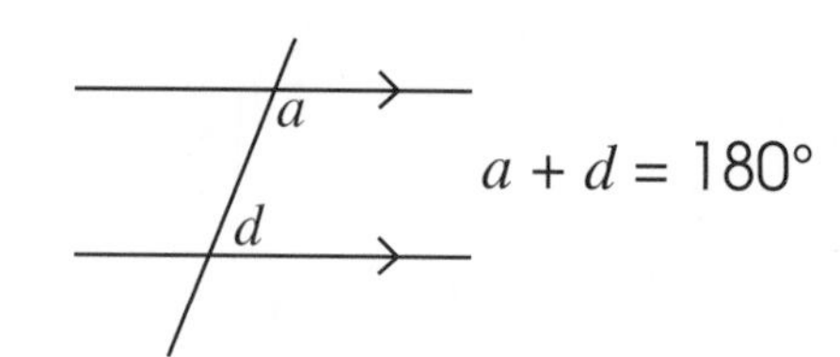

$a + d = 180°$

Grade 7

QR Code

QR Code – a quick way to access our fun-filled videos

Our QR code provides you with a quick and easy link to our fun-filled videos, which can help enrich your learning experience while you are working on the workbook. Below is a summary of the topics that the QR code brings you to. You may scan the QR code in each unit to learn about the topic or the QR code on the right to review all the topics you have learned in this book.

Scan this QR code or visit our Download Centre at ***www.popularbook.ca.***

The topics introduced with the QR code:

1. **What is a perfect square?** (p. 19)
Discover what makes a number a perfect square.

2. **How to Divide Fractions** (p. 27)
Learn how to divide fractions.

3. **How to Multiply Decimal Numbers** (p. 31)
Learn how to multiply decimal numbers.

4. **How to Divide Decimal Numbers** (p. 33)
Learn how to divide decimal numbers.

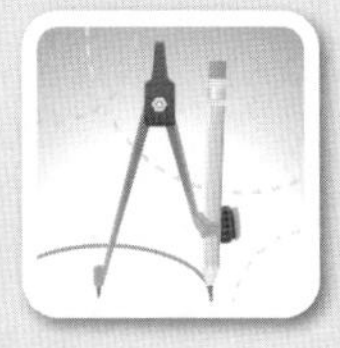

5. **Which compass?** (p. 51)
Distinguish the differences between the compass for directions and the compass for angles.

6

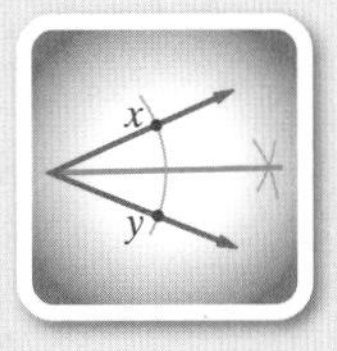

How to Draw an Angle Bisector (p. 53)
Learn how to draw an angle bisector.

7

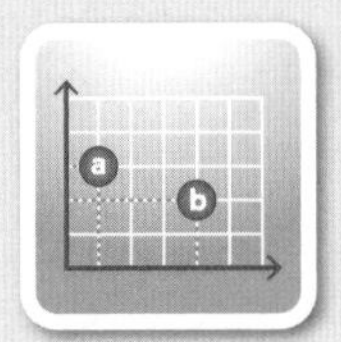

What is the Cartesian coordinate plane? (p. 59)
Discover how the Cartesian coordinate plane is used to describe locations.

8

Conversions among Fractions, Decimals, and Percents (p. 89)
Explore the relationships among fractions, decimals, and percents.

9

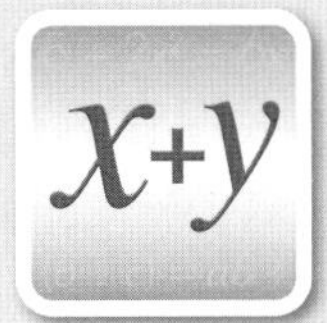

How to Solve an Equation (p. 109)
Learn how to solve an equation.

10

What is tax? (p. 147)
Explore the purpose of taxation and the general types of taxes.

11

What are Platonic solids? (p. 155)
Investigate Platonic solids and discover their properties.

12

Congruence and Similarity (p. 159)
Learn how to identify shapes that are congruent or similar.

Level 1

1 Multiples

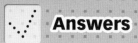

Try It
3 ; 6 ; 9 ; 12 ; 15
3, 6, 9, 12, 15

1. 4 ; 8 ; 12 ; 16 ; 20
   4, 8, 12, 16, 20
2. 5 ; 10 ; 15 ; 20 ; 25
   5, 10, 15, 20, 25
3. 6 ; 12 ; 18 ; 24 ; 30
   6, 12, 18, 24, 30
4. 7 ; 14 ; 21 ; 28 ; 35
   7, 14, 21, 28, 35
5. 8 ; 16 ; 24 ; 32 ; 40
   8, 16, 24, 32, 40
6. 9 ; 18 ; 27 ; 36 ; 45
   9, 18, 27, 36, 45
7.

1	2	3	4	5	6	7	8	9	10
11	12	13	14	15	16	17	18	19	20
21	22	23	24	25	26	27	28	29	30
31	32	33	34	35	36	37	38	39	40
41	42	43	44	45	46	47	48	49	50
51	52	53	54	55	56	57	58	59	60
61	62	63	64	65	66	67	68	69	70
71	72	73	74	75	76	77	78	79	80
81	82	83	84	85	86	87	88	89	90
91	92	93	94	95	96	97	98	99	100

a. 2, 4, 6, 8, 10, 12, 14, 16, 18, 20
3, 6, 9, 12, 15, 18, 21, 24, 27, 30
5, 10, 15, 20, 25, 30, 35, 40, 45, 50
7, 14, 21, 28, 35, 42, 49, 56, 63, 70

b. 2 and 3: 6, 12, 18, 24, 30, 36, 42, 48, 54, 60, 66, 72, 78, 84, 90, 96
2 and 5: 10, 20, 30, 40, 50, 60, 70, 80, 90, 100
3 and 5: 15, 30, 45, 60, 75, 90
3 and 7: 21, 42, 63, 84
5 and 7: 35, 70

8.

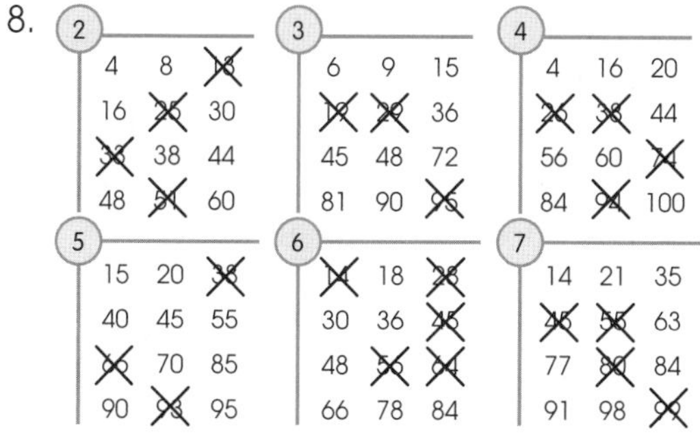

2
4 8
16 30
38 44
48 60
3
6 9 15
36
45 48 72
81 90
4
4 16 20
44
56 60
84 100
5
15 20
40 45 55
70 85
90 95
6
18
30 36
48
66 78 84
7
14 21 35
63
77 84
91 98

9a. 3 ; 5 ; 3 ; 5
b. 3 ; 4 ; 3 ; 4
c. 4 ; 5 ; 4 ; 5
d. 4 ; 8 ; 4 ; 8

10.
6, 12 | 10, 20, 30, 40, 50 | 15, 35, 45

11.
66 | 18, 36, 54, 72, 90 | 27, 45, 63, 81

12.

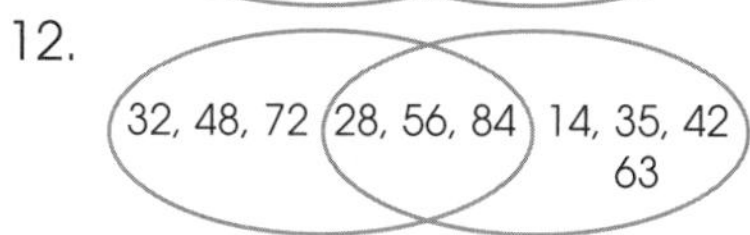

13.

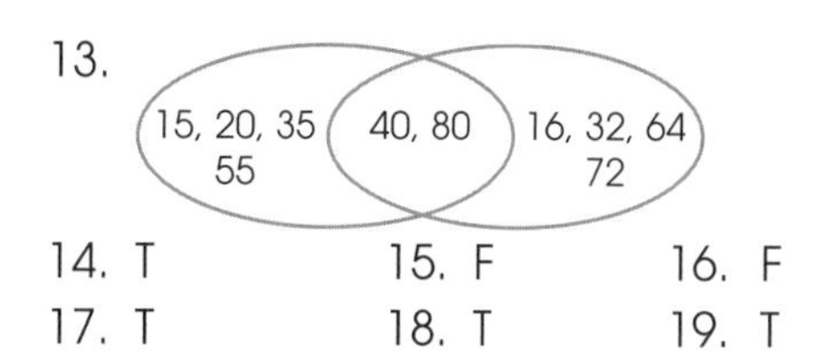

14. T
15. F
16. F
17. T
18. T
19. T

2 Factors

Try It
4 ; 2 ; 2
1 ; 2 ; 4

1. 3 ; 3
2. 1 ; 1 ; 5
3. 1 ; 7 ; 1 ; 7
4. 1 x 8
   2 x 4
   1, 2, 4, 8
5. 1 x 10
   2 x 5
   1, 2, 5, 10
6. 1 x 9
   3 x 3
   1, 3, 9
7. 1 x 12
   2 x 6
   3 x 4
   1, 2, 3, 4, 6, 12
8. 1 x 16
   2 x 8
   4 x 4
   1, 2, 4, 8, 16
9. 1 x 20
   2 x 10
   4 x 5
   1, 2, 4, 5, 10, 20
10. 1 x 24 = 24
    2 x 12 = 24
    3 x 8 = 24
    4 x 6 = 24
    1, 2, 3, 4, 6, 8, 12, 24
11. 1 x 27 = 27
    3 x 9 = 27
    1, 3, 9, 27
12. 1 x 30 = 30
    2 x 15 = 30
    3 x 10 = 30
    5 x 6 = 30
    1, 2, 3, 5, 6, 10, 15, 30
13. 1 x 32 = 32
    2 x 16 = 32
    4 x 8 = 32
    1, 2, 4, 8, 16, 32

14.

2	9	12	1
~~8~~	~~7~~	4	3
~~11~~	36	6	18

15.

3	1	~~5~~	18
~~12~~	9	2	~~8~~
6	~~7~~	~~16~~	~~4~~

16.

6	2	42	~~4~~
~~10~~	1	~~5~~	7
21	~~8~~	3	14

17.

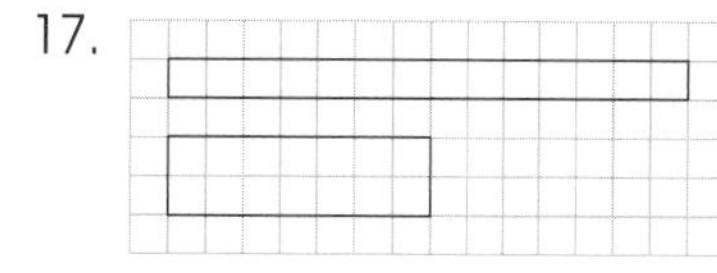

1, 2, 7, 14

18.

1, 3, 5, 15

19.

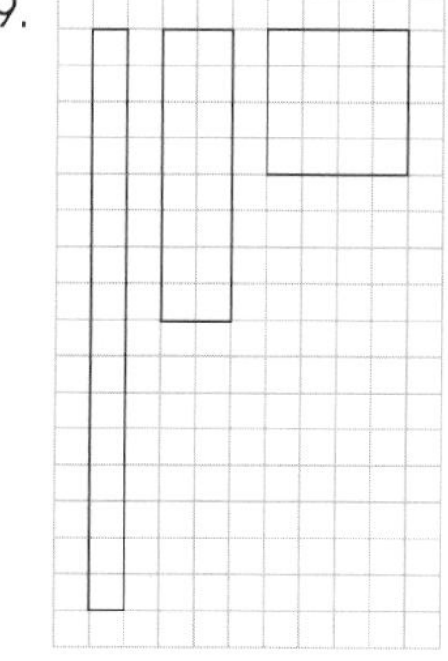

1, 2, 4, 8, 16

20. ① 3, ⑦ 21
① 5, ⑦ 35
1, 7

21. ①②③⑥ 7, 14, 21, 42
①②③ 4, 5, ⑥ 10, 12, 15, 20, 30, 60
1, 2, 3, 6

22. ①②④⑧ 16, 32
①②④ 5, ⑧ 10, 20, 40
1, 2, 4, 8

23. ①③⑨ 27
①③ 7, ⑨ 21, 63
1, 3, 9

24.
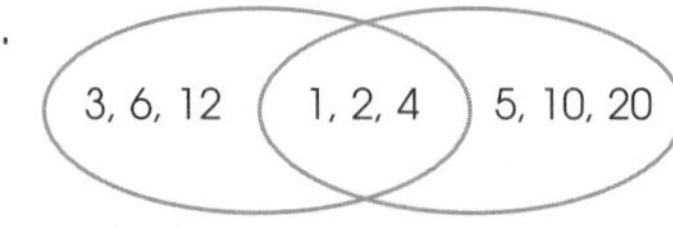

1, 2, 4

25.
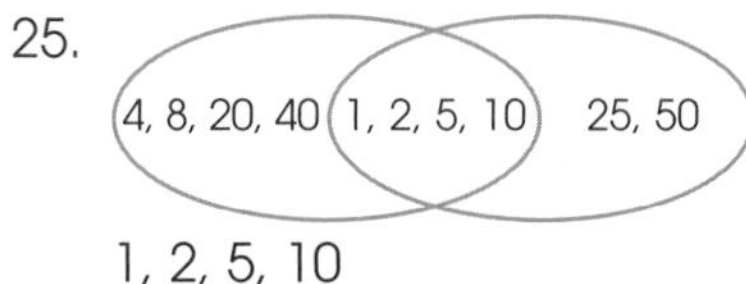

1, 2, 5, 10

3 Exponents

Try It
A

1. B 2. B 3. A 4. A 5. B
6. B 7. A 8. A 9. A

10.

11. 8^3 ; 8 ; 3
12. 9^4 ; 9 ; 4
13. 2^5 ; 2 ; 5
14. 4^5 ; 4 ; 5
15. 3^4 ; 3 ; 4
16. 6^3 ; 6 ; 3
17. 2 ; 2 ; 8
18. 3 x 3 x 3 ; 27
19. 5 x 5 x 5 ; 125
20. 6 x 6 x 6 ; 216
21. 4 x 4 x 4 x 4 ; 256
22. 2 x 2 x 2 x 2 x 2 ; 32
23. 4 x 4 x 4 ; 64
24. 5 x 5 x 5 x 5 ; 625
25. 8 x 8 ; 64
26. 1 x 1 x 1 x 1 x 1 x 1 ; 1
27. 7 x 7 x 7 ; 343
28. 12 x 12 ; 144

29. 16 30. 16 31. 1 32. 6
33. 49 34. 81 35. 10 000
36. 16 37. 1 38. 512 39. 45
40. 64 41. 119 42. 81 43. 37
44. 1 45. 533 46. 1

47. 48.

49. 50.

51a. $<$ b. $<$ c. $<$ d. $>$
52a. $<$ b. $>$ c. $<$ d. $>$
53a. $=$ b. $<$ c. $<$
d. $>$ e. $=$ f. $<$
54a. 6^3 b. 9^5 c. 5^5
55a. 7^0 b. 2^2 c. 3^4

56. 2^1, 2^2, 2^3, 2^4
57. 3^0, 3^1, 3^3, 3^9
58. 1^3, 4^3, 7^3, 8^3
59. 0^2, 1^2, 5^2, 8^2
60. 4^2, 5^2, 4^3, 5^3
61. 8^8, 9^8, 8^9, 9^9

4 Squares and Square Roots

Try It
5

1. 1 ; 4 ; 9 ; 16 ; 25 ; 36 ; 49 ; 64 ; 81 ; 100 ;
121 ; 144 ; 169 ; 196 ; 225 ; 256 ; 289 ; 324 ; 361 ; 400

2a. 0 b. 1 c. 4 d. 5 e. 6
f. 9

3. cannot 4. cannot 5. 5 6. 6
7. 9 8. 0 9. 1 10. 4
11. 4 12. 5 13. 1 14. 0
15. 6 16. 6

17. 18.

19. 20.

21. A: 289 B: $34^2 = 1156$
C: $15^2 = 225$ D: $13^2 = 169$
E: $19^2 = 361$

22. 1 ; 2 ; 3 ; 4 ; 5 ; 6 ; 7 ; 8 ; 9 ; 10 ;
11 ; 12 ; 13 ; 14 ; 15 ; 16 ; 17 ; 18 ; 19 ; 20

23. A: 9 B: $\sqrt{9} = 3$
C: $\sqrt{16} = 4$ D: $\sqrt{64} = 8$
E: $\sqrt{16 + 64 + 64} = \sqrt{144} = 12$

24. 5 ; 7 ; 11 ; 12 ; 15

25. 9 ; 36 ; 81 ; 100 ; 144 ; 169 ; 196 ; 225 ; 289 ;
324 ; 400 ; 529 ; 625

26. $\sqrt{4}$ $\sqrt{7}$ $\sqrt{9}$ $\sqrt{25}$ $\sqrt{30}$ $\sqrt{36}$ $\sqrt{49}$ $\sqrt{55}$ $\sqrt{64}$ $\sqrt{81}$ $\sqrt{90}$ $\sqrt{100}$
0 1 2 3 4 5 6 7 8 9 10

a. 3 b. 9 ; 10 c. 7 ; 8 d. 5 ; 6

27. $\sqrt{256}$ $\sqrt{275}$ $\sqrt{289}$ $\sqrt{324}$ $\sqrt{350}$ $\sqrt{361}$ $\sqrt{400}$ $\sqrt{425}$ $\sqrt{441}$ $\sqrt{484}$ $\sqrt{500}$ $\sqrt{529}$
15 16 17 18 19 20 21 22 23 24 25

a. 16 ; 17 b. 22 ; 23 c. 20 ; 21 d. 18 ; 19

5 Integers

Try It

-15	~~0.8~~
20	~~$1\frac{3}{4}$~~
~~-5.1~~	+9

1.

a.	b.	c.	d.
~~0.9~~	-15	0	$\sqrt{25}$
+7	~~0.04~~	-120	~~-3.3~~
-12	1	~~30%~~	4
e.	**f.**	**g.**	**h.**
10^2	~~15%~~	$\sqrt{100}$	$-\frac{16}{8}$
~~$-3\frac{3}{4}$~~	-5	~~-2.5~~	500%
5	-11	-9	~~$\frac{22}{5}$~~

2. +45 3. -200 4. -12 5. +140
6. +35 7. -2 8. -8 9. +8
10. +25 11. -5

12. Number line (-3 to 3) marked at -1 and 1 ; 1
13. Number line (-3 to 3) marked at -3 and 3 ; -3
14. Number line (-6 to 6) marked at -5 and 5 ; 5
15. Number line (-6 to 6) marked at -4 and 4 ; -4
16. < (-3, 0 marked on number line -4 to 2)
17. > (-1, 2 marked on number line -3 to 3)
18. < (-4, -1 marked on number line -6 to 0)
19. > (-9, -5 marked on number line -10 to -2)
20. Number line (-10 to 10) marked at -10, -9, -8, -7, -6, -5, -4, -3, -1, 0, 2, 3, 4, 8, 10

a. < b. > c. < d. < e. >
f. < g. > h. < i. <

21. Number line (-10 to 7) marked at -8, -2, 1, 6
-8, -2, 1, 6
22. Number line (-10 to 7) marked at -7, -5, -3, 5
5, -3, -5, -7
23. Number line (-10 to 7) marked at -6, -4, 2, 5
-6, -4, 2, 5
24. Number line (-10 to 7) marked at -10, -8, -4, -1
-1, -4, -8, -10

25. -6 ; -3 ; 0 ; 4 26. 5 ; -4 ; -7 ; -8
27. 9 ; 3 ; -2 ; -7 28. -6 ; -5 ; -1 ; 2
29. T 30. F 31. T 32. F 33. T
34a. -6 b. 5 c. -1, 0, 2, 5, 15
d. -0.5, 0, 2, $3\frac{1}{2}$, 5, 9.2, 15 e. -11, -9, -8, -6
f. -11, -10.2
g. -2, -1, 0
35. -7 ; -1°C ; 1°C ; -12°C ; -9°C
a. Wednesday b. Thursday
36. -3 ; 2°C ; -5°C ; -7°C ; -13°C
a. Friday b. Wednesday

6 Fractions

Try It
$2 ; 3 ; \frac{1}{2}$

1. $= \frac{10}{16} + \frac{1}{16} = \frac{11}{16}$
2. $= \frac{14}{24} + \frac{7}{24} = \frac{21}{24} = \frac{7}{8}$
3. $= \frac{3}{10} + \frac{8}{10} = \frac{11}{10} = 1\frac{1}{10}$
4. $= \frac{21}{35} + \frac{30}{35} = \frac{51}{35} = 1\frac{16}{35}$
5. $= 1\frac{8}{18} + 8\frac{1}{18} = 9\frac{9}{18} = 9\frac{1}{2}$
6. $= 7\frac{12}{15} + 2\frac{10}{15} = 9\frac{22}{15} = 10\frac{7}{15}$
7. $= \frac{5}{10} - \frac{1}{10} = \frac{4}{10} = \frac{2}{5}$
8. $= \frac{14}{15} - \frac{5}{15} = \frac{9}{15} = \frac{3}{5}$
9. $= 2\frac{3}{3} - 1\frac{1}{3} = 1\frac{2}{3}$
10. $= \frac{15}{18} - \frac{10}{18} = \frac{5}{18}$
11. $= 4\frac{7}{6} - 3\frac{4}{6} = 1\frac{3}{6} = 1\frac{1}{2}$
12. $= 6\frac{2}{4} - 5\frac{1}{4} = 1\frac{1}{4}$
13. $= \frac{\cancel{2}^{1}}{5} \times \frac{3}{\cancel{4}_{2}} = \frac{3}{10}$
14. $= \frac{5}{18}$
15. $= \frac{5}{\cancel{8}_{4}} \times \frac{\cancel{2}^{1}}{3} = \frac{5}{12}$
16. $= \frac{\cancel{7}^{1}}{3} \times \frac{2}{\cancel{14}_{2}} = \frac{2}{6} = \frac{1}{3}$
17. $= \frac{\cancel{11}^{1}}{\cancel{2}_{1}} \times \frac{\cancel{4}^{2}}{\cancel{11}_{1}} = 2$
18. $= \frac{17}{\cancel{6}_{2}} \times \frac{\cancel{15}^{5}}{2} = \frac{85}{4} = 21\frac{1}{4}$

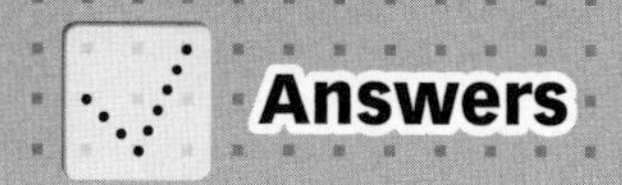

19. $= \frac{^{1}\cancel{4}}{1} \times \frac{29}{\cancel{8}_{2}}$
$= \frac{29}{2}$
$= 14\frac{1}{2}$

20. $= \frac{24}{_{1}\cancel{7}} \times \frac{\cancel{14}^{2}}{1}$
$= 48$

21a. $\frac{7}{12}$; 60 ; 35
b. $\frac{2}{3} \times 12 = 8$
c. $1\frac{1}{2} \times 15 = 22\frac{1}{2}$
d. $\frac{2}{5} \times 20 = 8$

22. $\frac{4}{3}$ 23. $\frac{2}{5}$ 24. $\frac{7}{6}$ 25. $\frac{3}{5}$

26. $\frac{5}{14}$ 27. $\frac{10}{17}$ 28. $\frac{7}{12}$ 29. $\frac{2}{7}$

30. $= \frac{2}{_{1}\cancel{3}} \times \frac{\cancel{15}^{5}}{1}$
$= 10$

31. $= \frac{5}{_{1}\cancel{6}} \times \frac{\cancel{12}^{2}}{3}$
$= \frac{10}{3}$
$= 3\frac{1}{3}$

32. $= {}^{2}\cancel{20} \times \frac{19}{\cancel{10}_{1}}$
$= 38$

33. $= \frac{^{1}\cancel{5}}{_{7}\cancel{14}} \times \frac{\cancel{6}^{3}}{\cancel{5}_{1}}$
$= \frac{3}{7}$

34. $= \frac{^{1}\cancel{3}}{_{2}\cancel{4}} \times \frac{\cancel{2}^{1}}{\cancel{9}_{3}}$
$= \frac{1}{6}$

35. $= 8 \times \frac{3}{11}$
$= \frac{24}{11}$
$= 2\frac{2}{11}$

36. $= \frac{25}{_{2}\cancel{4}} \times \frac{\cancel{2}^{1}}{3}$
$= \frac{25}{6}$
$= 4\frac{1}{6}$

37. $= \frac{11}{_{1}\cancel{5}} \times \frac{\cancel{10}^{2}}{1}$
$= 22$

38. $\frac{4}{5}$ 39. $\frac{4}{7}$ 40. 9 41. $\frac{7}{18}$

42. $2\frac{1}{3}$ 43. $1\frac{2}{3}$ 44. $\frac{23}{36}$ 45. $\frac{3}{16}$

46. $\frac{1}{14}$ 47. $\frac{2}{3}$ 48. $3\frac{9}{10}$ 49. $7\frac{1}{2}$

50. $4\frac{1}{2}$ 51. $4\frac{8}{15}$ 52. $\frac{37}{40}$ 53. 3

54. $\frac{2}{3}$ 55. $\frac{4}{9}$

56. $= 12\frac{2}{5} - 2$
$= 10\frac{2}{5}$

57. $= 3 - 3$
$= 0$

58. $= 3\frac{1}{2} + 6\frac{7}{8}$
$= 10\frac{3}{8}$

59. $= 1\frac{5}{12} + 1\frac{5}{6}$
$= 3\frac{1}{4}$

60. $= 10 - 4$
$= 6$

61. $4\frac{9}{10}$ 62. $14\frac{3}{5}$ 63. $1\frac{5}{6}$ 64. 7

65. $\frac{5}{8}$ 66. $19\frac{2}{3}$ 67. $2\frac{1}{10}$ 68. $1\frac{7}{9}$

7 Decimals

Try It
0.61

1. 77.93 2. 91.91 3. 40.07
4. 38.89 5. 40.39 6. 93.77
7. 101.12 8. 75.61 9. 34.32
10. 4.8 11. 70.33 12. 68.91

13. 76.28
69.56

14. 4.52 + 7.65
12.17

15. 4.18 + 49.12
53.3

16. 8.87 – 8.66
0.21

17. 104.03 – 20.98
83.05

18. 35.2 – 19.91
15.29

19. 21.39 20. 31.38
21. 2.461 22. 4.1328
23. 3.0375 24. 3.4030

25. $\begin{array}{r} 18.2 \\ \times\ \ 0.3 \\ \hline 5.46 \end{array}$

26. $\begin{array}{r} 0.096 \\ \times\ \ 9 \\ \hline 0.864 \end{array}$

27. $\begin{array}{r} 11.93 \\ \times\ \ 0.04 \\ \hline 0.4772 \end{array}$

28. $\begin{array}{r} 8.29 \\ \times\ \ 0.6 \\ \hline 4.974 \end{array}$

29. $\begin{array}{r} 7.11 \\ \times\ \ 1.5 \\ \hline 3555 \\ 7110 \\ \hline 10.665 \end{array}$

30. $\begin{array}{r} 0.05 \\ \times\ \ 84 \\ \hline 20 \\ 400 \\ \hline 4.20 \end{array}$

31. $\begin{array}{r} 3.08 \\ \times\ \ 0.26 \\ \hline 1848 \\ 6160 \\ \hline 0.8008 \end{array}$

32. $\begin{array}{r} 2.53 \\ \times\ \ 7.1 \\ \hline 253 \\ 17710 \\ \hline 17.963 \end{array}$

33. 2.48 34. 27.208 35. 3.888
36. 333.12 37. 0.672 38. 1.0944
39. 89.18 40. 46.516 41. 52.13
42. 133.515

43. A: 1.8
6
$\begin{array}{r} 1.8 \\ 6\overline{)10.8} \\ \underline{6} \\ 48 \\ \underline{48} \end{array}$

B: 5.2
41.6 ; 8
$\begin{array}{r} 5.2 \\ 8\overline{)41.6} \\ \underline{40} \\ 16 \\ \underline{16} \end{array}$

C: 9.6
76.8 ; 8
$\begin{array}{r} 9.6 \\ 8\overline{)76.8} \\ \underline{72} \\ 48 \\ \underline{48} \end{array}$

D: 250
6250 ; 25
$\begin{array}{r} 250 \\ 25\overline{)6250} \\ \underline{50} \\ 1250 \\ \underline{1250} \end{array}$

E: 201
804 ; 4
$\begin{array}{r} 201 \\ 4\overline{)804} \\ \underline{80} \\ 4 \\ \underline{4} \end{array}$

F: 32
224 ; 7
$\begin{array}{r} 32 \\ 7\overline{)224} \\ \underline{21} \\ 14 \\ \underline{14} \end{array}$

G: 45

405 ; 9

```
   45
9)405
  36
   45
   45
```

H: 2.5

42.5 ; 17

```
    2.5
17)42.5
   34
    8 5
    8 5
```

44. 28　45. 23　46. 63
47. 124　48. 0.305　49. 29.3
50. = 16.39 – 4.34
 = 12.05
51. = 4.2 + 8
 = 12.2
52. = 1350 ÷ 0.2
 = 6750
53. = 53.9 – 4
 = 49.9
54. = 20 – 9.02
 = 10.98
55. = 14.58 x 1.25
 = 18.225
56. 8.4 ; 3.7 ; 6.38 ; 6.16
57. (4.27 + 5.62 + 9.15 + 13.28) ÷ 4 ; 8.08
58. (5.49 + 4.91 + 6.5 + 3.7 + 1.35) ÷ 5 ; 4.39
59. 0.5 ; 0.5
 7.5 ; 0.4
 7.9
60. = (20 + 0.4) ÷ 0.2
 = 20 ÷ 0.2 + 0.4 ÷ 0.2
 = 100 + 2
 = 102
61. = (4 + 0.25) x 0.4
 = 4 x 0.4 + 0.25 x 0.4
 = 1.6 + 0.1
 = 1.7
62. = (2 + 0.98) ÷ 0.2
 = 2 ÷ 0.2 + 0.98 ÷ 0.2
 = 10 + 4.9
 = 14.9

8 Rates

Try It
2 ; 80

1. 140 ; 140　2. 0.25 ; 0.25
3. 196 ; 196　4. 0.5 ; 0.5
5. 0.35 ; 0.35　6. 1.25
7. 35 words/min　8. 93 pages/day
9. $0.68/mL　10. 2 claps/s
11. 0.3 push-ups/s　12. 23 rotations/s
13. 7.5 bracelets/h
14. A: $0.60/apple　B: $3.99/toothbrush
 C: $0.42/cup　D: $0.03/mL
 E: $1.79/kg　F: $0.21/m
15. $1.20/apple ; $1.10/apple ; B
16. $0.45/banana ; $0.46/banana ; A
17. $3.20/kg ; $2.80/kg ; B
18. $9.85/L ; $8.25/L ; B
19. $0.66/juice box ; $0.49/juice box ; B
20. $0.47/dog treat ; $0.53/dog treat ; A

21a. 127.5 ; 106.25 m/min ; 100 m/min
 b. 96 m/min ; 80 m/min ; 120 m/min
 c. 98 m/min ; 137.2 m/min ; 122.5 m/min
 d. 5100 ; 2400 ; 3430 ; 40 ; 25 ; 35 ; 109.3 m/min
22a. 75 x 6.2 = 465 (km)
 b. 200 x 125 = 25 000 (m)
23a. 280 ÷ 25 = 11.2 (min)
 b. 765 ÷ 85 = 9 (h)
24a. 285 x 2 = 570 (calories)
 b. 254 x 2 + 365 x 1 = 873 (calories)
 c. 285 x 3 + 150 x 2 = 1155 (calories)
 d. 117 x 3 + 150 x 2 = 651 (calories)
 e. 254 x 1 + 365 x 1 + 124 x 2 = 867 (calories)
 f. 285 x 1 + 117 x 2 + 150 x 2 = 819 (calories)
25a. Eggs: 155 ÷ 100 = 1.55 (calories/g)
 Bacon: 541 ÷100 = 5.41 (calories/g)
 180 x 1.55 + 250 x 5.41 = 1631.5 (calories)
 b. Pancakes: 227 ÷ 100 = 2.27 (calories/g)
 Orange slices: 47 ÷ 100 = 0.47 (calories/g)
 300 x 2.27 + 50 x 0.47 = 704.5 (calories)

9 Perimeter and Area

Try It
32 ; 48

1. 4 x 5 ; 20 (cm)
 5 x 5 ; 25 (cm^2)
2. 2 x (8 + 3) ; 22 (m)
 8 x 3 ; 24 (m^2)
3. 2 x (4.5 + 2) ; 13 (cm)
 2 x 4 ; 8 (cm^2)
4. 9 + 8 + 6 ; 23 (m)
 9 x 5 ÷ 2 ; 22.5 (m^2)
5. 6 + 4 + 3 ; 13 (m)
 3 x 3.6 ÷ 2 ; 5.4 (m^2)
6. 6 ; 8 ; 5 ; 35
7. (9 + 12) x 8 ÷ 2 = 84 (cm^2)
8. (2 + 2.5) x 2 ÷ 2 = 4.5 (m^2)
9. (10 + 20) x 9 ÷ 2 = 135 (m^2)
10. A: 42 cm ; 110.25 cm^2
 B: 26.6 m ; 28.8 m^2
 C: 54.4 m ; 171.72 m^2
 D: 18 cm ; 15.6 cm^2
 E: 27 m ; 31.5 m^2
 F: 18 cm ; 13.5 cm^2
 G: 94.1 m ; 441.75 m^2
11. P: 3.7 + 2.4 + 2.3 + 1.8 + 3 = 13.2
 A: (3 x 2.2) ÷ 2 + (1.8 + 2.4) x 2.2 ÷ 2 = 7.92
 13.2 m ; 7.92 m^2
12. P: 8 + 4.5 + 4 + 3 + 5 + 4.5 = 29
 A: 8 x 4.5 – 4 x 3 ÷ 2 = 30
 29 cm ; 30 cm^2
13. P: 6 x 4 + 8.5 = 32.5
 A: 6 x 6 + 6 x 6 ÷ 2 = 54
 32.5 cm ; 54 cm^2

14. P: $5.5 + 4.5 + (3.2 - 1.2) + 6.5 + 1.2 + 6.5 = 26.2$
A: $6.5 \times 1.2 + 4.5 \times 3.2 \div 2 = 15$
26.2 m ; 15 m^2
15. P: $15 + 7 + (15 - 8) + 5 + 8 + (5 + 7) = 54$
A: $15 \times (7 + 5) - 5 \times (15 - 8) = 145$
54 m ; 145 m^2
16. $(4.8 + 2.6) \times (1 + 0.8) \div 2 - 4.8 \times 1 \div 2 - 2.6 \times 0.8 \div 2 = 3.22$
3.22 m^2
17. $16 \times 12 - 12 \times 12 \div 2 - (12 - 9) \times 16 \div 2 = 96$
96 cm^2
18a. F b. T
19a. T b. F c. F
20. 140 000 21. 165 000 cm^2
22. 84 500 cm^2 23. 9 400 000 cm^2
24. 126 m^2 25. 4.732 m^2
26-28. (Suggested answers)
26. 14 ; 140 000 27. 12 ; 120 000
28. 15 ; 150 000

10 Volume

Try It
144
1. 5 ; 1 ; 7 ; 35
2. $7 \times 6 \times 6 = 252$ (cm^3)
3. $4 \times 12 \times 8 = 384$ (cm^3)
4. $11.2 \times 2.8 \times 8 = 250.88$ (m^3)
5. $8 \times 8 \times 8 = 512$ (m^3)
6. $5 \times 5 \times 16 = 400$ (cm^3)
7. $30 \times 5 \times 10 = 1500$ (m^3)
8. $13 \times 12 \div 2 \times 4 = 312$ (cm^3)
9. $8.6 \times 10.5 \div 2 \times 3 = 135.45$ (cm^3)
10. $5.6 \times 3 \times 1.8 = 30.24$ (m^3)
11. $(4.2 + 7.6) \times 3.9 \div 2 \times 2.5 = 57.525$ (m^3)
12. $(5.1 + 2) \times 9 \div 2 \times 2 = 63.9$ (m^3)
13. $4.4 \times 10 \times 3.5 = 154$ (m^3)
14. $(2.9 + 12.7) \times 11.4 \div 2 \times 3.2 = 284.544$ (cm^3)
15. $(13.2 + 18) \times 6.4 \div 2 \times 4 = 399.36$ (cm^3)
16. 16 cm 17. 5 m 18. 14 cm
19. 32 cm 20. 5 m 21. 12 cm
22a. 8 cm^3 b. 4 cm^3 c. 8 cm^3
A: $8 \times 3 + 4 + 8 = 36$ (cm^3)
B: $8 + 4 + 8 \times 3 = 36$ (cm^3)
C: $4 \times 4 + 8 \times 2 = 32$ (cm^3)
D: $8 \times 2 + 4 \times 2 + 8 \times 2 = 40$ (cm^3)
23. $12 \times (5 + 6) - 6 \times 4 = 108$ (cm^2)
$108 \times 3 = 324$ (cm^3)
24. $16.8 \times 9.5 + 3 \times 2 = 165.6$ (cm^2)
$165.6 \times 4 = 662.4$ (cm^3)
25. $5.4 \times 4 - 1.8 \times 1.8 = 18.36$ (cm^2)
$18.36 \times 3 = 55.08$ (cm^3)
26. $6 \times 5 \div 2 + 6 \times 2 = 27$ (cm^2)
$27 \times 9.2 = 248.4$ (cm^3)
27. $6 \times 4 - 2.5 \times 1.5 = 20.25$ (m^2)
$20.25 \times 3 = 60.75$ (m^3)

11 Surface Area

Try It
24
1. B ; 184 cm^2 2. A ; 600 cm^2
3. A ; 252 cm^2 4. A ; 480 cm^2
5. B ; 232 cm^2
6. $12 \times 12 \times 6 = 864$ (cm^2)
7. $16 \times 7 \times 2 + 16 \times 5 \times 2 + 7 \times 5 \times 2 = 454$ (cm^2)
8. $12.5 \times 5.2 \times 2 + 12.5 \times 4 \times 2 + 5.2 \times 4 \times 2 = 271.6$ (m^2)
9. $14 \times 5 \div 2 \times 2 + 14 \times 3.2 + 8.6 \times 3.2 \times 2 = 169.84$ (m^2)
10. $5 \times 5.2 \div 2 \times 2 + 11 \times 0.8 + 7 \times 0.8 + 5 \times 0.8 = 44.4$ (cm^2)
11. $(10.3 + 7.9) \times 6.8 \div 2 \times 2 + 10.3 \times 2.2 + 6.8 \times 2.2 + 7.9 \times 2.2 + 7.2 \times 2.2 = 194.6$ (m^2)
12. 4
a. 16 ; 64 cm^2 b. 18 ; 72 cm^2
c. 22 ; 88 cm^2 d. 20 ; 80 cm^2
e. 24 ; 96 cm^2
13a. $4 \times 3 \div 2 \times 2 + 5 \times 4 \times 2 + 5 \times 12 \times 3 + 3 \times 12 + 4 \times 12 = 316$ (cm^2)
b. $4 \times 3 \div 2 \times 2 + 5 \times 4 \times 2 + 4 \times 12 \times 3 + 3 \times 12 + 5 \times 12 = 292$ (cm^2)
14. $8.4 \times 4 \times 2 + 8.5 \times 3.5 \times 2 + (8.4 + 4 + 4.9 + 8.5 + 3.5 + 12.5) \times 5 = 335.7$ (cm^2)
15. $(10 + 14) \times 5 \div 2 \times 2 \times 2 + (10 + 5.4 \times 2) \times 2 \times 3 = 364.8$ (cm^2)
16. $7 \times 8 \div 2 \times 2 + 6 \times 5 \div 2 \times 2 + (7 + 10.6 + 3 + 7.8 + 6) \times 3.4 = 202.96$ (cm^2)
17. $20.3 \times 9.7 \times 2 + 5.1 \times 3.8 \times 2 + (20.3 + 9.7 + 20.3 + 3.8 + 3.8 + 9.7) \times 4.9 = 763.82$ (cm^2)
18. $(18.5 + 15.7) \times 12.4 \div 2 \times 2 + 16.2 \times 18.5 \div 2 \times 2 + (15.7 + 12.7 + 24.6 + 16.2 + 12.4) \times 7.5 = 1335.78$ (cm^2)
19. $(38.2 \times 38.2 - 19.1 \times 12.7) \times 2 + (38.2 \times 4 + 19.1 \times 2) \times 10.8 = 4496.14$ (cm^2)

12 Lines and Angles

Try It
parallel lines
intersecting lines

1. Intersecting Lines: B, E, F
Parallel Lines: A, C, D
2. A ; C
3. B ; not perpendicular
4. A
5. B ; perpendicular
6. B ; perpendicular
7.

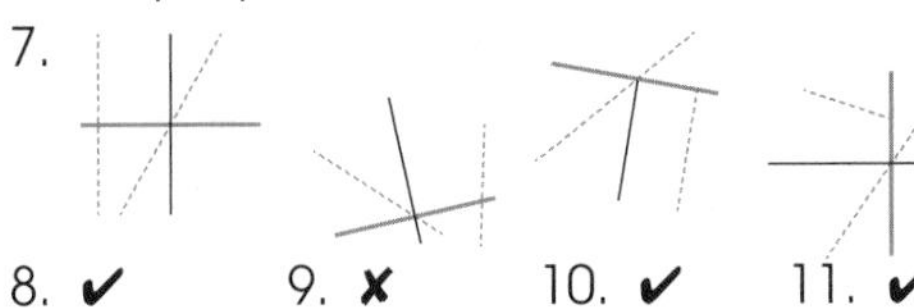

8. ✔ 9. ✘ 10. ✔ 11. ✔

12. 25°
13. 55°
14. 20°
15. 50°
16. 45°
17. 15°
18.
19.

13 Angles and Shapes

Try It
acute triangle
equilateral triangle

1. obtuse triangle
scalene triangle
2. right triangle
isosceles triangle
3. acute triangle
equilateral triangle
4. T
5. T
6. F
7. F
8. T
9. F
10. 95°
11. 29°
12. 16°
13. 84°
14. 40°
15. 25°
16. 48°
17. 71°
18. 3 cm, 4 cm, 5 cm
19. 2 cm, 3 cm, 4 cm
20. 4 cm, 4 cm, 3 cm

21-23. (Suggested drawings)

21.

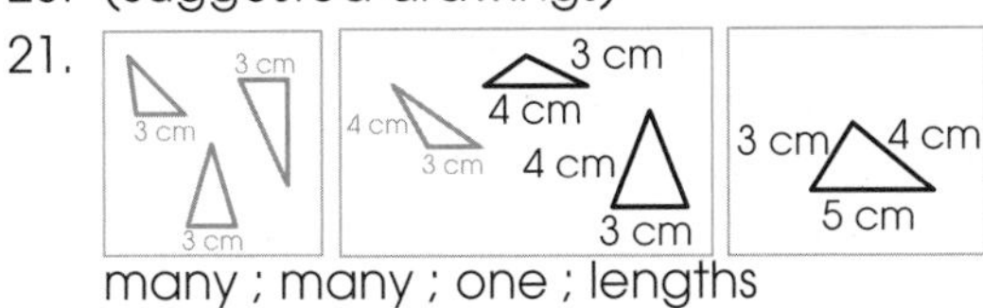

many ; many ; one ; lengths

22.

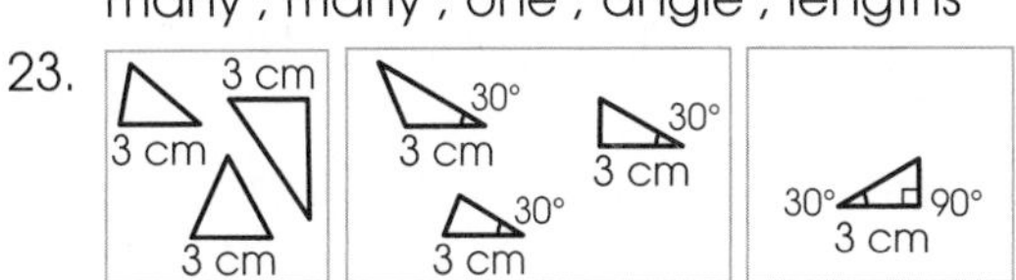

many ; many ; one ; angle ; lengths

23.

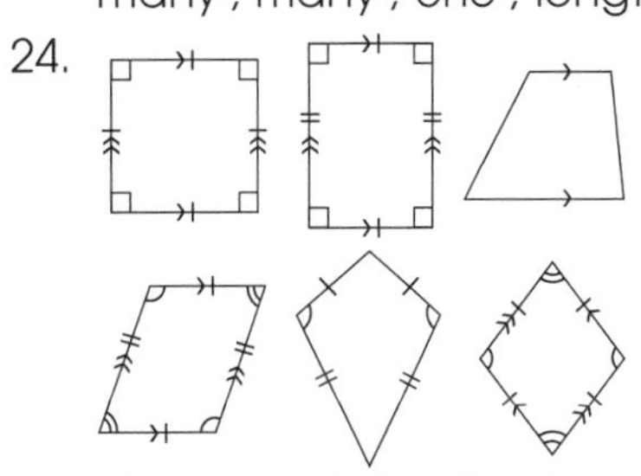

many ; many ; one ; length ; angles

24.

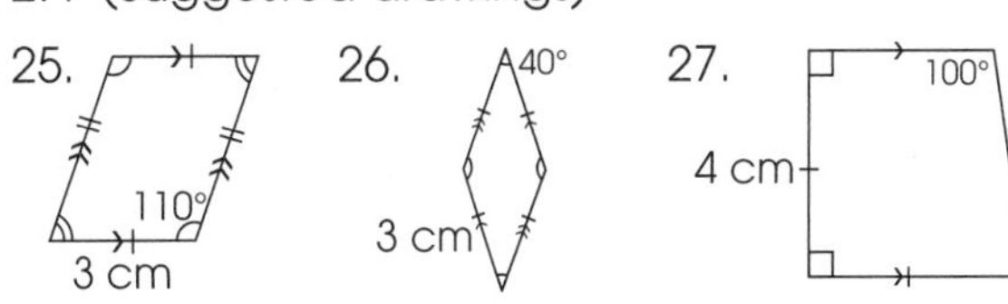

25-27. (Suggested drawings)

25. 110°, 3 cm
26. 40°, 3 cm
27. 100°, 4 cm

14 Coordinates

Try It
II ; x ; y

1a. 2 ; 5
-4 ; 8
-1 ; -7
2 ; -6
5 ; -3
-5 ; -5
6 ; 3
-6 ; 2

b. A, G, N
D, E, P

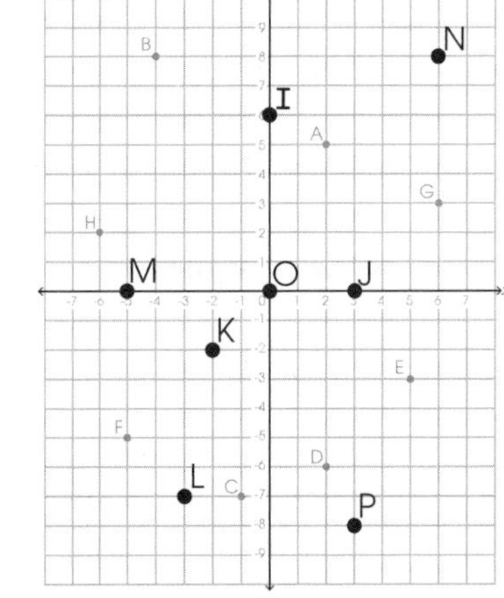

2a.

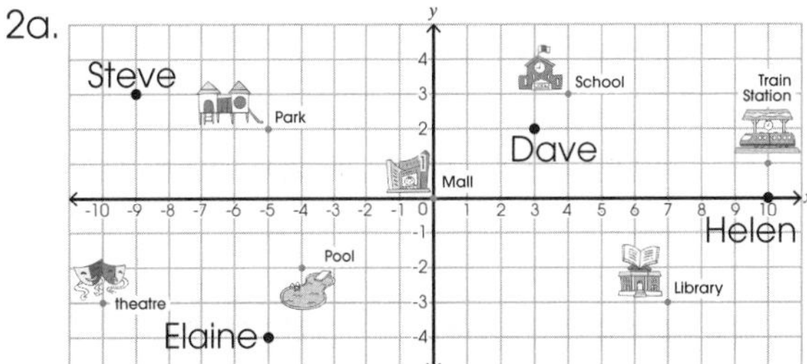

b. the mall
c. 6 units
d. 8 units
e. the park
f. (-5,2)
g. the library
h. She should move 7 units up and 9 units to the right.
i. He should move 6 units down and 1 unit to the left.

3. Rectangle: A(-8,3) ; B(-6,3) ; C(-6,-5) ; D(-8,-5)
   2 ; 8 ; 16
   Trapezoid: E(2,4) ; F(5,4) ; G(7,1) ; H(2,1)
   3 ; 5 ; 12
   Triangle: L(-3,-1) ; M(-3,-5) ; N(1,-5)
   4 ; 4 ; 8
4. the trapezoid 5. the triangle
6.

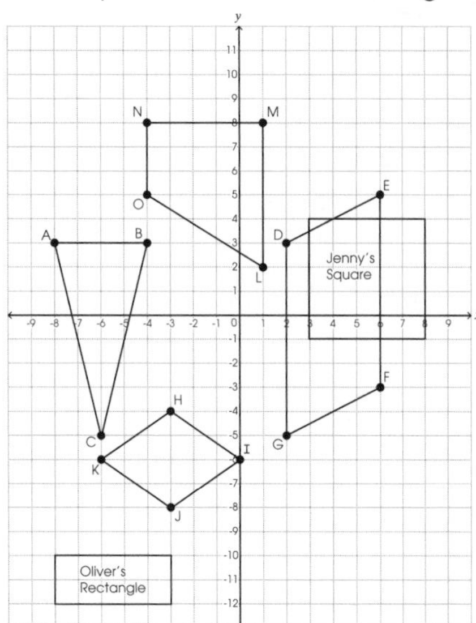

a. triangle b. parallelogram
c. rhombus d. trapezoid
e. Quadrants **II** and **III**
   Quadrants **I** and **IV**
   Quadrant **III**
   Quadrants **I** and **II**
f. (Refer to the coordinate plane.)

15 Transformations

Try It
dilatation

1. reflection 2. translation 3. rotation
4. dilatation, enlargement
5. dilatation, reduction
6. rotation
7. reflection/rotation

8a. 3 ; 3
b. a translation of 2 units up and 3 units to the right

9a. 2
b. an enlargement by a scale factor of 3

10a. L
b. a reflection in Line M

11a. P
b. a $\frac{1}{2}$ rotation about Point P

12. (Suggested answers)
a. 3 ; 1 ; 180
b. Rotate it 90° clockwise about Point Q and then reflect it in Line L.
c. Enlarge it by a scale factor of 2 and then translate it 1 unit up and 7 units to the right.
d. Reflect it in Line L and then translate it 8 units down and 2 units to the right.

13.

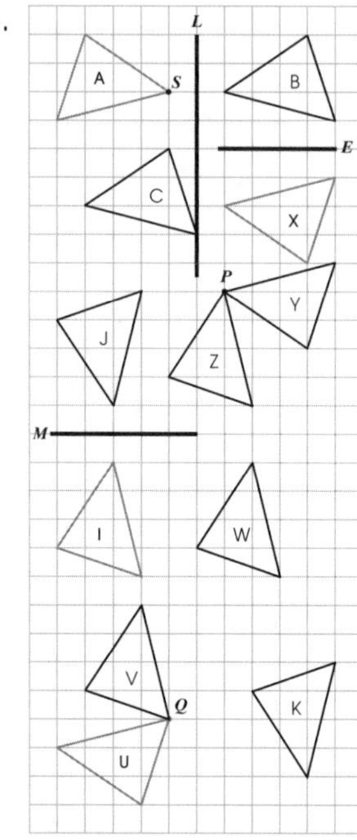

14. (Suggested answers)
    Rotate A 180° about Point S and translate it 4 units down and 2 units to the right.
    Reflect A in Line L and then reflect it in Line E.

16 Algebraic Expressions

Try It
B

1. A 2. A 3. A 4. B
5. D ; C ; A ; B ; E ; G ; F
6. $4x$ 7. $x \div 8$ 8. $x \div 2$
9. $(x - 7) \div 2$ 10. $2x + 3$ 11. $4 \div x - 5$
12. $x \div 4 - 2$ 13. $3x + 7$
14. + ; x ; + ; ÷ ; x ; – ; – ; ÷ ; –
15. the product of a number and 5
16. divide 9 by a number
17. the sum of a number and 1
18. subtract a number from 10
19. the sum of 8 and a number
20. half of a number
21. the sum of 1 and four times a number
22. the product of 6 and 2 less than a number

23. the difference of a quarter of a number and 9

24. 2 ; 7

25. = (5 – 1) ÷ 2
= 2

26. = 9 ÷ 3 – 2
= 1

27. = 10 – 4 x 2
= 2

28. = 8 x (2 + 5)
= 56

29. = 5 x 4 ÷ 2
= 10

30. = 30 ÷ (6 ÷ 2)
= 10

31a. 4 x 1 – 3
= 1

b. 4 x 2 – 3
= 5

c. 4 x 3 – 3
= 9

32a. (3 + 1) ÷ 2
= 2

b. (5 + 1) ÷ 2
= 3

c. (7 + 1) ÷ 2
= 4

33a. B

b. 3 x 4 + 5
= 17
17

3 x 7 + 5
= 26
26 trees

34a. B

b. 8(3 – 1)
= 16
16 cookies

8(5 – 1)
= 32
32 cookies

35a. A

b. 0.5(5 – 1)
= 2
2 kg

0.5(9 – 1)
= 4
4 kg

Level 2

1 Multiples and Factors

Try It
6, 12, 18, 24, 30, 36, 42, 48, 54, 60
10, 20, 30, 40, 50, 60, 70, 80, 90, 100
30, 60

1. Multiples:
4, 8, 12, 16, 20
5, 10, 15, 20, 25
20
Factors:
1, 2, 4
1, 5
1

2. Multiples:
8, 16, 24, 32, 40
12, 24, 36, 48, 60
24
Factors:
1, 2, 4, 8
1, 2, 3, 4, 6, 12
1, 2, 4

3. Multiples:
7, 14, 21, 28, 35
14, 28, 42, 56, 70
14, 28
Factors:
1, 7
1, 2, 7, 14
1, 7

4. Multiples:
9, 18, 27, 36, 45
15, 30, 45, 60, 75
45
Factors:
1, 3, 9
1, 3, 5, 15
1, 3

5.

6.

7.

8.

9.

10.

11a. ✗ ; 30
b. ✔
c. ✔
d. ✗ ; 140
e. ✗ ; 225
f. ✔

12a. ✗ ; 4
b. ✔
c. ✔
d. ✗ ; 2
e. ✔
f. ✗ ; 2

13. 1

14. no

15a. 30 ; 36
b. 16 ; 40
c. 30 ; 45
d. 30 ; 40

16a. 45 ; 52
b. 39 ; 52
c. 30 ; 60
d. 30 ; 39

17. 5, 10, 15, 20, 25, 30, 35, (40)
8, 16, 24, 32, (40)
40

18. 1, 2, (4), 8
1, 2, 3, (4), 6, 12
4

19. multiples
multiples of 2: 2, 4, (6)
multiples of 3: 3, (6)
6

20. factors
factors of 14: 1, 2, (7), 14
factors of 21: 1, 3, (7), 21
7

21. multiples
multiples of 10: 10, 20, 30, 40, 50, (60)
multiples of 12: 12, 24, 36, 48, (60)
60

22. multiples
multiples of 5: 5, 10, 15, 20, 25, 30, 35, (40)
multiples of 8: 8, 16, 24, 32, (40)
40

2 Exponents

Try It
A ; C

1. A: 5^3
B: ✗
C: ✗
D: 10^3
E: ✗
F: 11^4
G: 200^2
H: ✗
I: 8^4
J: 5^5
K: ✗
L: ✗
M: 6^3
N: 15^5

2. 2   3. 8   4. 6   5. 9   6. 3
7. 2   8. 4   9. 2   10. 4   11. 5^3
12. 2^4 13. 4^4 14. 3^5 15. 9^4 16. 8^5
17. 6^6 18. 2^8 19. 3^7 20. 5^6 21. 8^5
22. 4^9 23. 10^{12} 24. 25^{15} 25. 18^8 26. 100^{11}
27. A   28. B   29. A   30. B
31. $3^3 \times 5^4$ 32. $4^3 \times 7^2$
33. $2 \times 3^2 \times 7^3$ 34. $4^3 \times 5^2 \times 6$
35. $2^2 \times 3^3 \times 7$ 36. $6^2 \times 8^3 \times 9$
37. $5^4 \times 6 \times 7^2$ 38. $2 \times 5^2 \times 7^2 \times 9^3$
39. $2 \times 3^2 \times 5^3$
40. 1
10
100
1000
10 000
100 000
1 000 000
10 000 000
100 000 000
1 000 000 000
10 000 000 000
41. 10^3 42. 10^5 43. 10^3 44. 10^7
45. 10^4 46. 10^6 47. 10^5
48. 500   49. 2000   50. 90 000
51. 700   52. 9   53. 300
54. 50 000   55. 12 000   56. 8000
57. 200 000   58. 4000   59. 6 000 000
60. 80   61. 110 000   62. 7
63. 150 000 000
64.   65.

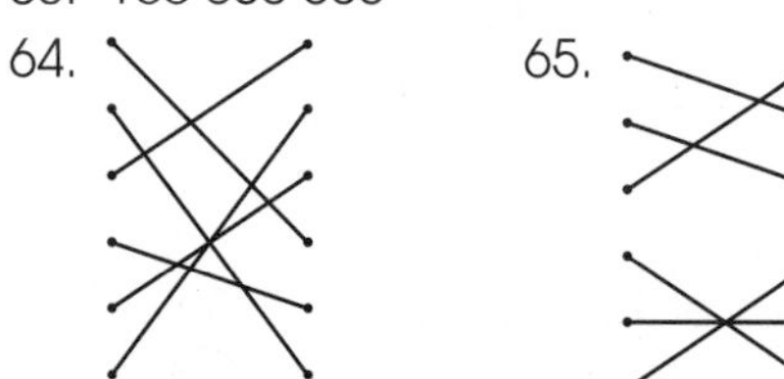

66. 9 ; 10^2 67. 8×10^3 68. 2×10^5
69. 7×10^3 70. 5×10^2 71. 6×10^4
72. 5×10^4 73. 6×10^3 74. 7×10^4
75. 6×10^1 76. 9×10^3 77. 9×10^5
78. 2 ; 1 ; 3   79. 2 ; 3 ; 1   80. 1 ; 3 ; 2
81. 1 ; 3 ; 2   82. 2 ; 3 ; 1   83. 2 ; 1 ; 3

3 Squares and Square Roots

Try It
9.5 ; 5.1
1. 9 ; 8.7   2. 9 ; 10 ; 9.2
3. 10 ; 11 ; 10.7   4. 5 ; 6 ; 5.2
5. 12 ; 13 ; 12.2   6. 9 ; 10 ; 9.7
7. 3 ; 4 ; 3.9   8. 14 ; 15 ; 14.1

9. $>$ 10. $>$ 11. =   12. $<$ 13. =
14. $>$ 15. =   16. $>$ 17. =   18. $>$
19. =   20. $<$
21. 6, $\sqrt{42}$, 7, $\sqrt{50}$ 22. 2^2, 5, $\sqrt{30}$, $\sqrt{40}$
23. $\sqrt{50}$, $\sqrt{60}$, 8, 3^2 24. $\sqrt{10}$, $\sqrt{15}$, 2^2, 4^2
25. 4 ; 4 ; 4   26. 5 ; 5 ; 5
27. $\sqrt{9 \times 9}$ $= 9$ 28. $\sqrt{11} \times \sqrt{11}$ $= 11$
29. $\sqrt{7} \times \sqrt{7}$ $= 7$ 30. $\sqrt{6 \times 6}$ $= 6$
31. 2, 4, 4, 2, 4   32. 8, 8, 64, 64, 8
33. 100 10   34. $= \sqrt{81}$ $= 9$ 35. $= \sqrt{16}$ $= 4$
36. $= \sqrt{16}$ $= 4$ 37. $= \sqrt{16}$ $= 4$ 38. $= \sqrt{144}$ $= 12$
39. $= \sqrt{100}$ $= 10$ 40. $= 5 + 2 + 3$ $= 10$ 41. $= \sqrt{16}$ $= 4$
42. $= \sqrt{25} \times \sqrt{9}$ $= \sqrt{225}$ $= 15$ 43. $= \sqrt{64}$ $= 8$ 44. $= \sqrt{100}$ $= 10$
45. $= 3 + \sqrt{25}$ $= 3 + 5$ $= 8$ 46. $= \sqrt{16} + \sqrt{16}$ $= 4 + 4$ $= 8$
47. $= \sqrt{12} \times \sqrt{3}$ $= \sqrt{36}$ $= 6$ 48. $= \sqrt{25} \times \sqrt{4}$ $= \sqrt{100}$ $= 10$
49. $= \sqrt{18} \times \sqrt{2}$ $= \sqrt{36}$ $= 6$ 50. $= \sqrt{4} + \sqrt{25}$ $= 2 + 5$ $= 7$
51. F   52. T   53. T   54. F
55. T   56. F   57. T

4 Integers

Try It
5 ;
1 ;
1. -1 ;
2. -5 ;
3. -2 ;
4. -6 ;
5. 4 ;

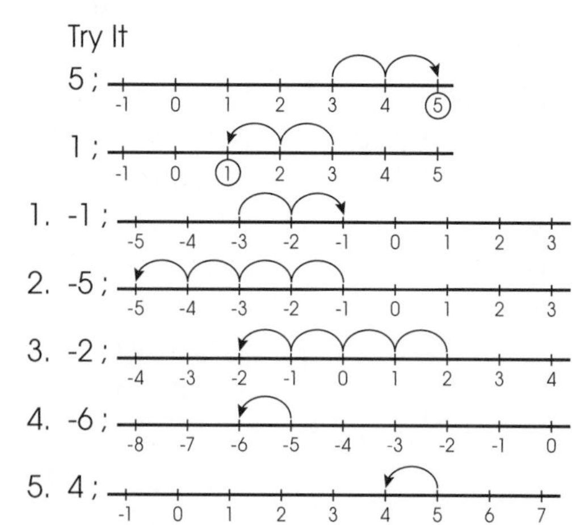

6. -3 ; 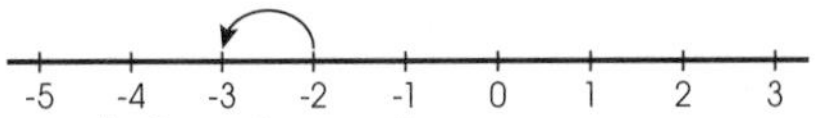 forward ; backward
7. -2 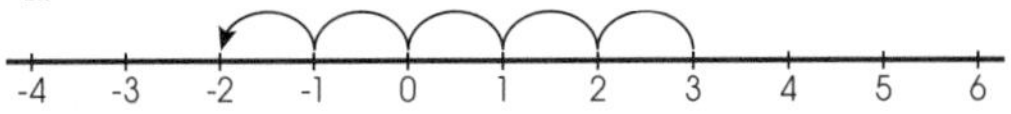
8. 3 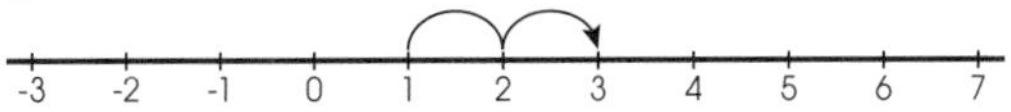
9. -5 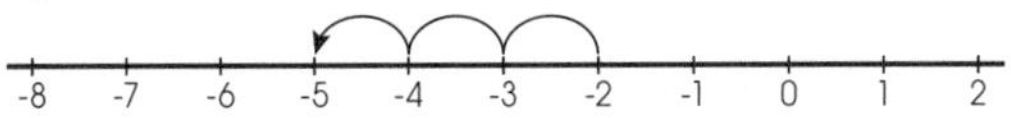
10. -3 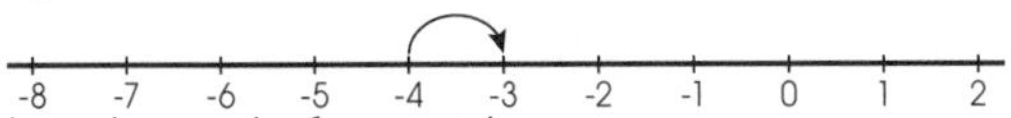 ; backward ; forward
11. -1 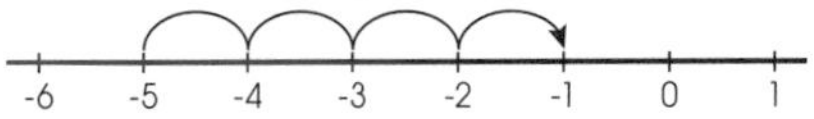
12. -4 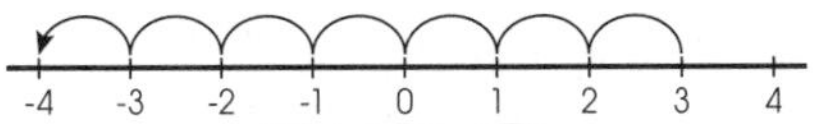
13. -5 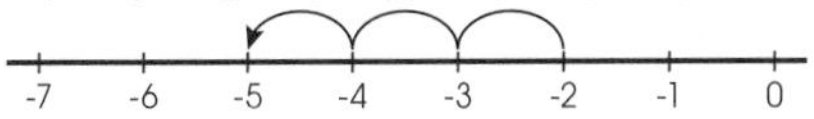
14. 6 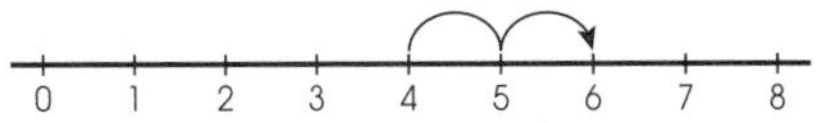
15. 10 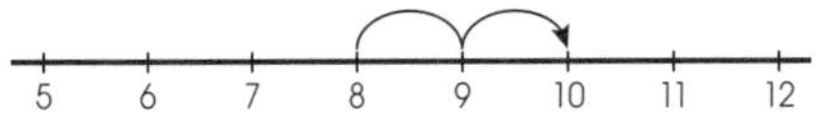
16. -10 
17. -3 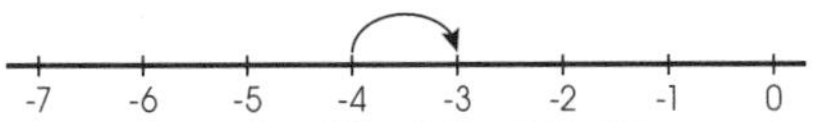
18. 2 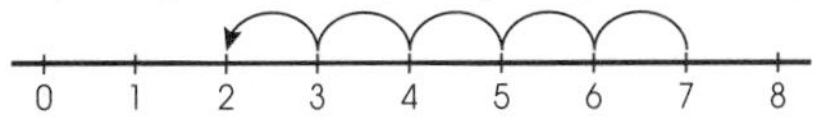
19. 3 
20. 0

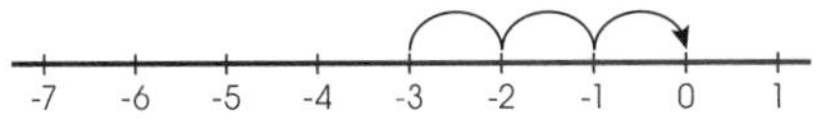

21. -2
22. = 8 + 1
 = 9
23. = 3 + 2
 = 5
24. = 3
25. = -3
26. = 4 + 2
 = 6
27. = -2 + 6
 = 4
28. = -2
29. = 5 - 5
 = 0
30. = -6 + 3
 = -3
31. = -6
32. -1 - (-2) ; -2 + 1
 -2 - 1 ; -1 + (-2)
33. -4 - 3 ; -3 - (-4)
 -3 + (-4) ; -4 + 3
34. -6
35. 4 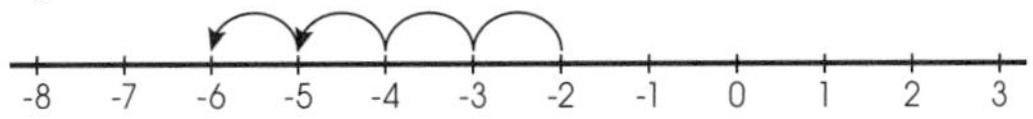
36. -1 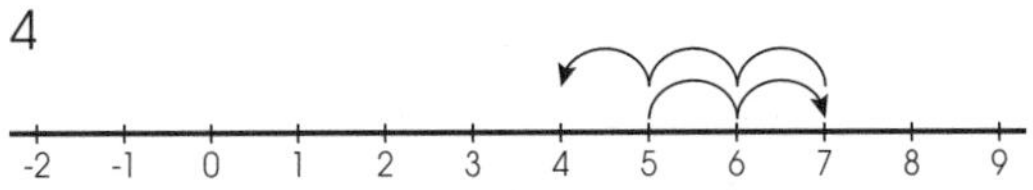
37. 6 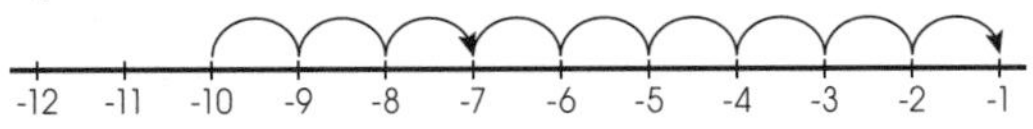
38. -9 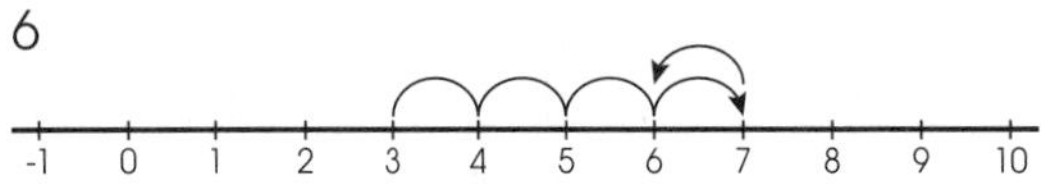
39. 7 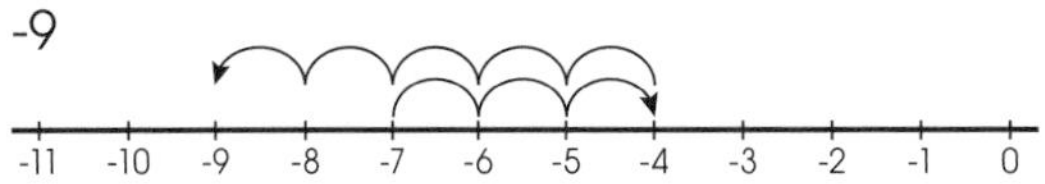
40. 4 ; 5 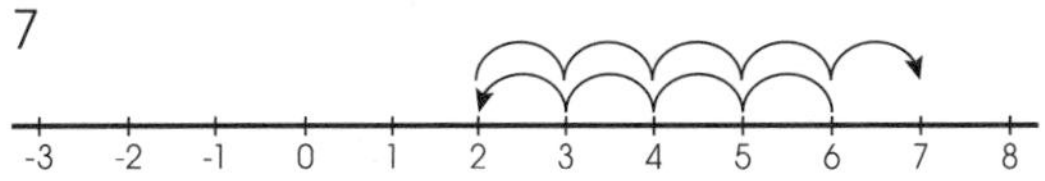
 -2
41. = -1 + 6 + 8
 = 13
42. = 8 - 12 - 3
 = -7
43. = -7 + 8 + 9
 = 10
44. = -10 - 6 - 4
 = -20
45. = 4 + 8 - 3
 = 9
46. = -3 + 2 + 6
 = 5
47. = 9 + 4 + 1
 = 14
48. = -2 - 3 - 4
 = -9
49. = 2 + 2 - 5
 = -1
50. = -15 - 4 + 2
 = -17
51. = 8 - 1 + 6
 = 13

5 Fractions, Decimals, and Percents

Try It
$\frac{15}{100}$; 0.15 ; 15%

1a. 0.6 b. 0.25 c. 0.4 d. 0.75
e. 0.375 f. 0.8 g. 0.7 h. 1.5
i. 1.2 j. 0.625 k. 0.5 l. 0.3
m. 0.875 n. 1.4

2a. 20% b. 25% c. 90% d. 60%
e. 80% f. 175% g. 30% h. 40%
i. 125% j. 15% k. 120% l. 60%
m. 80% n. 75%

3a. $\frac{3}{100}$ b. $\frac{1}{2}$ c. $\frac{7}{25}$ d. $\frac{17}{20}$
e. $1\frac{1}{5}$ f. $\frac{7}{10}$ g. $2\frac{9}{20}$ h. $1\frac{1}{20}$

4a. 35% b. 18% c. 40% d. 95%
e. 210% f. 17% g. 350% h. 103%

5. $\frac{11}{20}$; 55% ; A
 $\frac{3}{4}$; 75% ; F
 $\frac{2}{5}$; 40% ; C
 $\frac{3}{5}$; 60% ; E
 $\frac{1}{20}$; 5% ; B
 $\frac{5}{8}$; 62.5% ; G
 $\frac{3}{8}$; 37.5% ; D

6a. $\frac{1}{2}$
b. $\frac{1}{5}$
c. $\frac{7}{20}$
d. $\frac{21}{50}$
e. $\frac{39}{50}$
f. $\frac{16}{25}$
g. $\frac{41}{50}$
h. $\frac{7}{25}$

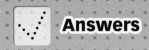

7a. 0.1 b. 0.7 c. 0.14 d. 0.29
e. 0.59 f. 0.06 g. 0.08 h. 0.24

8-13. (Suggested drawings)

8. $\frac{3}{10}$; 0.3

9. $\frac{1}{4}$; 0.25

10. $\frac{3}{5}$; 0.6

11. $\frac{7}{10}$; 0.7

12. $\frac{7}{20}$; 0.35

13. $\frac{17}{20}$; 0.85

14. 0.36 ; 36%
15. 0.24 ; $\frac{6}{25}$
16. 18% ; $\frac{9}{50}$
17. 15% ; $\frac{3}{20}$
18. 0.64 ; 64%
19. 0.35 ; 35%
20. 1.5 ; 150%
21. 250% ; $2\frac{1}{2}$
22. 1.8 ; 180%
23. $\frac{8}{25}$; 0.53 ; 85%
24. $70.5\% < \frac{29}{40} < 0.73$
25. $1.15 < 123\% < 1\frac{1}{2}$
26. $60\% < 0.65 < \frac{14}{20}$
27. $110\% < 1\frac{2}{5} < 1.5$
28. $200\% < 2\frac{1}{4} < 2.5$
29. ✘ ; $\frac{1}{4}$
30. ✘ ; 10%
31. ✔
32. ✘ ; 0.25
33. ✘ ; 0.5
34. ✘ ; 3.75
35. ✔
36. ✔
37. ✘ ; 85%

6 Ratios and Rates

Try It
3 to 2 ; 3:2 ; $\frac{3}{2}$

1. 1 to 2 ; 1:2 ; $\frac{1}{2}$
   1 to 1 ; 1:1 ; 1
   2 to 5 ; 2:5 ; $\frac{2}{5}$
2. 1 to 3 ; 1:3 ; $\frac{1}{3}$
   1 to 4 ; 1:4 ; $\frac{1}{4}$
   3 to 4 ; 3:4 ; $\frac{3}{4}$
3. 1 to 2 ; 1:2 ; $\frac{1}{2}$
   3 to 2 ; 3:2 ; $\frac{3}{2}$
   1 to 3 ; 1:3 ; $\frac{1}{3}$
4. 1 to 3 ; 1:3 ; $\frac{1}{3}$
   1 to 3 ; 1:3 ; $\frac{1}{3}$
   1 to 2 ; 1:2 ; $\frac{1}{2}$
5. A ; B
6. A ; C
7. B ; D
8. B ; C
9. B ; C

10.–15. (Drawings)

16. 17 ; 11:17 ; 4:17 ; 2:17
    18 ; 2:3 ; 1:6 ; 1:6
    15 ; 2:3 ; 2:15 ; 1:5
    17 ; 15:17 ; 1:17 ; 1:17
17. The Bears and the Raiders have the same ratio of wins to games.

18a. 5:6 b. 1:9

19. ✔ ; 20.73
    $23.94/kg
20. ✔ ; $3.33/cup
    $3.89/cup
21. $8.59/bag
    ✔ ; $6.72/bag
22. ✔ ; 1.15
    1.54
23. ✔ ; 0.03
    0.05
24. 15
    ✔ ; $25/h
    $18/h
25. $10.67/h
    ✔ ; $17.50/h
    $16/h
26. ✔ ; $20/h
    $12/h
    $18/h
27. 20% to 80% = 1:4
28. 3 to 7 = 3:7
29. 2 to (5 – 2) = 2:3
30. (4 + 4) to 4 = 2:1
    (4 + 8) to (4 + 4) = 3:2
31. 3 min – 20 s = 180 s – 20 s = 160 s
    120 m ÷ 160 s = 0.75 m/s = 45 m/min
32. Discount: $30.99 x 85% = $26.34
    $26.34 ÷ 24 cartons = $1.10/carton

7 Volume and Surface Area

Try It
120 ; 158

1. A: 143.5 cm^3 ; 174.4 cm^2
   B: 63 cm^3 ; 105 cm^2
   C: 57 cm^3 ; 126 cm^2
   D: 775.5 m^3 ; 569.2 m^2
   E: 33.06 m^3 ; 66.46 m^2
2. 27 cm^3 ; 54 cm^2
   a. V: 108 cm^3
   S.A.: 162 cm^2
   b. V: 108 cm^3
   S.A.: 144 cm^2

c. V: 162 cm^3 d. V: 135 cm^3
S.A.: 198 cm^2 S.A.: 198 cm^2

3a. F b. F c. F d. T e. T

4. V: 10.4 x 8.5 ÷ 2 x 5 + 10.4 x 3 x 5 = 377 (m^3)
S.A.: 10.4 x 8.5 ÷ 2 x 2 + 10.4 x 3 x 2 + (10.4 + 3 + 8.5 + 13.4 + 3) x 5 = 342.3 (m^2)

5. V: 19 x 4 x 7 + 16 x 4.5 x 7 = 1036 (m^3)
S.A.: 19 x 4 x 2 + 16 x 4.5 x 2 + (4 + 19 + 4 + 19 + 16 + 16) x 7 = 842 (m^2)

6. V: (14 + 23) x 10 ÷ 2 x 8 + (14 + 24) x 12 ÷ 2 x 8 = 3304 (cm^3)
S.A.: (14 + 23) x 10 ÷ 2 x 2 + (14 + 24) x 12 ÷ 2 x 2 + (11 + 23 + 11 + 13 + 24 + 13) x 8 = 1586 (cm^2)

7. V: (34 x 34 – (34 + 10) x 14.3 ÷ 2) x 6 = 5048.4 (cm^3)
S.A.: (34 x 34 – (34 + 10) x 14.3 ÷ 2) x 2 + (34 x 3 + 18.7 + 10 + 18.7) x 6 = 2579.2 (cm^2)

8a. 4 cm b. 64 cm^3

9a. 180 cm^2 b. 20 cm

10a. 24 cm b. 6 cm c. 648 cm^2

11. 4 12. 2.356 13. 1.73

14. 1 500 000 15. 7000 16. 600 000

17. > 18. > 19. < 20. <

21. < 22. = 23. = 24. <

8 Coordinates and Transformations

Try It

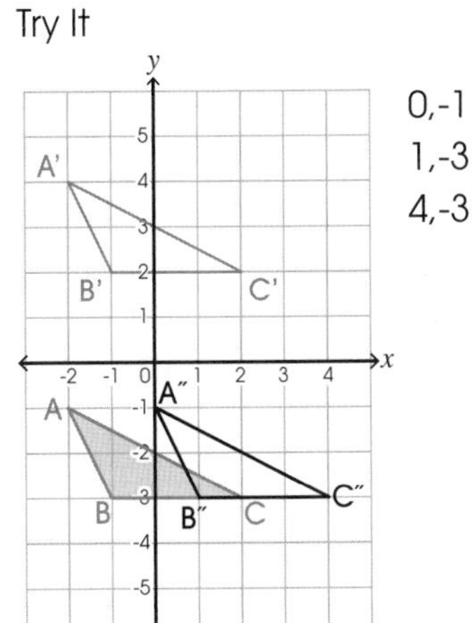

0,-1
1,-3
4,-3

1.

A(3,5) → A′(1,2)
B(1,4) → B′(-1,1)
C(3,1) → C′(1,-2)

2.

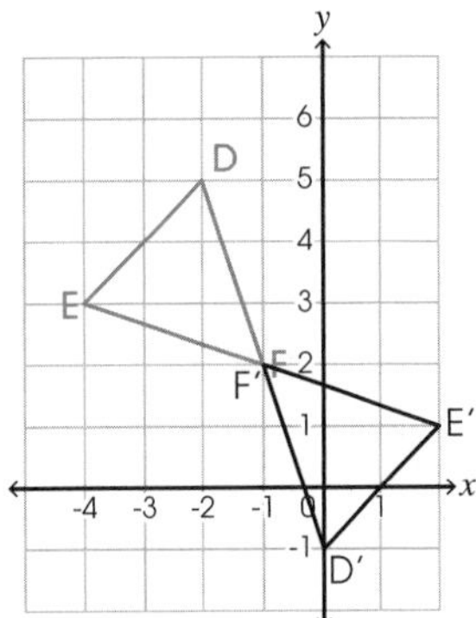

D(-2,5) → D′(0,-1)
E(-4,3) → E′(2,1)
F(-1,2) → F′(-1,2)

3.

G(-5,2) → G′(-5,-2)
H(-1,1) → H′(-1,-1)
I(-4,0) → I′(-4,0)

4.

a. (-8,5) → (-3,0)
(-5,1) → (-7,-3)
(-8,0) → (-8,0)

b. (1,5) → (-1,3)
(3,5) → (-3,3)
(1,1) → (-1,-1)

c. (3,-2) → (6,2)
(5,-2) → (8,2)
(6,-5) → (9,-1)
(1,-5) → (4,-1)

d. (-10,-3) → (10,-3)
(-8,-3) → (10,-5)
(-7,-6) → (13,-6)
(-9,-6) → (13,-4)

5.

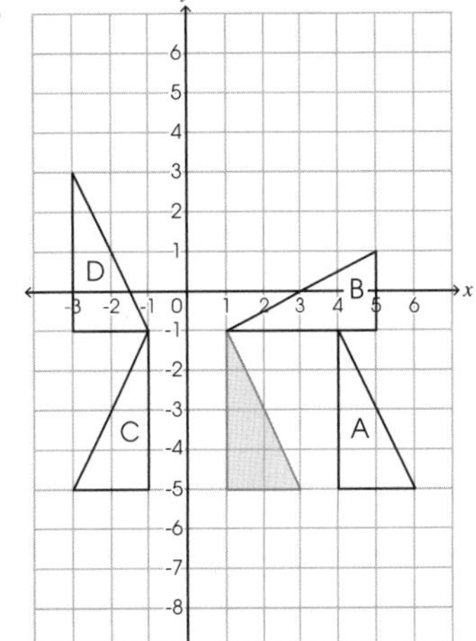

C ; A ; B ; D

6.

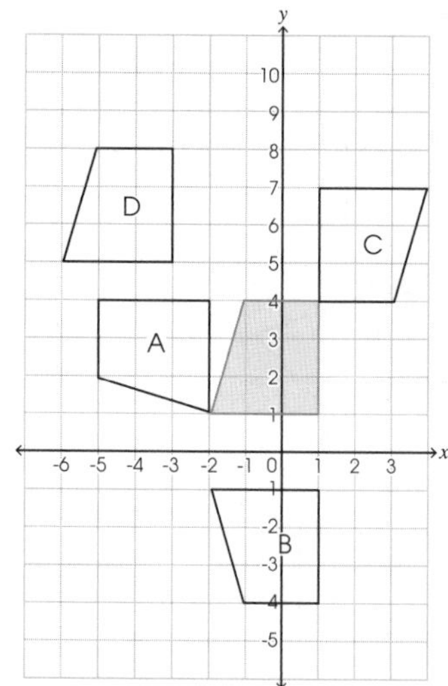

A ; B ; D ; C

7.

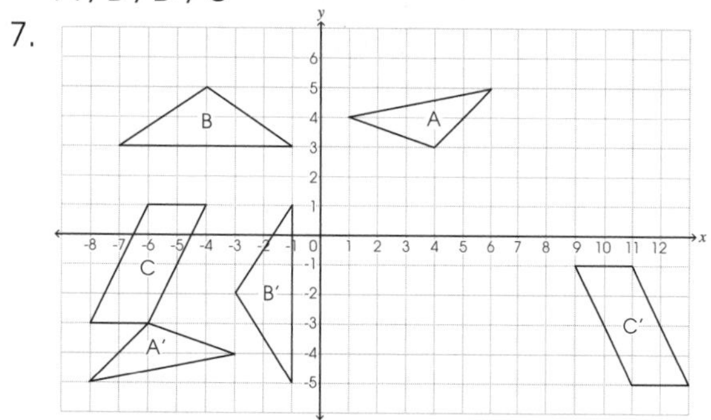

(Suggested answers)

a. rotate by a $\frac{1}{2}$ turn about (0,0) and then translate 2 units to the left

b. rotate $\frac{1}{4}$ counterclockwise about (0,2)

c. reflect in the y-axis and then translate 5 units to the right and 2 units down

8.

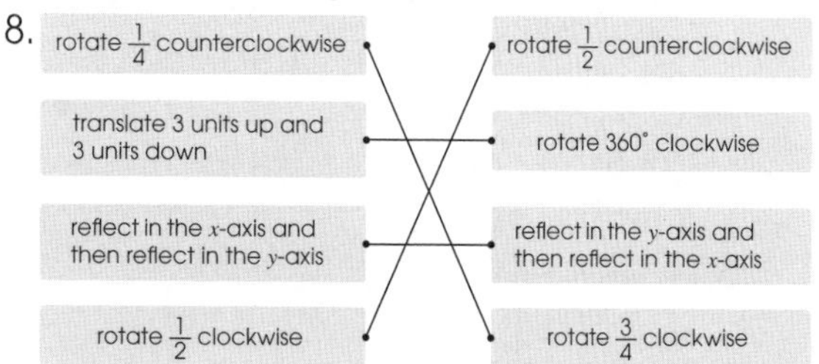

9 Algebraic Expressions

Try It

3 ; 5 ; 5 ; 7

1. 2 x 3 + 5
   = 11
   2 x 2 + 5
   = 9
2. (3 – 1) ÷ 2
   = 1
   (8 – 1) ÷ 2
   = 3.5
3. 9 x (6 – 2)
   = 36
   9 x (12 – 2)
   = 90
4. 10 – 2 x (1 + 1)
   = 6
   10 – 2 x (2 + 1)
   = 4
5. 4 + (6 – 2) ÷ 2
   = 6
   4 + (6 – 4) ÷ 2
   = 5
   4 + (6 – 6) ÷ 2
   = 4
   4 + (6 – 0) ÷ 2
   = 7
6. 3 x 2 + 10 ÷ 2
   = 11
   3 x 5 + 10 ÷ 5
   = 17
   3 x 1 + 10 ÷ 1
   = 13
   3 x 10 + 10 ÷ 10
   = 31

7a. = 6 x 2 – 2 x 5
    = 2
  b. = 5 x (3 – 1)
    = 10
  c. = 9 – 3 x 3 + 2 x 4
    = 8

8a. = 2 x (3 + 5)
    = 16
  b. = 6 + 2 x 5
    = 16
  c. = 8 x 1 + 4 x (3 – 2)
    = 12

9a. = 9 ÷ (10 – 1)
    = 1
  b. = 9 x 1 + 10
    = 19
  c. = 3 x (3 – 1) + 2 x (2 – 1)
    = 8

10. $2m + 3n$
  a. 2 x 8 + 3 x 5
    = 31
  b. 2 x 6 + 3 x 2
    = 18

11. $6(x - y)$
  a. 6 x (8 – 5)
    = 18
  b. 6 x (6 – 1)
    = 30

12a. B b. A

13a. A b. A

14a. B b. A

15a. A b. B

16a. B b. A

17a. $a + b$
  b. 20 + 5 = 25 ; 25
    15 + 10 = 25
    25 cookies

18a. $m(n - 1)$
  b. 4 x (1 – 1) = 0
    0 trees
    5 x (2 – 1) = 5
    5 trees

19a. $q(p \div 2)$
  b. 5 x (2 ÷ 2) = 5
    5 laps
    3 x (4 ÷ 2) = 6
    6 laps

20a. $(i \div 3) \times (j - 1)$
  b. (3 ÷ 3) x (4 – 1) = 3
    3 quilts
    (6 ÷ 3) x (5 – 1) = 8
    8 quilts

10 Equations

Try It

1 ; 1 ; 4

1. $a + 2 - 2 = 7 - 2$
   $a = 5$
2. $d - 4 + 4 = 11 + 4$
   $d = 15$
3. $3y \div 3 = 18 \div 3$
   $y = 6$
4. $x \div 6 \times 6 = 2 \times 6$
   $x = 12$
5. $n + 4 - 4 = 24 - 4$
   $n = 20$
6. $s - 16 + 16 = 2 + 16$
   $s = 18$

7. $4i \div 4 = 12 \div 4$
$i = 3$

8. $\frac{k}{9} \times 9 = 10 \times 9$
$k = 90$

9. $14n \div 14 = 18 \div 14$
$n = 1\frac{2}{7}$

10. $\frac{3}{4}y \div \frac{3}{4} = 24 \div \frac{3}{4}$
$y = 32$

11. $2x \div 2 = 16 \div 2$
$x = 8$

12. $20 - d - 20 = 11 - 20$
$-d = -9$
$d = 9$

13. $11 - e - 11 = 4 - 11$
$-e = -7$
$e = 7$

14. A
$a + 8 = 15$
$a + 8 - 8 = 15 - 8$
$a = 7$

15. B
$d - 15 = 9$
$d - 15 + 15 = 9 + 15$
$d = 24$

16. A
$3i = 15$
$3i \div 3 = 15 \div 3$
$i = 5$

17. B
$x \div 2 = 8$
$x \div 2 \times 2 = 8 \times 2$
$x = 16$

18-21. (Individual guess-and-check)

18. 2 19. 24 20. 3 21. 2

22. $b + 4 - 4 = 13 - 4$
$b = 9$
$9 + 4 = 13$
$10 + 3 = 13$

23. $m - 2 + 2 = 7 + 2$
$m = 9$
$9 - 2 = 7$
$8 - 1 = 7$

24. $k \div 2 \times 2 = 3 \times 2$
$k = 6$
$6 \div 2 = 3$
$6 - 3 = 3$

25. $p \times 4 \div 4 = 8 \div 4$
$p = 2$
$2 \times 4 = 8$
$16 \div 2 = 8$

26. $q - 5 + 5 = 4 + 5$
$q = 9$
$9 - 5 = 4$
$12 \div (2 + 1) = 4$

27. A 28. B 29. A 30. B

31. $x + 5 = 11$
$x + 5 - 5 = 11 - 5$
$x = 6$
6

32. $2x = 18$
$2x \div 2 = 18 \div 2$
$x = 9$
9

33. $x \div 5 = 6$
$x \div 5 \times 5 = 6 \times 5$
$x = 30$
30

34. $(x - 4) \div 2 = 7$
$(x - 4) \div 2 \times 2 = 7 \times 2$
$x - 4 = 14$
$x - 4 + 4 = 14 + 4$
$x = 18$
18

35. $3x + 5 = 35$
$3x + 5 - 5 = 35 - 5$
$3x = 30$
$3x \div 3 = 30 \div 3$
$x = 10$
10

11 Data Management

Try It
300 kg ; 200 kg

1. 100 kg 2. 50 kg 3. Friday
4. Monday 5. Sunny Town
6. A
a. Rain Town
b. Bridge Town and Silver Town

7.

Material	Percent	Angle
Paper	30%	108°
Glass	20%	72°
Plastic	25%	90°
Metal	15%	54°
Others	10%	36°

Recycling Materials

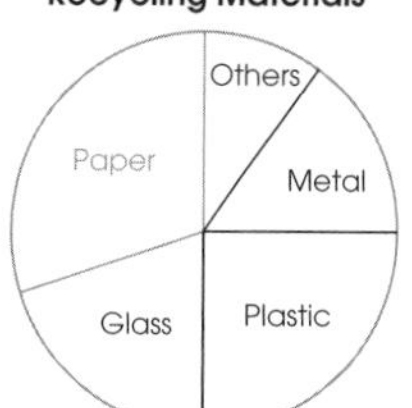

8a. 1500 kg b. 1000 kg
c. 1250 kg d. 750 kg

9a. histogram
b. (Suggested answers)
A bar graph contains space between bars. A histogram's bars represent a range of values and have no space between them.
c. 30 boxes d. 25 boxes e. 10 boxes
f. 80 boxes g. 10 boxes h. 10 boxes
i. 0 boxes j. 6 kg

10.

Weight (kg)	Frequency
56 – 60	||
61 – 65	|||
66 – 70	~~||||~~ ||
71 – 75	~~||||~~
76 – 80	~~||||~~ ~~||||~~ |||

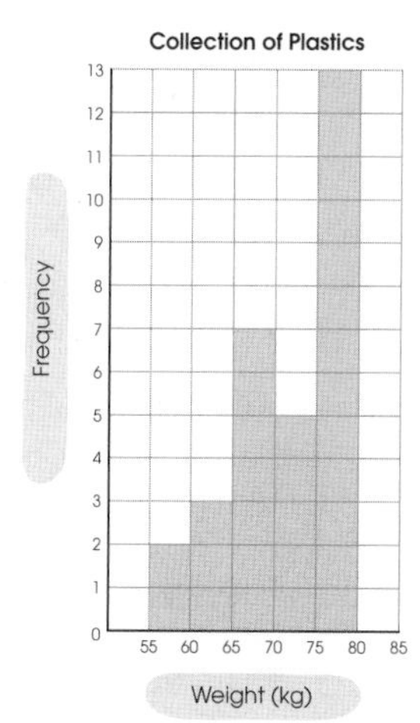

a. frequency
b. weight (kg)
c. 30 weeks
d. 74 kg
e. 76 – 80 kg is the most frequent. This implies that collecting 76 – 80 kg of plastics is more likely than any other weight range.

12 Probability

Try It

60% ; 50%

1. theoretical probability
2. experimental probability
3. theoretical probability
4. experimental probability
5. theoretical probability

6a. $\frac{1}{8}$; $\frac{3}{8}$

$\frac{1}{4}$; $\frac{1}{8}$

$\frac{7}{8}$; $\frac{1}{4}$

b. two 2s and one 3

c. any three cards that are not 2

7a. $\frac{1}{3}$; $\frac{1}{3}$

$\frac{1}{3}$; $\frac{2}{3}$

$\frac{2}{3}$

b. A

$\frac{1}{6}$; $\frac{1}{6}$; $\frac{1}{6}$; 0

8a. $\frac{1}{6}$

b. $\frac{1}{6}$

c. $\frac{5}{6}$

d. $\frac{1}{2}$

9a.

+	1	2	3	4	5	6
1	2	3	4	5	6	7
2	3	4	5	6	7	8
3	4	5	6	7	8	9
4	5	6	7	8	9	10
5	6	7	8	9	10	11
6	7	8	9	10	11	12

b. $\frac{1}{9}$; $\frac{1}{12}$; $\frac{31}{36}$

c. No, some sums are more likely to occur than others. For example, there is only one way to get a sum of 2 (i.e. 1 + 1) but there are multiple ways to get a sum of 5 (1 + 4, 2 + 3, etc.).

d. 7 is the most likely. The probability is $\frac{1}{6}$.

e. The probability is $\frac{1}{36}$.

f. The probability is $\frac{11}{36}$.

10a. The theoretical probability is $\frac{1}{6}$.

b. The experimental probability is $\frac{1}{4}$.

c. Yes, Rex has been lucky because his experimental probability of rolling doubles is greater than the theoretical probability.

11a. The probability is $\frac{1}{4}$ on each spinner.

b. He wants his mom to spin Spinner B.

c. No, the probability does not change. Each spin is independent of any previous spins.

Level 3

1 Exponents

Try It

10 000 000 ; 20 000 000 ; 20 000 000

1. 5^4 is 5 to the power of 4 or 5 x 5 x 5 x 5 and it involves repeated multiplication unlike 5 x 4.
2. $2^3 = 2 \times 2 \times 2 = 8$ ← must be a multiple of 2 (the base)

   Yes, any power is a multiple of its base since it is a product of its base.
3. $3^4 = 3 \times 3 \times 3 \times 3 = 81$

   $4^3 = 4 \times 4 \times 4 = 64$

   3^4 is greater.
4. $16 = 2^4 = 4^2$

   Yes, it is possible.
5. Yes. From inspection, $3^4 > 3^3$ since they have the same base and 4 > 3. Then $5^4 > 3^4$ since they have the same exponent and 5 > 3. So, 5^4 is the greatest.
6. A ; $100 \times 2^4 = 100 \times 16 = 1600$; 1600
7. A ; $2^9 = 512$; 512

8a. B ; $100\,000 \times 2^3 = 100\,000 \times 8 = 800\,000$; 800 000

b. B ; $100\,000 \div 2 = 50\,000$; 50 000

c. A ; $100\,000 \div 2^3 = 100\,000 \div 8 = 12\,500$; 12 500

9a. $3 \times 10^3 = 3 \times 1000 = 3000$; 3000

b. $3 \times 10^5 = 3 \times 100\,000 = 300\,000$; 300 000

c. $3\,000\,000 = 3 \times 1\,000\,000 = 3 \times 10^6$; 6

10a. $200 \times 3^3 = 200 \times 27 = 5400$; 5400

b. $5400 \div 2^3 = 5400 \div 8 = 675$; 675

c. $5400 \div 1350 = 4 = 2^2$; 2:00 p.m.

11. Drew's cube: $6^3 = 216$ (cm^3)

    Andy's cone: $2 \times 10^2 = 2 \times 100 = 200$ (cm^3)

    Drew's cube has a greater volume.
12. Cube A: $10 \times 10 \times 10 = 1000$ (cm^3)

    Cube B: $30 \times 30 \times 30 = 27\,000$ (cm^3)

    Cube C: $7 \times 7 \times 7 = 343$ (cm^3)

    Grace's cube: $27 \times 10^3 = 27\,000$ (cm^3)

    It is Cube B.

13a. Program A: $1 \times 2^9 = 512$
Program B: $1 \times 3^6 = 729$
Program C: $1 \times 4^5 = 1024$
It will take Program A 9 minutes, Program B 6 minutes, and Program C 5 minutes.
b. $1 \times 4^8 - 1 \times 3^8 = 65\,536 - 6561 = 58\,975$
Program C can send 58 975 more e-mails.
14. $\$1000 \times 2^{10} = \$1000 \times 1024 = \$1\,024\,000$.
The value would be $1 024 000.

2 Order of Operations

Try It
5 ; 5 ; 9 ; 45 ; 57

1. = 25 x 4
= 100
2. = 60 ÷ 12
= 5
3. = 21 ÷ (4 – 1)
= 21 ÷ 3
= 7
4. = 25 x 4 – 81 + 6
= 100 – 81 + 6
= 25
5. = 16 ÷ (2 x 9 – 10)
= 16 ÷ 8
= 2
6. = 81 ÷ 9 x 3
= 9 x 3
= 27
7. = 7 + 7 x 4
= 7 + 28
= 35
8. = 30 + (18 ÷ 9)
= 30 + 2
= 32
9. = 24 ÷ 6 x (2 + 16)
= 4 x 18
= 72
10. = 4 – 2 + 5 x 16 ÷ 8
= 4 – 2 + 10
= 12
11. = 19 – 2 + (36 ÷ 3) + 12
= 19 – 2 + 12 + 12
= 41
12. = 65 ÷ (9 – 4) + (25 x 3)
= 65 ÷ 5 + 75
= 88
13. = (7 – 4) x 16 ÷ (60 ÷ 5)
= 3 x 16 ÷ 12
= 4

14a. 10 + 12 ÷ 2 → 22 ÷ 2 → 11
It gave 11 as the answer.
b. 10 + 12 ÷ 2
= 10 + 6
= 16
No, it was incorrect. The correct answer is 16.
15. He is correct. The brackets are not necessary because multiplication comes before addition in the order of operations.
16. $2^2 + 3^2 = 4 + 9 = 13$
$(2 + 3)^2 = 5^2 = 25$
No, he is incorrect.
17. Way 1: (10 + 3) x 2 = 13 x 2 = 26
Way 2: 10 x 2 + 3 x 2 = 26
18a. No, it is incorrect because the answer is not 65.
b. (35 + 3) ÷ 19 + 7 x (22 – 13) = 65
19. A
(6 – 1) x 3
= 5 x 3
= 15
15
20. A
100 – (23 x 4)
= 100 – 92
= 8
8
21. B
(19 x 3 – 5) ÷ 2
= (57 – 5) ÷ 2
= 52 ÷ 2
= 26
26
22. A
17 x (6 + 5) + 20 x 4
= 17 x 11 + 20 x 4
= 187 + 80
= 267
267
23. B
Shorter call: (46 – 10) ÷ 2 = 36 ÷ 2 = 18
Longer call: 46 – 18 = 28
18 ; 28
24. (200 ÷ 2) – 90 = 100 – 90 = 10
10 brown-eyed Grade 7 students do not have brown hair.
25. ($100 – $25 – $15) ÷ 2 = $60 ÷ 2 = $30
The price of the less expensive pair was $30.
26. ($106 – $55) ÷ 3 + $5
= $51 ÷ 3 + $5
= $17 + $5
= $22
The price of each T-shirt before the sale was $22.
27. (4000 – 1200) – 1800 = 2800 – 1800 = 1000
The difference is 1000 m.
28a. $6 x 5 + $7 x 8 – $50 = $30 + $56 – $50 = $36
Her profit was $36.
b. New profit: $7 x (5 + 8) – $50 = $7 x 13 – $50
= $91 – $50 = $41
Difference: $41 – $36 = $5
She would have earned $5 more.

3 Squares and Square Roots

Try It
15 ; 15

1. $30^2 = 900$
The area of the frame is 900 cm^2.
2. Side length: $\sqrt{9} = 3$
Perimeter: 3 x 4 = 12
He needs 12 m of lumber.

3. Side length: $100 \div 4 = 25$
   Area: $25^2 = 625$
   The area of the largest square is 625 cm^2.
4. Area of 1 square: $3200 \div 2 = 1600$
   Side length: $\sqrt{1600} = 40$
   40 cm 40 cm
   40 cm
   The dimensions are 80 cm by 40 cm.
5. $\sqrt{289} - \sqrt{121} = 17 - 11 = 6$
   The difference is 6 m.
6. $(2 + 3)^2 - 3^2 = 5^2 - 3^2 = 25 - 9 = 16$
   The area is 16 m^2.
7. $20^2 - 18^2 = 400 - 324 = 76$
   The area of the frame is 76 cm^2.

8a. Side length of big square: $\sqrt{576} = 24$
   Side length of small square: $24 \div 2 = 12$
   Area of small square: $12^2 = 144$
   The area is 144 cm^2.
b. $576 \times 2 + 144 \times 7 = 2160$
   The area is 2160 cm^2.
c. $24 \times 3 + 12 \times 10 = 192$
   The perimeter is 192 cm.

9. Area of a square: $48 \div 3 = 16$
   Side length of a square: $\sqrt{16} = 4$

   The dimensions are 12 km by 4 km.
10. Side length of large square: $\sqrt{49} = 7$
   Side length of small square: $\sqrt{36} = 6$
   Perimeter: $7 \times 3 + 6 \times 3 + (7 - 6) = 40$
   The perimeter is 40 m.
11. $(6 + 1 + 1)^2 - (1^2 \times 4) = 8^2 - (1 \times 4) = 64 - 4 = 60$
   The area is 60 m^2.
12. $\sqrt{680} \times 4 = 26 \times 4 = 104$
   A player needs to run 104 m.
13. $\sqrt{53\,000} = 230$
   Each side is about 230 m long.
14. $\sqrt{48\,500} \times 4 = 220 \times 4 = 880$
   The perimeter is approximately 880 m.

15a. 1, 4, 9, 16, 25, 36, 49, 64, 81, 100
b. 3, 5, 7, 9, 11, 13, 15, 17, 19
c. The pattern starts at 3 and increases by 2 each time.

16. (Suggested examples)
a. $4 \times 9 = 36$
   Yes, the answer is always a perfect square.
b. $4 \times 10 = 40$
   40 is not a perfect square. She is not correct.
c. $10 \times 40 = 400$
   400 is a perfect square. He is correct.
d. $\sqrt{9 + 16} = \sqrt{25} = 5$
   She is not correct.

4 Integers

Try It
-22 ; -12 ; -10 ; 10

1a. always ; 2 + 3 = 5 b. never ; -2 + (-3) = -5
c. never ; 2 > -3
d. sometimes ; 2 + (-3) = -1
e. always ; -2 > -3

2. A ; 8
3. A ; lost ; 3

4a.

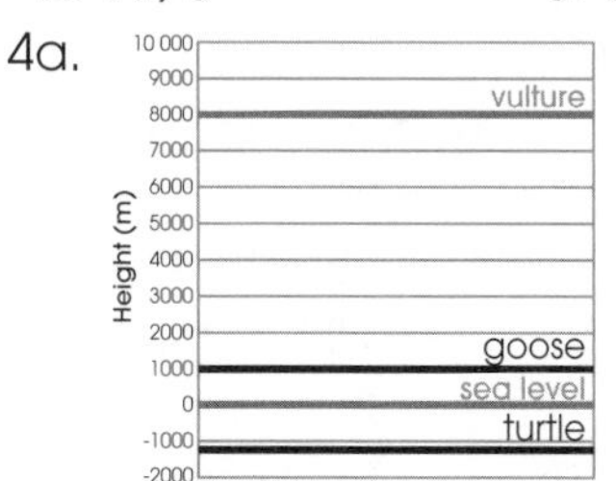

b. $8000 - 1000 = 7000$
   It is 7000 m higher.
c. $8000 - (-1200) = 9200$
   It is 9200 m higher.
d. $8000 + (-1500) = 6500$
   The prey was 6500 m high.

5a. Mount Logan: +5900 ;
   Mariana Trench: -11 000
b. (-11 000) + (+5900) = -5100
   It would not be seen as an island because its height would be negative, meaning its peak would be below sea level.

6a. (-2) + (+2) + (-3) + (+2) + (-3) = -4
   The overall price change was -\$4.
b. Monday: $23 \times 1000 = 23\,000$
   Friday: $(23 + (-4)) \times 1000 = 19 \times 1000 = 19\,000$
   Change: $19\,000 - 23\,000 = -4000$
   He lost \$4000.

7a. Mercury can get the coldest.
b. (-90) - (-123) = 33
   The difference is 33°C.
c. -90°C > -123°C > -184°C

8a. (-\$35) - (-\$19) = -\$16
   He has -\$16 in Account B.
b. (-\$16) - (-\$19) = \$3
   The difference is \$3.

9a. (-4) - (-5) = 1
   The difference was 1°C.

b. $(-4) - 7 = -11$
The temperature was -11°C.

10a. $2 - (-16) = 18$
He got 18 more points in Round 2.

b. $(-16) + 15 + (-9) = -10$
They got -10 points in Round 1 in all.

c. $29 - 2 - 31 = -4$
His score was -4 in Round 2.

11a. $0 + (+2) + (-1) + (-2) + 0 + (+1) + (+1) + (-1) + (+2) = +2$
His score was over par.

b. $36 + (+2) = 38$
His score was 38.

12. $(-2) + (-7) = -9$
$(-2) - (-7) = 5$
The integers are -2 and -7.

5 Fractions

Try It
$\frac{1}{2}$; 10 ; 5 ; 5

1a. $\frac{1}{4} \times 48 = 12$; 12 b. $\frac{3}{8} \times 48 = 18$; 18

2a. $\frac{2}{5} \times 80 = 32$; 32 b. $\frac{7}{10} \times 80 = 56$; 56

3a. $\frac{1}{2} \times 1000 = 500$; 500

b. $\frac{18}{20} \times 1000 = 900$; 900

4a. $1\frac{1}{2} + 1\frac{4}{5} = 3\frac{3}{10}$
He spent $3\frac{3}{10}$ h playing soccer altogether.

b. $3\frac{3}{10} - 2\frac{3}{4} = \frac{11}{20}$
He spent $\frac{11}{20}$ h less on soccer.

5. $\frac{5}{12} \times 8 = 3\frac{1}{3}$
It costs $\$3\frac{1}{3}$.

6a. $\frac{2}{3} \times 12 = 8$
There are 8 apples.

b. $\frac{2}{3} \times 1500 = 1000$
They weigh 1000 g altogether.

7. $12 \div 1\frac{1}{4} = 9\frac{3}{5}$
She needs $9\frac{3}{5}$ planks.

8. $13 \div 3\frac{1}{4} = 4$
She can make 4 dresses.

9. $90 \div 4\frac{1}{4} = 21\frac{3}{17}$
21 textbooks will fit on the shelf.

10. $1 - \frac{1}{3} - \frac{1}{4} - \frac{1}{5} = \frac{13}{60}$
$\frac{13}{60}$ of the cars are neither white, black, nor red.

11. $(\frac{1}{4} + \frac{3}{4}) \div 2 = \frac{1}{2}$
The mean height is $\frac{1}{2}$ m.

12. $\frac{3}{4} \times 240 + \frac{2}{3} \times 210 = 320$
She ate 320 g of chocolate.

13. $10 \times 2 \times \frac{3}{5} = 12$
12 people are in the classroom.

14. $\frac{3}{5} \times 20 + \frac{5}{6} \times 30 = 37$
She got 37 correct answers.

15a. $\frac{20 + 15}{66} = \frac{35}{66}$
$\frac{35}{66}$ of the Grade 7 students got an "A" on the Math test.

b. $66 \div \frac{2}{17} = 561$
There are 561 students in the school.

c. $\frac{1}{2} \times 30 = 15$
15 students in Class A are boys.

d. $(30 - \frac{1}{2} \times 30) + (\frac{1}{3} \times 36) = 27$
27 Grade 7 students are girls in total.

e. $\frac{1}{3} \times 60 = 20$
A Grade 7 student spends 20 minutes on Math on average.

f. $(\frac{1}{5} + \frac{1}{4} + \frac{1}{3} + \frac{1}{4}) \times 60 = 62$
A Grade 7 student spends 62 minutes on homework each day on average.

6 Decimals

Try It
\$5.99 ; \$5.99

1a. \$7.98 x 3 = \$23.94
She pays \$23.94.

b. \$8.95 x 2 = \$17.90
She pays \$17.90.

c. \$23.94 + \$17.90 = \$41.84
They pay \$41.84 altogether.

d. \$50 - \$41.84 = \$8.16
They will get \$8.16 in change.

e. \$41.84 - \$35 = \$6.84
They need \$6.84 more.

2. \$12.30 ÷ \$2.50 = 4.92
You can buy 4 comic books.

3. 320 km ÷ 100 km x 7 L x $1.02/L = $22.848
   His weekly gas expense is $22.85.
4. $100 – $15.49 x 5 = $22.55
   He had $22.55 left.

5a. Cost before tax: $16.99 x 3 + $24.25 x 2 = $99.47
   Tax: $99.47 x 0.15 = $14.92
   Total: $99.47 + $14.92 = $114.39
   He spent $114.39.

b. $150 – $114.39 = $35.61
   He got $35.61 in change.

6. $4.75 ÷ 8 = $0.59375
   Each of them got $0.59.
7. $3.48 ÷ $0.05 = 69.6
   She can have a maximum of 69 nickels.
8. 3.73 x 150 = 559.5
   They are 559.5 km apart.
9. 365.3 x 1.88 = 686.764
   A Mars year is 686.764 days long.

10a. 70.83 ÷ 48 = 1.48
   Each orbit took about 1.48 h.

b. 10 ÷ 1.48 = 6.8
   She made about 6.8 orbits in 10 hours.

11. $4.5 \div 1\frac{1}{4} = 4.5 \div 1.25 = 3.6$
   The original number is 3.6.
12. 0.357
   4 ÷ 13 = 0.308
   4 ÷ 12 = 0.333
   5 ÷ 16 = 0.313
13. 5 ÷ 17 = 0.294
   His new batting average was 0.294.
14. 0.375 x 16 = 6
   6 – 4 = 2
   He got 2 more hits.
15. 5 ÷ 0.333 = 15
   15 – 12 = 3
   He was at bat 3 times that day.
16. (5 + 4 + 4 + 5) ÷ (14 + 13 + 12 + 16) = 18 ÷ 55 = 0.327
   The batting average of the four batters was 0.327.
17. (35.2 + 27.4 + 48.7) ÷ 65.5 = 111.3 ÷ 65.5 = 1.699
   It will take him 1.699 h.
18. (12.5 + 28.3 + 27.4 + 48.7) ÷ 2.2 = 116.9 ÷ 2.2 = 53.136
   His average speed is 53.136 km/h.
19. Distance: 70 x 2.9 = 203
   52.8 + 48.7 + 48.7 + 52.8 = 203
   He took the route from Huntsville to Greenpark via Brownsville and back.
20. (52.8 + 12.5 + 28.3 + 27.4) ÷ 1.7 = 121 ÷ 1.7 = 71.176
   Her average speed was 71.176 km/h.
21. $35.2 \times \frac{3}{5} \div 65 + 35.2 \times \frac{2}{5} \div 40 = 0.325 + 0.352 = 0.677$
   His trip was 0.677 h long.

7 Percents

Try It
0.1 or $\frac{1}{10}$; 5 ; 5

1a. 50 x 20% = 10 ; 10

b. 50 x 10% = 5 ; 5

c. $40 x 80% = $32 ; 32

d. $40 x 15% = $6 ; 6

2. 170 x 30% = 51 ; 51

3a. $36 – $36 x 25% = $36 – $9 = $27 ; 27

b. $27 + $27 x 10% = $27 + $2.70 = $29.70 ; 29.70

4a. $80 – $80 x 20% = $80 – $16 = $64 ; 64

b. $64 – $64 x 25% = $64 – $16 = $48 ; 48

5a. $40 + ($40 – $40 x 50%) = $40 + $20 = $60 ; 60

b. $60 + $60 x 5% = $60 + $3 = $63 ; 63

6. Sale: 12 ; 24.23 ; 24.15 ; 21.39
   Cap: 80% ; 80% ; 12
   T-shirt: 100% – 15% = 85%
   $28.50 x 85% = $24.23
   Shorts: 100% – 30% = 70%
   $34.50 x 70% = $24.15
   Soccer ball: 100% – 35% = 65%
   $32.90 x 65% = $21.39

7a. $42.59 +$42.59 x 15% = $48.98

b. $22.99 + $22.99 x 15% = $26.44

c. $45.50 + $45.50 x 15% = $52.33

d. $65.50 + $65.50 x 15% = $75.33

8. $45 + $45 x 9% = $49.05
   She paid $49.05.
9. $9 ÷ 15% = $60
   She paid $60 before tax.
10. $8.50 ÷ 20% = $42.50
   The price was $42.50 before tax.
11. $2 ÷ 10% = $20
   His old allowance was $20.
12. 34 ÷ 40% = 85
   There are 85 national parks in North America.
13. 210 ÷ 70% = 300
   There are 300 calories in a piece of regular cheese.

14. Tom: $64 \times 75\% = 48$
John: $20 + 64 \times 25\% = 36$
Tom will have 48 cards and John will have 36 cards.

8 Perimeter and Area

Try It
3×2 ; 2×2 ; 10 ; 10

1. $4 + 3 + 2 + 1 + 2 + 2 = 14$
14 m of fencing is needed.

2a. $(3 + 1.4) \times 2 = 8.8$
The perimeter is 8.8 m.

b. Square: $1.4 \times 1.4 = 1.96$
Rectangle: $(3 \times 1.4 - 1.96) \div 2 = 1.12$
The white square has an area of 1.96 m^2 and each red rectangle has an area of 1.12 m^2.

3. Perimeter: $12 + 13 + 3 + 4 = 32$
Area: $12 \times 5 \div 2 + 3 \times 4 \div 2 = 36$
The perimeter is 32 cm and the area is 36 cm^2.

4a. Length: $50 \times 3 = 150$
Width: $50 \times 2 = 100$
The length is 150 cm and the width is 100 cm.

b. Perimeter: $(150 + 100) \times 2 = 500$
Area: $150 \times 100 = 15\,000$
The perimeter is 500 cm and the area is 15 000 cm^2.

5a. $50 \times 6 = 300$
Its perimeter is 300 cm.

b. $88 \times 50 + (88 \times 24 \div 2) \times 2 = 6512$
The area is 6512 cm^2.

6. Base: $5 \div 2 = 2.5$
Height: $6 \div 2 = 3$
Total area: $2.5 \times 3 \div 2 \times 3 = 11.25$
The total area is 11.25 cm^2.

7. Area of small triangle:
$(2 - 1.2) \div 2 \times (2 \div 2) \div 2 = 0.2$
Area of hexagon: $2 \times 2 - 0.2 \times 4 = 3.2$
The area of the shaded hexagon is 3.2 m^2.

8a. $(36 + 30) \times 30 \div 2 \times 2 = 1980$
The total area of its front and back is 1980 cm^2.

b. $(48 + 44) \times 30 \div 2 \times 2 = 2760$
The total area of its sides is 2760 cm^2.

9a. $10 \times 2 \times 8 = 160$
160 cm^2 of fabric is needed.

b. $(10 + 3) \times 8 = 104$
104 cm of lace trimming is needed.

10. 10 cm 5 cm
Area of square: $5 \times 5 = 25$
Height of triangle: $25 \times 2 \div 10 = 5$
The height of the triangle is 5 cm.

11a. Fencing: $16 + 10 + 16 + 1 + (10 - 1 - 7) = 45$
Cost: $\$15 \times 45 = \$675$
It cost \$675.

b. $(7 \times 9) - (5 \times 7) = 28$
The area is 28 m^2.

c. $\$12 \times (16 \times 10 - 8 \times 1 - 7 \times 9) = \$1068$
The sod cost \$1068.

12a. $\$70 \times 5 \times 4 = \$1400$
It will cost \$1400 in total.

b. Area to be painted:
$4 \times 2.5 \times 2 + 5 \times 2.5 \times 2 + 5 \times 4 - 7 = 58$
Cost: $\$8.99 \times 58 \div 5 = \$104.28$
The paint will cost \$104.28.

9 Volume and Surface Area

Try It
12 ; 22 ; 7 ; 1848 ; 1848

1. $30 \times 30 \times 6 = 5400$
5400 cm^2 of cardboard is needed.

2. Area of 1 face: $54 \div 6 = 9$
Side length: $\sqrt{9} = 3$
The side length of the cube is 3 cm.

3. $12 \times 12 \times 12 = 1728$
1728 cm^3 = 1728 mL = 1.728 L
It can hold 1.728 L of water.

4. $5 \times 5 \times 2 + 5 \times 35 \times 4 = 750$
The area to be painted is 750 cm^2.

5.

a. $12 \div 4 = 3$
$12 - 3 - 3 = 6$
The dimensions of the prism are 3 cm by 3 cm by 6 cm.

b. $3 \times 3 \times 2 + 3 \times 6 \times 4 = 90$
The surface area is 90 cm^2.

c. $3 \times 3 \times 6 = 54$
The volume is 54 cm^3.

6. $20 \times 30 \times 2 + 20 \times 8 \times 2 + 30 \times 8 \times 2 = 2000$ (cm^2)
1 m^2 = 10 000 cm^2
Yes, she has enough paper.

7a. The surface area is 4 times that of the original cube.

b. The volume is 8 times that of the original cube.

8. Side length: $\sqrt{216 \div 6} = 6$
   Volume: 6 x 6 x 6 = 216
   The volume of the cube is 216 m^3.

9a. 1 x 1.2 x 2 + 1 x 1.5 x 2 + 1.2 x 1.5 x 2 = 9
   9 m^2 of material was required.

b. 1 x 1.2 x 1.5 = 1.8
   It can hold 1.8 m^3 of compost.

10. Area painted: 3 x 5 x 2 + 3 x 4 x 2 – 3 = 51
    Paint needed: 51 x 2 ÷ 36 = 2.83
    3 cans of paint are needed.

11. Volume of dice: 2 x 2 x 2 = 8
    Volume of box: 10 x 10 x 10 = 1000
    No. of dice: 1000 ÷ 8 =125
    125 dice can be placed in the box.

12. 16 x 23 x 3 x 30 + 16 x 23 x 1.5 x 20 = 44 160
    The minimum volume of the container is 44 160 cm^3.

13. Box A: \$11.99 ÷ (17 x 30 x 30)
    = \$0.000784/cm^3
    Box B: \$6.99 ÷ (15 x 25 x 25)
    = \$0.000746/cm^3
    Box B is a better buy because it is cheaper for the same amount of detergent.

14. Volume: 31 x 20 x 7 = 4340
    No. of servings: 4340 ÷ 175 = 24.8
    It contains 24.8 servings.

15. Rate of a worker: 3 x 3 x 3 ÷ 1 = 27 (m^3/h)
    Time needed: 6 x 6 x 6 ÷ 27 ÷ 2 = 4
    It will take 4 hours.

16. Volume of water: 5 x 5 x 12 = 300
    Volume of pyramid: 300 x $\frac{1}{3}$ = 100
    Rise in water level: 100 ÷ (5 x 5) = 4
    New water level: 12 + 4 = 16
    The new water level will be 16 cm high.

17. 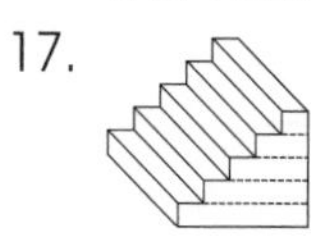

    1.5 x 1.5 x 0.2 + 1.5 x 1.2 x 0.2 + 1.5 x 0.9 x 0.2 + 1.5 x 0.6 x 0.2 + 1.5 x 0.3 x 0.2 = 1.35
    The volume of cement needed is 1.35 m^3.

18. Side length of square: $\sqrt{25} = 5$
    Length of box: 5 – 1 – 1 = 3
    Volume of box: 3 x 3 x 1 = 9
    Capacity of box: 9 cm^3 = 9 mL
    The capacity of the box is 9 mL.

10 Congruence and Similarity

Try It
congruent ; similar

1. neither
2. congruent
3. similar
4. similar
5. neither
6. congruent
7. 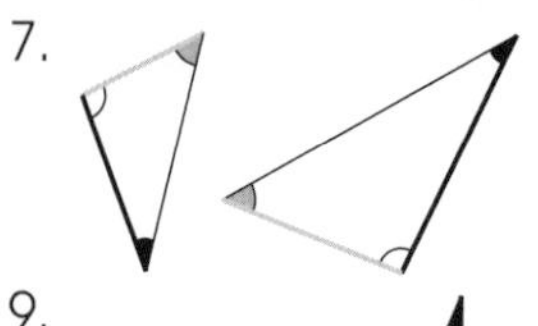
8. 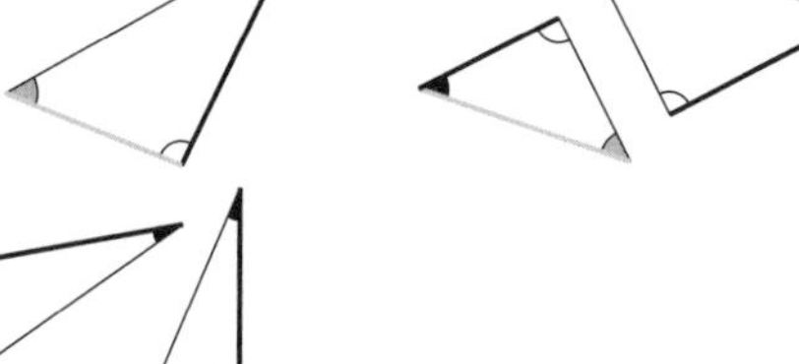
9.

10. $\frac{2}{3}$; $\frac{5}{7.5} = \frac{2}{3}$; $\frac{4}{6} = \frac{2}{3}$; is
11. $\frac{3}{6} = \frac{1}{2}$; $\frac{2}{4} = \frac{1}{2}$; $\frac{4}{8} = \frac{1}{2}$; is
12. $\frac{1}{2}$; $\frac{2}{3}$; $\frac{2.5}{4.5} = \frac{5}{9}$; is not
13. F
14. T
15. F
16. T
17. T
18. T
19. F
20. T
21. Yes, the triangles are congruent. There is only one way to draw a triangle with 3 given sides.
22. No, the triangles are probably not congruent. The sides of the triangles can be any length with the given set of angles.
23. Yes, the triangles are similar. All equilateral triangles are similar because their angles are all the same size (60°).
24. $\frac{3}{3.6} = \frac{5}{6}$ $\frac{7}{8.4} = \frac{5}{6}$
    Yes, the pools are similar because the sides are proportional and their angles are the same.
25. Ratio of length: $\frac{37}{26}$
    Ratio of width: $\frac{27}{21} = \frac{9}{7}$
    The pictures are not similar. The lengths and widths are not proportional.
26. Way 1: 3 ÷ 2 x 1.2 = 1.8
    Way 2: 3 ÷ 1.2 x 2 = 5
    The lengths of the other 2 sides can be 1.8 cm or 5 cm.
27. 5 ÷ 7.2 x 10.8 = 7.5
    The side length is 7.5 cm.
28. $\frac{6}{9} = \frac{2}{3}$ $\frac{4}{6} = \frac{2}{3}$
    The pictures are similar. The corresponding sides are proportional.

11 Transformations and Tiling

Try It

A ; B

1.

2.

3.

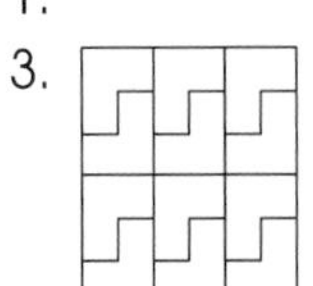

4.

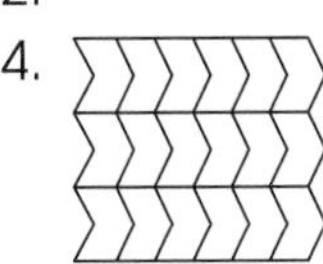

5.

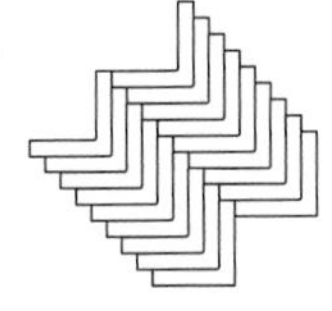

6.

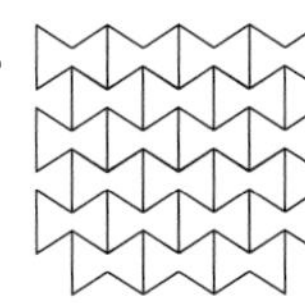

7.

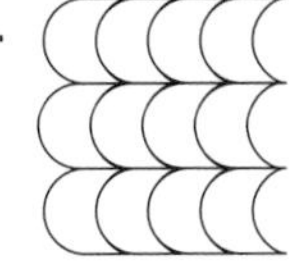

8.

9.

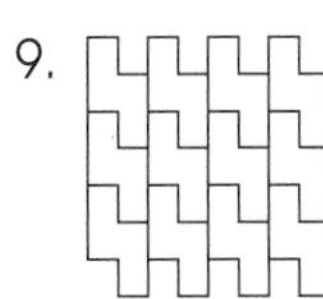

10. translation ; rotation ; reflection

11.

a. reflection b. rotation c. reflection

12.

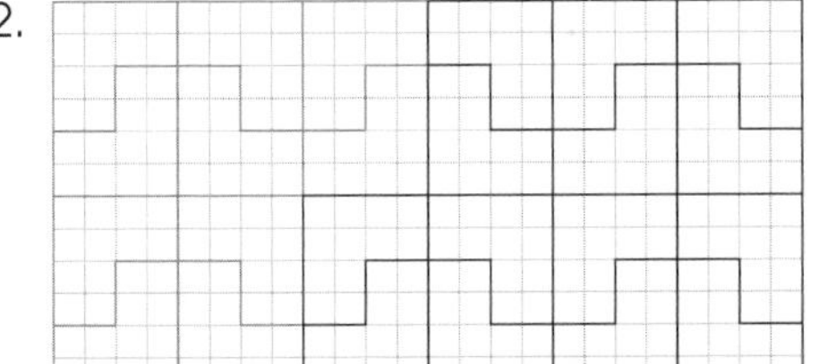

a. translation b. reflection/rotation

c. rotation

13. (Suggested drawing)

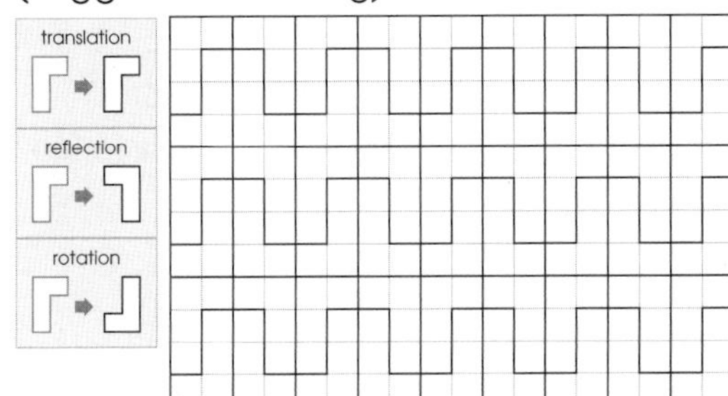

14. It is a translation, a reflection, and then a rotation.

15. (Suggested answer)

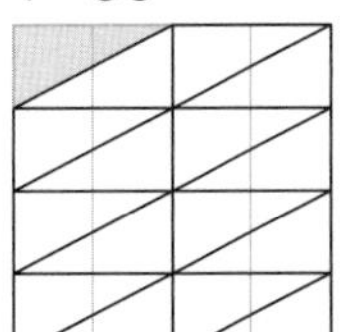

The triangle is translated and rotated.

16. (Suggested answer)
The tiles are reflected horizontally and vertically.

17. Shape D will not be considered because it will leave gaps in the tiling pattern.

18. Figures A and E can cover a flat surface without gaps or overlaps.

19. He should add square bricks to fill the gaps among the octagons.

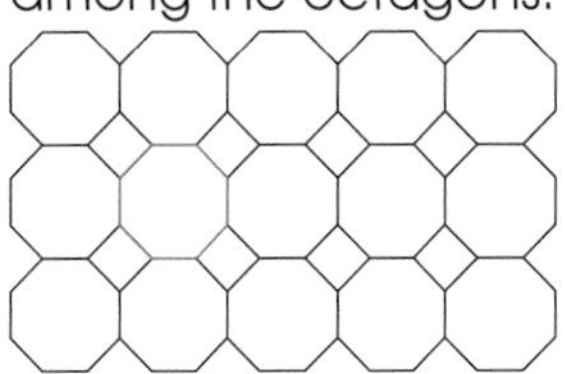

12 Patterning

Try It

$2n$; 20

1. 7 ; 8 ; 3 ; 3
2. 12 ; 15 ; 3 ; 3
3. 7 ; 9 ; Multiply the term number by 2 and subtract 1. ; $2n - 1$
4. 3 ; 4 ; Subtract 1 from the term number. ; $n - 1$
5. 16 ; 20 ; Multiply the term number by 4. ; $4n$
6. 10 ; 11 ; Add 6 to the term number. ; $n + 6$
7. 9 ; 11 ; $2n + 1$
8. 7 ; 5 ; $15 - 2n$

7. a.

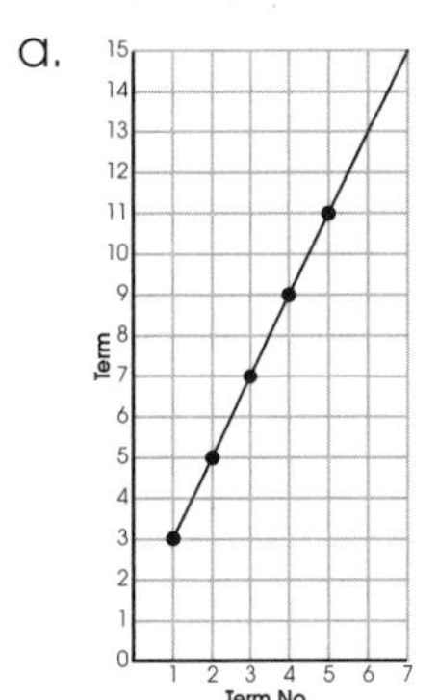

8. a.

7. b. 15

8. b. 1

9. 400 ; 500

a. 400 g ; 500 g

b. Week 6 ; Week 8

10. 10 ; 12

a. \$12 ; \$14

b. 7 km ; 9 km

11. 17 ; 21

a. 33 g ; 41 g

b. Day 7 ; Day 11

12a. Value: $n \times 3 + 2$
$n = 3$; Value: $3 \times 3 + 2 = 11$
The card will be worth $11 after the third month.
b. $n = 6$; Value: $6 \times 3 + 2 = 20$
The card will be worth $20 after the sixth month.
13a. No. of stamps: $20(n - 1) + 19$
$n = 5$;
No. of stamps: $20(5 - 1) + 19 = 99$
She will collect 99 stamps in the fifth year.
b. $20(n - 1) + 19 = 139$
$n = 7$
She will collect 139 stamps in the seventh year.
14a. Value: $x + 3n$
$45 = x + 3 \times 6$
$45 = x + 18$
$x = 27$
The teapot was $27.
b. Value: $27 + 3n$
Double the value it was sold: $45 \times 2 = 90$
$90 = 27 + 3n$
$63 = 3n$
$n = 21$
No. of years: $21 - 6 = 15$
It will be worth double that value in 15 years.

13 Equations

Try It
$2x + 1 = 11$
$2x = 10$
$x = 5$
5

1. x
$x + 6 + 8 = 24$
$x = 10$
The length of the third side is 10 cm.

2. x
$3x - 15 = 21$
$3x = 36$
$x = 12$
Her number is 12.

3. x be the length of the service
$30 + 25x = 80$
$25x = 50$
$x = 2$
The service was 2 hours long.

4. x be the snowfall in Sunville
$5x - 10 = 360$
$5x = 370$
$x = 74$
Sunville's snowfall was 74 cm.

5. Let x be the distance travelled.
$20 + 1x = 50$
$x = 30$
They travelled 30 km.

6. Let x be the original number.
$(x - 15) \times 2 = 22$
$x - 15 = 11$
$x = 26$
His original number was 26.

7. Let x be the time needed.
$(-80) + 2x = -14$
$2x = 66$
$x = 33$
It will take 33 seconds.

8. Let x be the number of marbles in each box.
$2x + 5 = 39$
$2x = 34$
$x = 17$
There are 17 marbles in each box.

9. Let x be the entrance fee per person.
$4x + 22 = 78$
$4x = 56$
$x = 14$
The entrance fee per person was $14.

10. Let x be the cost of a poster.
$3x + 2 \times 21 + 13 = 100$
$3x + 55 = 100$
$3x = 45$
$x = 15$
A poster was $15.

11. Let x be the distance Zoe jogs.
$2x + 2 = 8$
$2x = 6$
$x = 3$
She jogs 3 km.

12. Let x be the number of yoga classes.
$2x + 24 = 40$
$2x = 16$
$x = 8$
She went to 8 yoga classes.

13. Let x be the number of hours worked.
$16x + 50 = 370$
$16x = 320$
$x = 20$
She worked 20 hours.

14. Let x be the number of nickels.
$4x + 500 = 700$
$4x = 200$
$x = 50$
He has saved 50 nickels.

15. Let x be the wattage of each surrounding bulb.

$8x + 120 = 600$
$8x = 480$
$x = 60$

The maximum allowable wattage of each surrounding bulb is 60 W.

16. Let x be the bigger number.

$x + x - 2 = 40$
$2x - 2 = 40$
$2x = 42$
$x = 21$
$x - 2 = 21 - 2 = 19$

The numbers are 19 and 21.

17. Let x be the number of months.

$2x + 20 = 50$
$2x = 30$
$x = 15$

It will take 15 months.

18. Let x be the number of girls.

$x + 2x - 31 = 83$
$3x - 31 = 83$
$3x = 114$
$x = 38$
$2x - 31 = 2 \times 38 - 31 = 45$

There are 45 boys and 38 girls.

19a. Let x be the cost of each tea bag.

$24x + 1 = 4$
$24x = 3$
$x = 0.125$

Each tea bag costs \$0.125 to produce.

b. 75 x 0.125 + 1 = 10.375
It costs \$10.375.

14 Data Analysis

Try It

Stem	Leaf
14	4, 7, 8
15	0, 3, 4, 5, 5, 6, 8
16	0, 2, 3, 5
17	0

156 ; 155 ; 155

1a. 25 h ; 24 h ; 35 h
b. This is a sample because it does not examine all 13-year-old children in Calgary.
2a. The mean is 19 shots. ; The median is 18 shots. ; The mode is 18 shots.
b. This set is a census because it includes data from all the games played last month.
c. This is a set of primary data because the data was collected first-hand.
d. Yes, the mean would increase because 29 is greater than the mean.
Yes, the median would increase because the middle value would shift.
No, the mode would remain the same because 18 would still be the most frequent data value.

3-8. (Individual answers)

3. 2, 2, 5, 5, 5
4. 2, 2, 2, 5, 5
5. 1, 1, 1, 1, 5
6. 1, 1, 5, 5, 5
7. 1, 3, 3, 3, 5
8. 4, 4, 7, 10, 10

9a. The mean is \$39 750, the median is \$30 000, and the mode is \$30 000.
b. The mean would be used because it is the greatest value.
c. This would affect the mean only.
d. (Suggested answer)
The mode best represents the salaries because most employees earn \$30 000.

10. 78 – 68 = 10
The range is 10.

11. (25 x 70% + 30 x 60%) ÷ 55 = 64.55%
The average score was 64.55%.

12. 3 x 20 – 15 = 45
The sum of the other 2 numbers is 45.

13a. The number of hard drives produced had been increasing.
b. (20 + 18 + 26 + 32 + 46 + 50 + 47 + 57) ÷ 8 = 37
The mean is approximately 37 000 hard drives.
c. 57 – 18 = 39
The range is 39 000 hard drives.
d. (Suggested answer)
About 65 000 hard drives will be produced in September.

15 Data Management

Try It

C

1. B
2. C
3. B
4. B
5. A
6. B

7a. There were 15 employees in 2010.
b. There were 43 employees in 2015.

8. There were 3 more female employees in 2012, 7 more in 2015, and 3 more in 2016.

9a. The sum is 100%.
b. 2010: About 65% of the employees were male and about 35% were female.
2016: About 48% of the employees were male and about 52% were female.

10a. Employees in a Factory
b. Percent of Employees
11. No, a circle graph cannot be used because neither question is about finding parts of a whole.
12. Tyler ; Tavia
a. Tavia spent more time watching TV than playing sports.
b. Tyler spent 3.2 h and Tavia spent 2.5 h doing homework.
Tyler spent 2 h and Tavia spent 2.8 h eating.
c. The children spent the same amount of time at school. They spent 6 hours.
13.

School and homework took up about half of Terrance's day.
14a. (Suggested answers)
Using Graph C would emphasize the increase in profits, encouraging investors to invest in the company because it appears to be doing well.
Using Graph C would emphasize the profits made, making the profits appear excessive.
Using Graph B would make the increase of the company's profits appear smaller. This would help one argue that the increase in profit is not significant.
b. The profit had been increasing and would be about $9 million in 2020.
c. No, a bar graph is not more appropriate because the data is continuous. A line graph is more suitable to represent data over time.

16 Probability

Try It

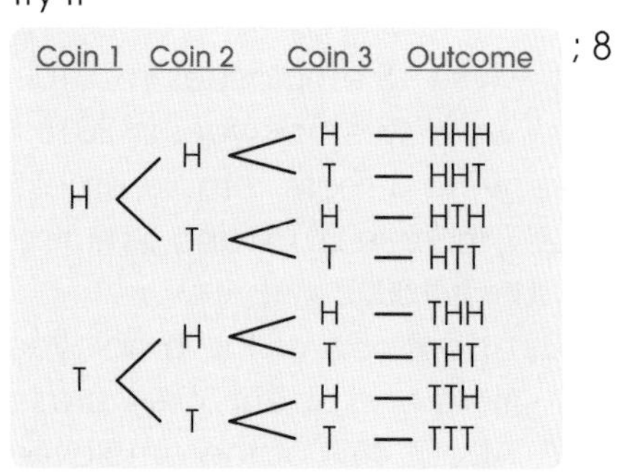

; 8

1a. The probability is $\frac{3}{8}$.
b. The probability is $\frac{3}{8}$.
c. The probability is $\frac{1}{2}$.
d. The probability is $\frac{1}{2}$.
e. The probability is $\frac{1}{8}$.
f. The probability is $\frac{1}{8}$.
g. The probability is $\frac{1}{4}$.
h. The probability is 0.
2a. He has 6 options.
b. The probability is $\frac{1}{3}$.
The probability is $\frac{1}{2}$.
The probability is $\frac{1}{6}$.
3a. $\frac{12}{25}$ of the flips were heads. This is close to $\frac{1}{2}$, which would have been expected.
b. She should expect to flip heads 250 times.
c. The probability is $\frac{1}{2}$ because each flip is independent and has the same probability of getting heads.
d. Yes, it is possible but very unlikely to happen.
4a. The probability is $\frac{1}{3}$ for Spinner A, $\frac{1}{2}$ for Spinner B, and $\frac{1}{6}$ for Spinner C.
b. He should expect 10 times for Spinner A, 15 for Spinner B, and 5 for Spinner C.
c. No, the probability is independent of the number of spins.
d. Spinner A is fair.
5a. The last contestant has a better chance to win because he/she would have the most information about the colours of the balls to be drawn before guessing.
b. The probability is $\frac{1}{3}$.
c. A black ball is most likely to be drawn.

ASSESSMENT TESTS 1 AND 2

Test-taking Tips

Writing tests can be stressful for many students. The best way to prepare for a test is by practising! In addition to practising, the test-taking tips below will also help you prepare for tests.

Multiple-choice Questions

- Read the question twice before finding the answer.
- Skip the difficult questions and do the easy ones first.
- Come up with an answer before looking at the choices.
- Read all four choices before deciding which is the correct answer.
- Eliminate the choices that you know are incorrect.
- Read and follow the instructions carefully:
  - Use a pencil only.
  - Fill one circle only for each question.
  - Fill the circle completely.
  - Cleanly erase any answer you wish to change.

  e.g.

  ● correct

  ⊗ ✓ ◉ ◒ incorrect

Open-response Questions

- Read the question carefully.
- Highlight (i.e. underline/circle) important information in the question.
- Use drawings to help you better understand the question if needed.
- Find out what needs to be included in the solution.
- Estimate the answer.
- Organize your thoughts before writing the solution.
- Write in the space provided.
- Always write a concluding sentence for your solution.
- Check if your answer is reasonable.
- Never leave a question blank. Show your work or write down your reasoning. Even if you do not get the correct answer, you might get some marks for showing your work.

Multiple-choice Questions

① Which set lists all the common factors of 36 and 60?

- O 2, 3, 6
- O 2, 3, 4, 6
- O 1, 2, 3, 4, 6, 12
- O 1, 2, 3, 4, 6, 9, 12

② If there is an 8% tax on the sale price, what is the total cost of the book?

- O $12.58
- O $13.59
- O $15.98
- O $17.02

③ Which number is greater than 4?

- O $\sqrt{20}$
- O 2.4
- O -7
- O $\frac{18}{5}$

④ What percent of the flowers are red?

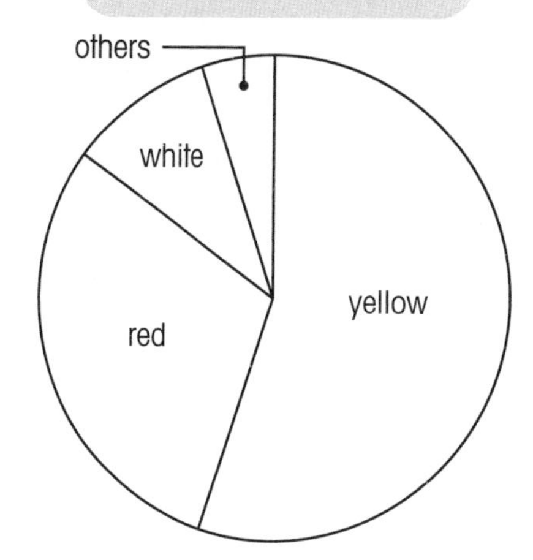

- O 18%
- O 30%
- O 42%
- O 55%

⑤ A hockey team has played 20 games. No games were tied. The ratio of wins to losses is 2:3. If the team plays 2 more games and wins them both, what will the new ratio be?

- O 4:7
- O 5:6
- O 4:3
- O 2:5

⑥ What is the volume of the solid?

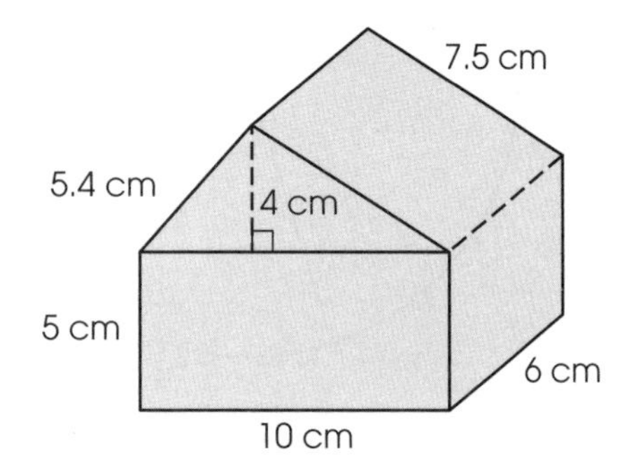

- O 337.4 cm^3
- O 340 cm^3
- O 420 cm^3
- O 421.5 cm^3

⑦ Evaluate $\sqrt{17-5} \times \sqrt{9 \times 3}$.

- O 8.66
- O 18
- O 36
- O 324

⑧ Which term cannot be used to describe the lines below?

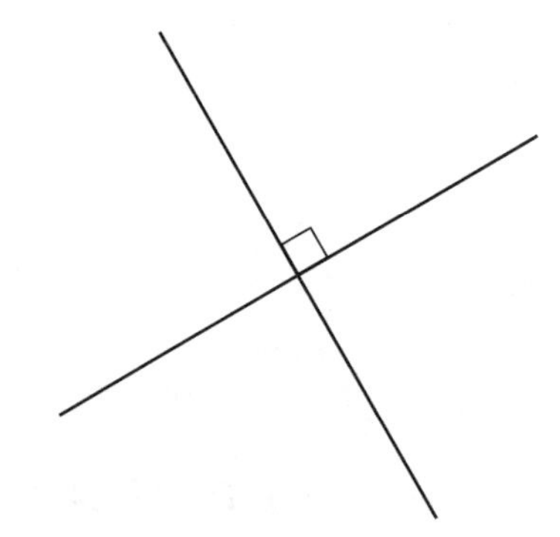

- o perpendicular lines
- o intersecting lines
- o parallel lines
- o bisectors

⑨ Which shape has the greatest area?

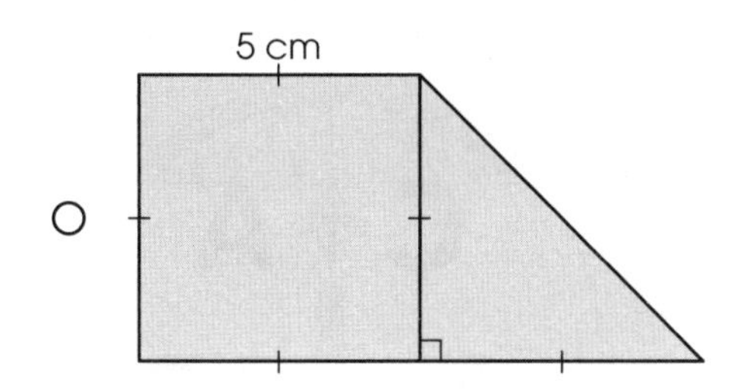

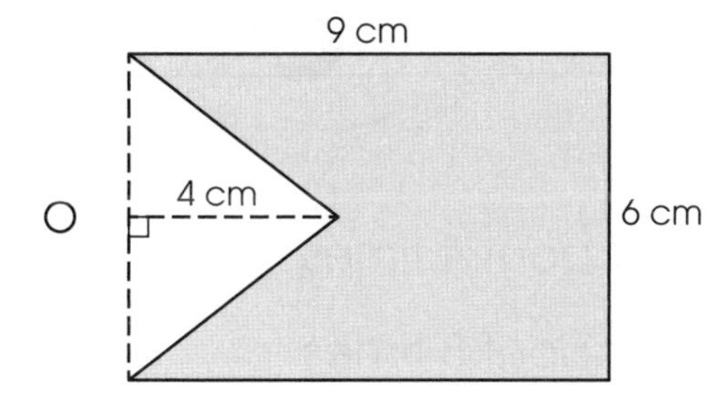

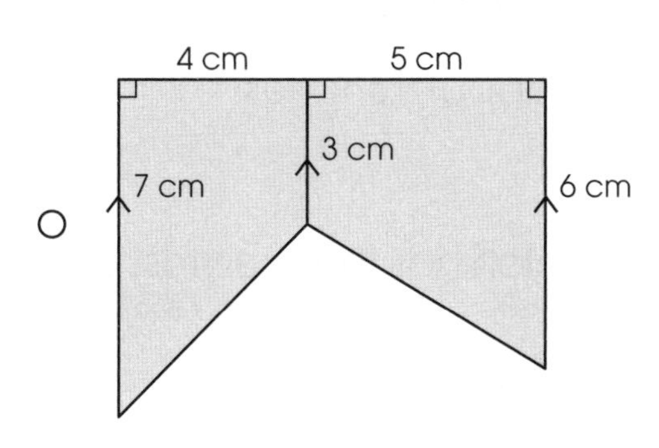

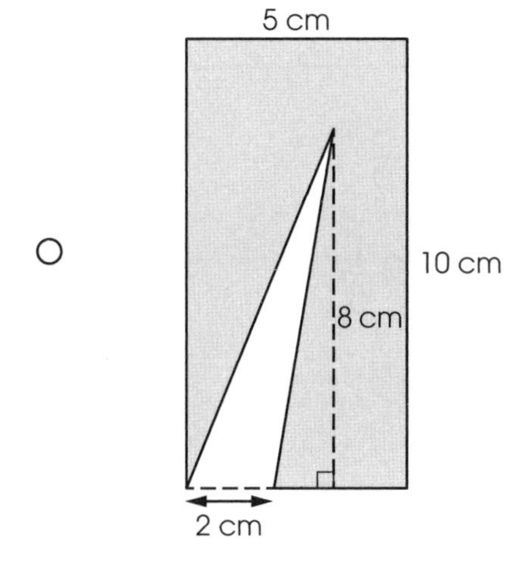

⑩ Which is the correct algebraic expression for "the difference of one third of a number and 8"?

- $(n - 8) \div 3$
- $\frac{1}{3} \times (8 - n)$
- $n \div 3 - 8$
- $3n - 8$

⑪ Which shape cannot form a tiling pattern by itself on a flat surface?

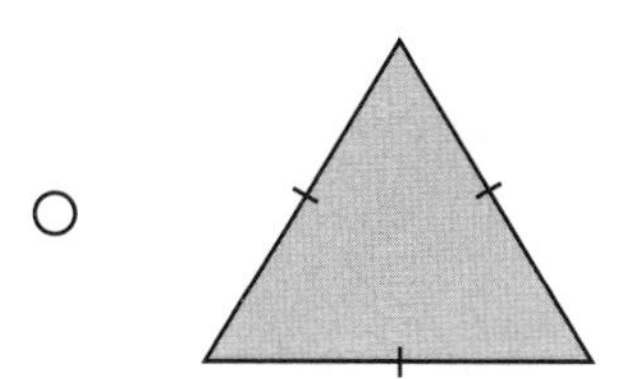

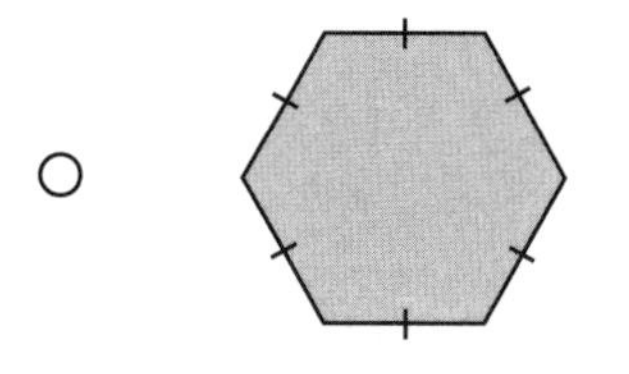

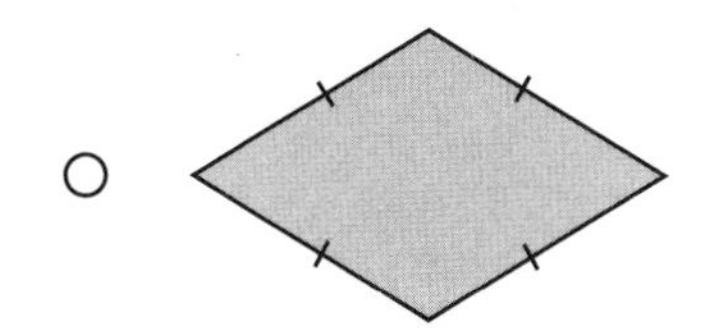

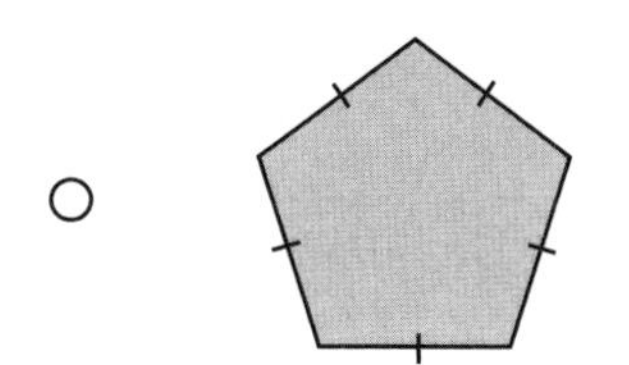

⑫ Which of the following correctly describes the set of data?

19	25	30	25	17

- O The median is smaller than the mean.
- O The mode is greater than the median.
- O The range is 15.
- O The mean is smaller than the mode.

⑬ Which of the following coordinates lie in Quadrant III?

- O (1,5)
- O (3,-1)
- O (-2,-4)
- O (-5,7)

⑭ If the spinner is spun 20 times, about how many times will A be spun?

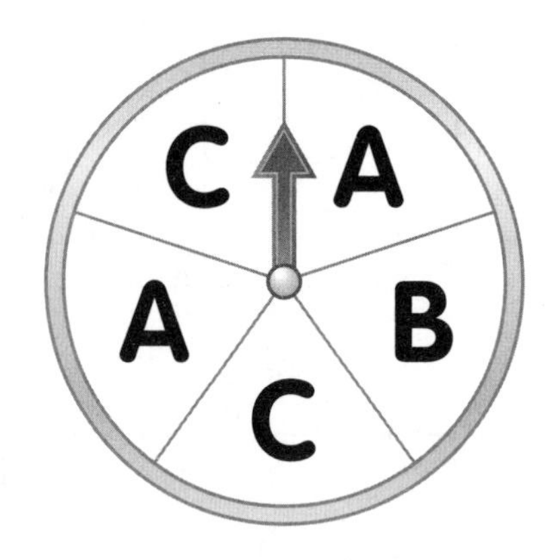

- O about 4 times
- O about 8 times
- O about 12 times
- O about 16 times

⑮ Which has the greatest answer?

- O $3\frac{1}{2} \div 1\frac{2}{5}$
- O $4 - \frac{15}{8}$
- O $3\frac{4}{5} \times \frac{5}{9}$
- O $4\frac{1}{3} - 2\frac{1}{6}$

Open-response Questions

⑯ Look at △ABC and △DEF on the grid.

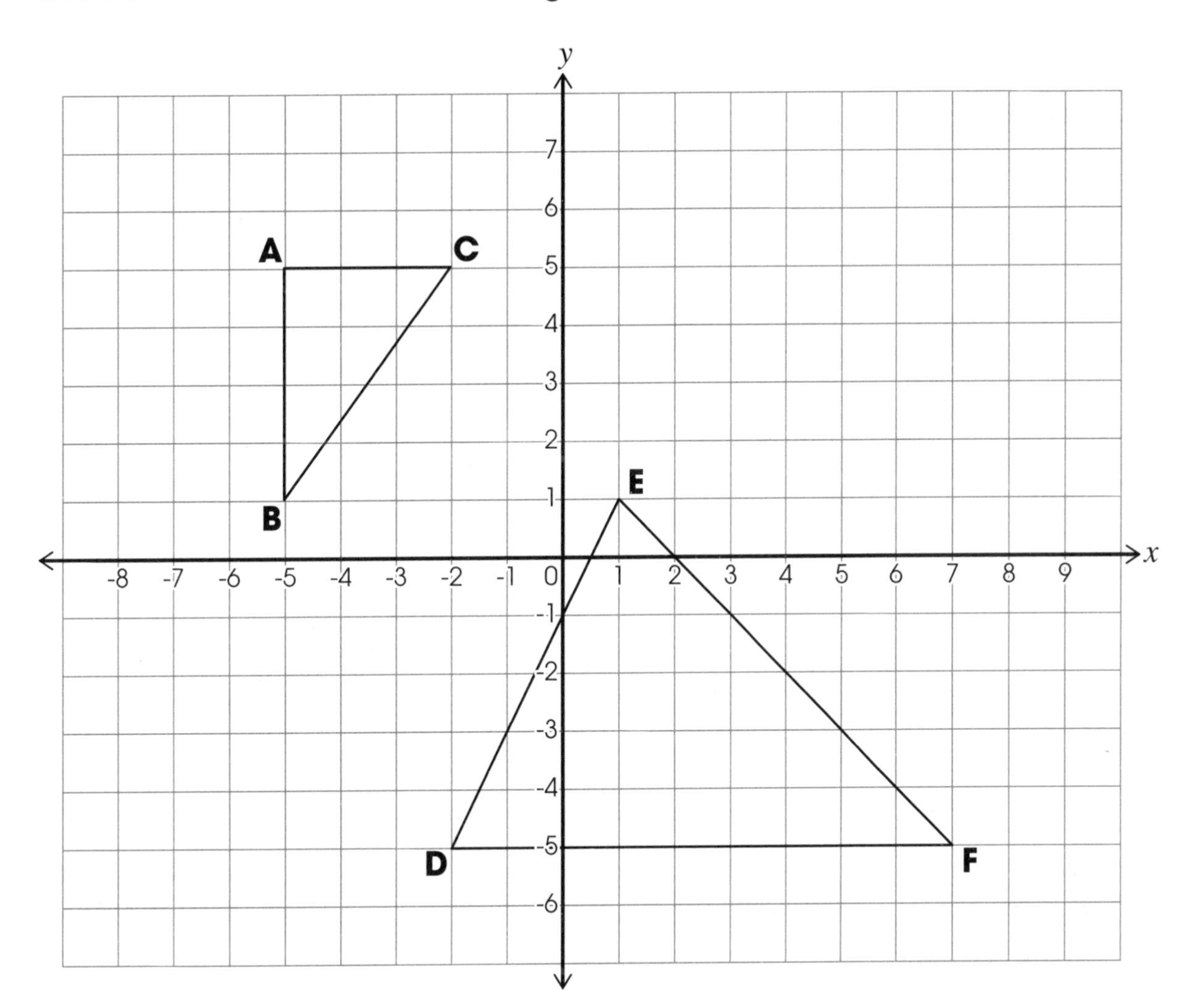

Draw to complete the dilatations and find the missing coordinates.

<u>enlargement by a factor of 2</u>	<u>reduction by a factor of 3</u>
A(-5,5) ⟶ A′(-6,7)	D(-2,-5) ⟶ D′(1,-4)
B(-5,1) ⟶ B′(,)	E(1,1) ⟶ E′(,)
C(-2,5) ⟶ C′(,)	F(7,-5) ⟶ F′(,)

Which has a greater area, △A′B′C′ or △DEF?

____

⑰ Write an algebraic expression for each pattern and extend it. Then find the 8th term for each pattern.

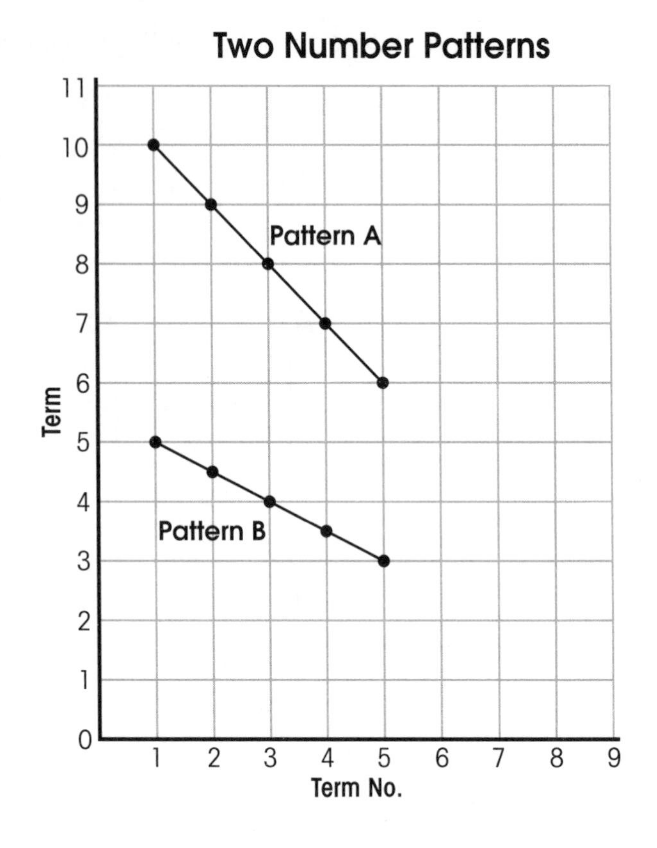

⑱ Bianca made a prediction about the revenue of her store from 2019 to 2025. However, some of the data are missing.

If the median of the predicted revenue is $30 000 and the mode is $60 000, what is the predicted mean revenue?

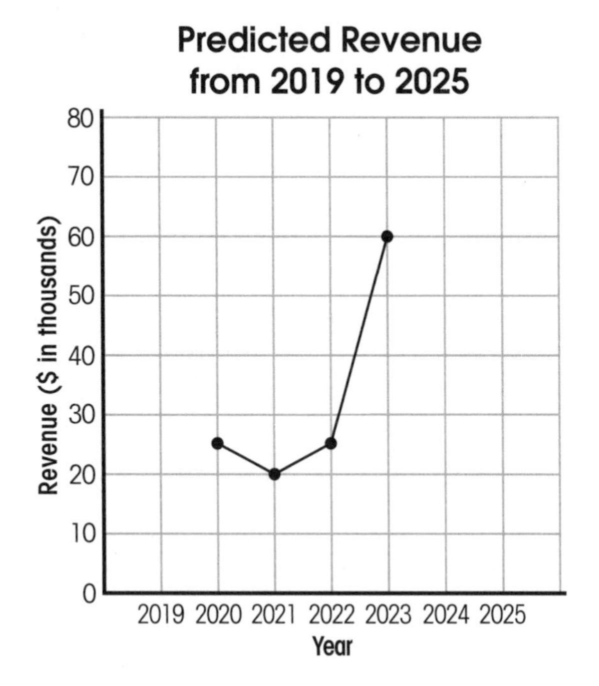

Multiple-choice Questions

① What is the least common multiple of 6, 8, and 10?

- 2
- 60
- 120
- 240

② Which of the following is incorrect?

- $$\begin{array}{r} 1.65 \\ \times \quad 4.2 \\ \hline 330 \\ 660 \\ \hline 0.990 \end{array}$$
- $$\begin{array}{r} 16.24 \\ + \ 23.76 \\ \hline 40.00 \end{array}$$
- $$\begin{array}{r} 65.00 \\ - \ 19.83 \\ \hline 45.17 \end{array}$$
- $$\begin{array}{r} 2.14 \\ 13 \overline{)27.82} \\ 26 \\ \hline 18 \\ 13 \\ \hline 52 \\ 52 \\ \hline \end{array}$$

③ Which has the greatest value?

- 4^2
- 2^4
- 2×10^4
- 4×10^2

④ Which of the following costs the most?

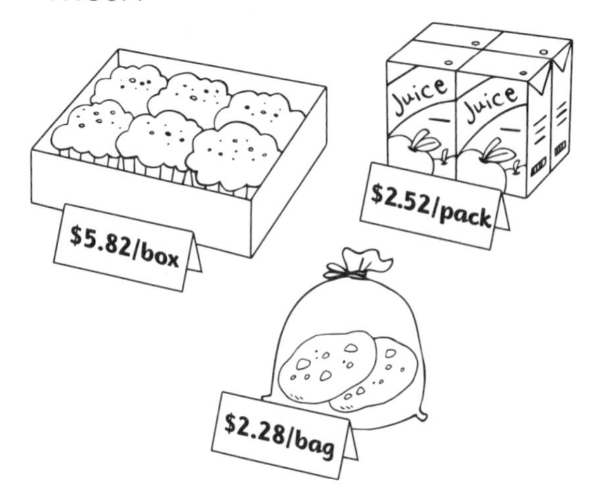

- 1 muffin and 5 juice boxes
- 3 muffins and 2 juice boxes
- 3 juice boxes and 3 cookies
- 4 muffins and 1 cookie

⑤ Look at the number of points each person got in each round. The score is the sum of the points in all 3 rounds.

Round	1	2	3
Alex	4	-2	-6
Bailey	8	-7	2
Casey	9	-1	-4
Dee	-5	11	-3

Who has the highest score?

- O Alex
- O Bailey
- O Casey
- O Dee

⑥ Which of the following is not in the correct order?

- O $\frac{7}{4} > 1.33 > 125\%$
- O $2.08 < \frac{8}{3} < 241\%$
- O $68\% > 0.55 > \frac{5}{13}$
- O $46\% < \frac{12}{25} < 0.49$

⑦ Four rectangles are connected to create a shape. What is the total perimeter of the shape including both the outer and inner perimeters?

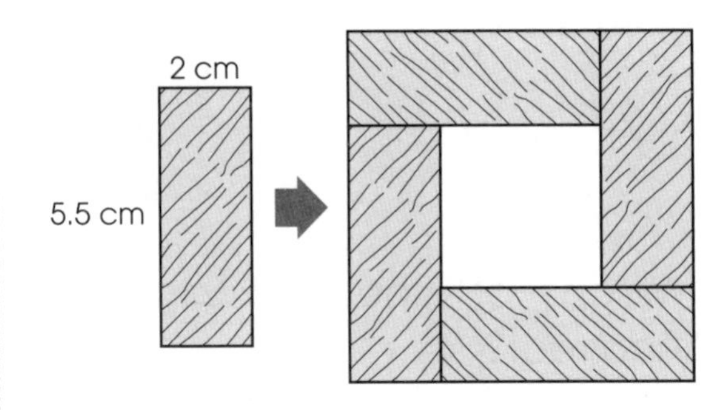

- O 30 cm
- O 38 cm
- O 44 cm
- O 60 cm

⑧ Which is the correct way to name this triangle by its angles?

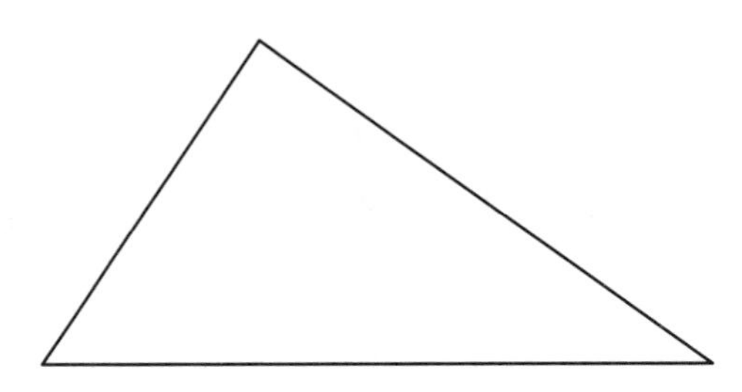

- O obtuse triangle
- O scalene triangle
- O right triangle
- O acute triangle

⑨ The net of a square-based pyramid is cut from a piece of cardboard. What is the surface area of the resulting pyramid?

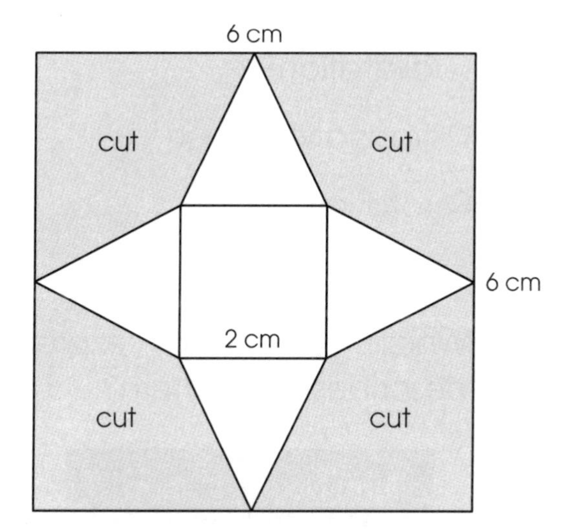

- O 8 cm^2
- O 12 cm^2
- O 15 cm^2
- O 18 cm^2

⑩ If 3 coins are flipped, what is the probability that all three coins land on the same side?

- O $\frac{1}{8}$
- O $\frac{1}{4}$
- O $\frac{3}{8}$
- O $\frac{1}{2}$

⑪ Look at the steps to solve the equation.

$$\triangle x + 6 = \heartsuit$$
$$\heartsuit - 6 = 28$$

$$\triangle x + 6 = \heartsuit$$
$$\triangle x + 6 - 6 = \heartsuit - 6$$
$$\triangle x = 28$$
$$\triangle x \div 4 = 28 \div 4$$
$$x = 7$$

What are the values of the symbols?

- O $\triangle = 6$ $\heartsuit = 35$
- O $\triangle = 6$ $\heartsuit = 28$
- O $\triangle = 4$ $\heartsuit = 34$
- O $\triangle = 4$ $\heartsuit = 25$

⑫ Which of the following shapes are not always similar?

- O right triangles
- O circles
- O equilateral triangles
- O squares

⑬ The triangle is translated 3 units down and 1 unit to the right. What are the new coordinates?

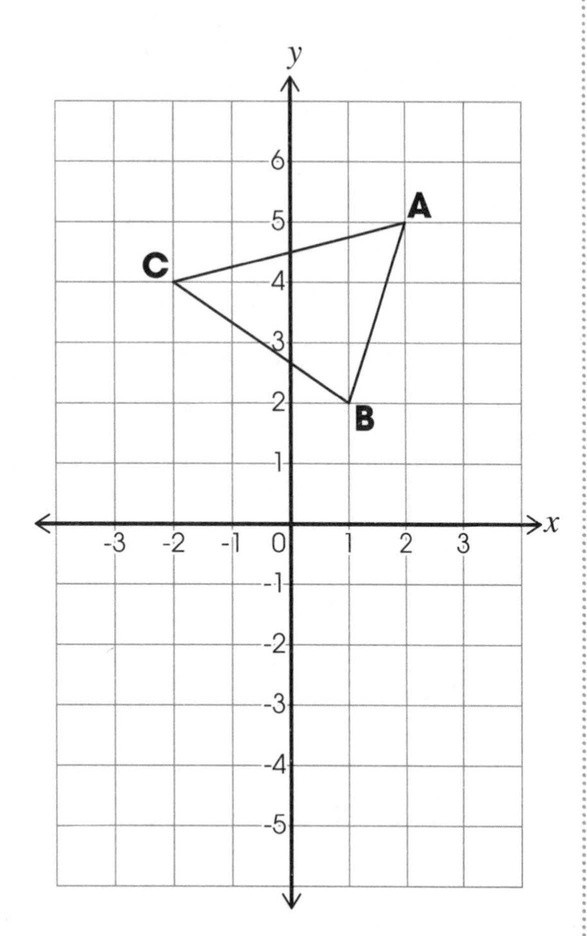

- O A′(-1,4) B′(-2,1) C′(-5,3)
- O A′(1,2) B′(0,-1) C′(-3,1)
- O A′(2,5) B′(1,2) C′(-2,4)
- O A′(3,2) B′(2,-1) C′(-1,1)

⑭ Which type of transformation does not result in a congruent image?

- O reflection
- O dilatation
- O translation
- O rotation

⑮ Which algebraic expression describes the pattern?

Term No.	Term
1	4
2	7
3	10
4	13
5	16

- O $4n - 2$
- O $(n + 2) \times 2$
- O $3n + 1$
- O $2n + 3$

Open-response Questions

⑯ On Earth Day, Ms. Lowy presented the following double line graph on how her Grade 7 students got to school each year.

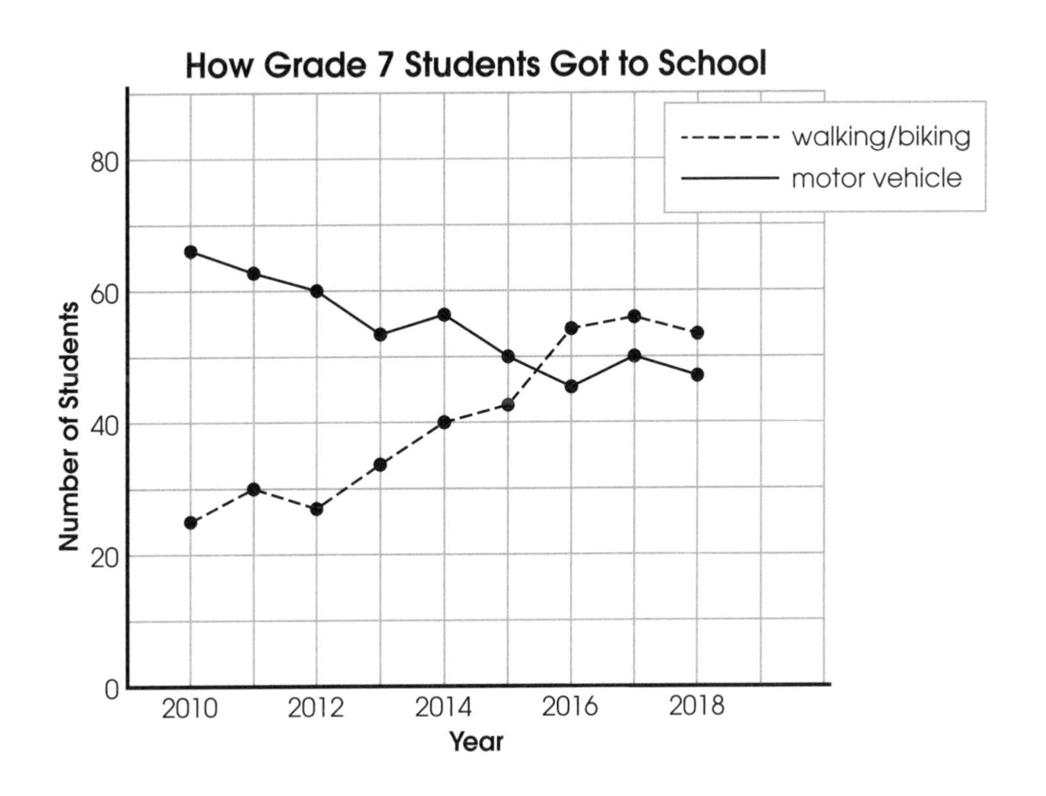

Check the correct answers and answer the question.

Set of Data	◯ census ◯ sample

Type of Data	◯ primary data ◯ secondary data

Describe the trend in the graph. Predict how many students would get to school by each method in 2020. Give a possible reason for this trend.

⑰ A garden is to be reduced in size proportionally. What is the new length of x? Find the percent change from the area of the old garden to the area of the new one.

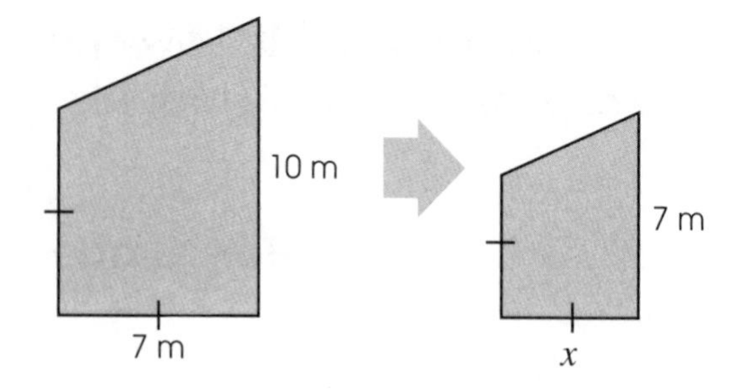

⑱ A trapezoidal prism has a volume of 330 cm^3. The trapezoidal base has a longer side of 15 cm and a height of 5 cm. The prism has a height of 6 cm. Use an equation to find the shorter side of the trapezoid. Is it possible to find the surface area of the prism? Explain.

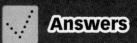

Assessment Test 1

1. 1, 2, 3, 4, 6, 12
2. $13.59
3. $\sqrt{20}$
4. 30%
5. 5:6
6. 420 cm^3
7. 18
8. parallel lines
9. 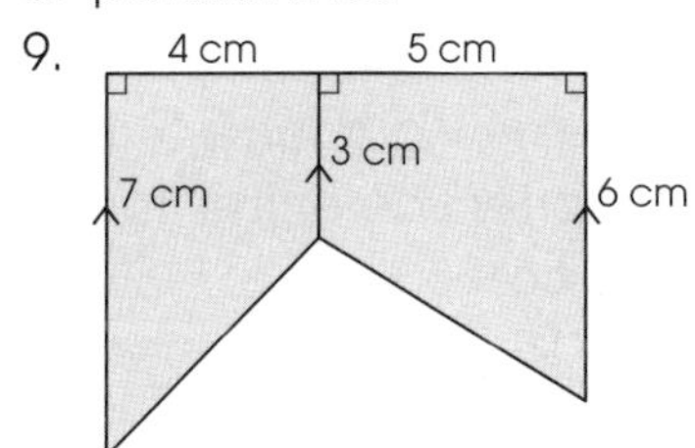

10. $n \div 3 - 8$
11. 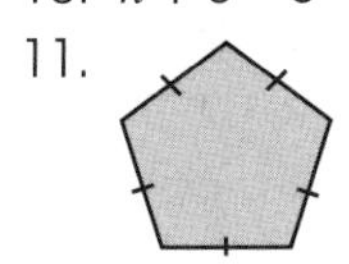
12. The mean is smaller than the mode.
13. (-2,-4)
14. about 8 times
15. $3\frac{1}{2} \div 1\frac{2}{5}$
16. 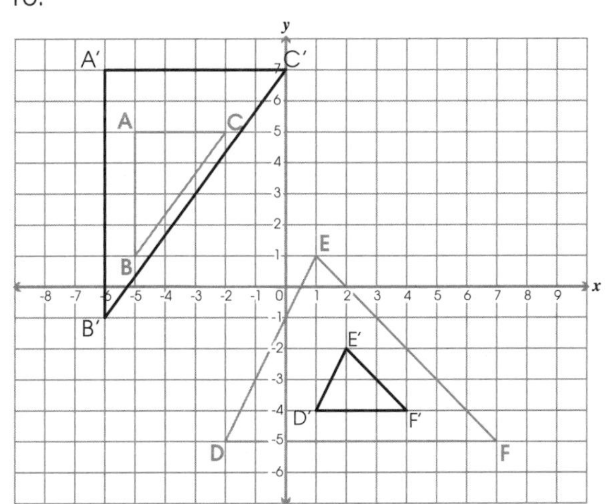

B'(-6,-1) E'(2,-2)
C'(0,7) F'(4,-4)
Area of △A'B'C': 6 x 8 ÷ 2 = 24
Area of △DEF: 9 x 6 ÷ 2 = 27
△DEF has a greater area.

17. Pattern A: $11 - n$
Pattern B: $5 - (n - 1) \div 2$
The 8th term of Pattern A is 3 and the 8th term of Pattern B is 1.5.

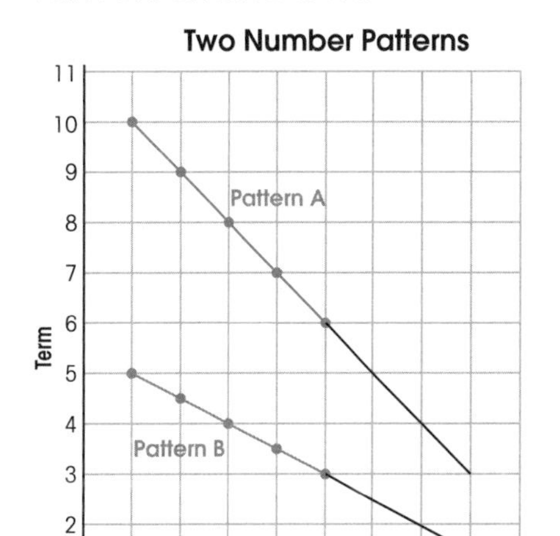

18. There are 3 missing values.
The data on the graph:
$20 000, $25 000, $25 000, $60 000
Since $60 000 is the mode, 2 of the missing values must be $60 000.
Since the median is $30 000, the last missing value must be $30 000.
Mean: (25 000 + 20 000 + 25 000 + 60 000 + 60 000 + 60 000 + 30 000) ÷ 7 = 40 000
The predicted mean revenue is $40 000.

Assessment Test 2

1. 120
2. $$\begin{array}{r} 1.65 \\ \times \quad 4.2 \\ \hline 330 \\ 660 \\ \hline 0.990 \end{array}$$
3. 2×10^4
4. 3 juice boxes and 3 cookies
5. Casey
6. $2.08 < \frac{8}{3} < 241\%$
7. 44 cm
8. right triangle
9. 12 cm^2
10. $\frac{1}{4}$
11. △ = 4 ♡ = 34
12. right triangles
13. A′(3,2) B′(2,-1) C′(-1,1)
14. dilatation
15. $3n + 1$
16. census ; primary data
    The number of students walking or biking to school had been increasing while the number of students taking a motor vehicle had been decreasing.
    In 2020, about 60 students would get to school by walking or biking and about 40 students would get to school by motor vehicle. A possible reason for this trend is increased promotion of walking and biking as a way to reduce the environmental impact of motor vehicles.
17. $\frac{7}{10} = \frac{x}{7}$
    $10x = 49$
    $x = 4.9$
    The new length is 4.9 m.
    Area of old garden: $(7 + 10) \times 7 \div 2 = 59.5$
    Area of new garden:
    $(4.9 + 7) \times 4.9 \div 2 = 29.155$
    Percent change:
    $\frac{29.155 - 59.5}{59.5} \times 100\% = -51\%$
    The area of the new garden is 51% less.
18. Let x be the shorter side of the trapezoid.
    $(x + 15) \times 5 \div 2 \times 6 = 330$
    $x + 15 = 22$
    $x = 7$
    The shorter side of the trapezoid is 7 cm. It is not possible to find the surface area of the prism because the lengths of the other two sides of the trapezoid are not given.